Modern Assembly Language Programming
with the ARM Processor

Modern Assembly Language Programming with the ARM Processor

Larry D. Pyeatt

ELSEVIER

AMSTERDAM • BOSTON • HEIDELBERG • LONDON
NEW YORK • OXFORD • PARIS • SAN DIEGO
SAN FRANCISCO • SINGAPORE • SYDNEY • TOKYO
Newnes is an imprint of Elsevier

Newnes

Newnes is an imprint of Elsevier
The Boulevard, Langford Lane, Kidlington, Oxford OX5 1GB, UK
50 Hampshire Street, 5th Floor, Cambridge, MA 02139, USA

Library of Congress Cataloging-in-Publication Data
A catalog record for this book is available from the Library of Congress

British Library Cataloging-in-Publication Data
A catalogue record for this book is available from the British Library

ISBN: 978-0-12-803698-3

For information on all Newnes publications
visit our website at https://www.elsevier.com/

Working together
to grow libraries in
developing countries

www.elsevier.com • www.bookaid.org

Publisher: Joe Hayton
Acquisition Editor: Tim Pitts
Editorial Project Manager: Charlotte Kent
Production Project Manager: Julie-Ann Stansfield
Designer: Mark Rogers

Typeset by SPi Global, India

Contents

List of Tables

List of Figures

List of Listings

Preface

This book is intended to be used in a first course in assembly language programming for Computer Science (CS) and Computer Engineering (CE) students. It is assumed that students using this book have already taken courses in programming and data structures, and are competent programmers in at least one high-level language. Many of the code examples in the book are written in C, with an assembly implementation following. The assembly examples can stand on their own, but students who are familiar with C, C++, or Java should find the C examples helpful.

Computer Science and Computer Engineering are very large fields. It is impossible to cover everything that a student may eventually need to know. There are a limited number of course hours available, so educators must strive to deliver degree programs that make a compromise between the number of concepts and skills that the students learn and the depth at which they learn those concepts and skills. Obviously, with these competing goals it is difficult to reach consensus on exactly what courses should be included in a CS or CE curriculum.

Traditionally, assembly language courses have consisted of a mechanistic learning of a set of instructions, registers, and syntax. Partially because of this approach, over the years, assembly language courses have been marginalized in, or removed altogether from, many CS and CE curricula. The author feels that this is unfortunate, because a solid understanding of assembly language leads to better understanding of higher-level languages, compilers, interpreters, architecture, operating systems, and other important CS an CE concepts.

One of the goals of this book is to make a course in assembly language more valuable by introducing methods (and a bit of theory) that are not covered in any other CS or CE courses, while using assembly language to implement the methods. In this way, the course in assembly language goes far beyond the traditional assembly language course, and can once again play an important role in the overall CS and CE curricula.

Choice of Processor Family

Because of their ubiquity, x86 based systems have been the platforms of choice for most assembly language courses over the last two decades. The author believes that this is unfortunate, because in every respect other than ubiquity, the x86 architecture is the worst possible choice for learning and teaching assembly language. The newer chips in the family have hundreds of instructions, and irregular rules govern how those instructions can be used. In an attempt to make it possible for students to succeed, typical courses use antiquated assemblers and interface with the antiquated IBM PC BIOS, using only a small subset of the modern x86 instruction set. The programming environment has little or no relevance to modern computing.

Partially because of this tendency to use x86 platforms, and the resulting unnecessary burden placed on students and instructors, as well as the reliance on antiquated and irrelevant development environments, assembly language is often viewed by students as very difficult and lacking in value. The author hopes that this textbook helps students to realize the value of knowing assembly language. The relatively simple ARM processor family was chosen in hopes that the students also learn that although assembly language programming may be more difficult than high-level languages, it can be mastered.

The recent development of very low-cost ARM based Linux computers has caused a surge of interest in the ARM architecture as an alternative to the x86 architecture, which has become increasingly complex over the years. This book should provide a solution for a growing need.

Many students have difficulty with the concept that a register can hold variable x at one point in the program, and hold variable y at some other point. They also often have difficulty with the concept that, before it can be involved in any computation, data has to be moved from memory into the CPU. Using a load-store architecture helps the students to more readily grasp these concepts.

Another common difficulty that students have is in relating the concepts of an address and a pointer variable. You can almost see the little light bulbs light up over their heads, when they have the "eureka!" moment and realize that pointers *are* just variables that hold an address. The author hopes that the approach taken in this book will make it easier for students to have that "eureka!" moment. The author believes that load-store architectures make that realization easier.

Many students also struggle with the concept of recursion, regardless of what language is used. In assembly, the mechanisms involved are exposed and directly manipulated by the programmer. Examples of recursion are scattered throughout this textbook. Again, the clean architecture of the ARM makes it much easier for the students to understand what is going on.

Some students have difficulty understanding the flow of a program, and tend to put many unnecessary branches into their code. Many assembly language courses spend so much time and space on learning the instruction set that they never have time to teach good programming practices. This textbook puts strong emphasis on using structured programming concepts. The relative simplicity of the ARM architecture makes this possible.

One of the major reasons to learn and use assembly language is that it allows the programmer to create very efficient mathematical routines. The concepts introduced in this book will enable students to perform efficient non-integral math on any processor. These techniques are rarely taught because of the time that it takes to cover the x86 instruction set. With the ARM processor, less time is spent on the instruction set, and more time can be spent teaching how to optimize the code.

The combination of the ARM processor and the Linux operating system provides the least costly hardware platform and development environment available. A cluster of 10 Raspberry Pis, or similar hosts, with power supplies and networking, can be assembled for 500 US dollars or less. This cluster can support up to 50 students logging in through ssh. If their client platform supports the X window system, then they can run GUI enabled applications. Alternatively, most low-cost ARM systems can directly drive a display and take input from a keyboard and mouse. With the addition of an NFS server (which itself could be a low-cost ARM system and a hard drive), an entire Linux ARM based laboratory of 20 workstations could be built for 250 US dollars per seat or less. Admittedly, it would not be a high-performance laboratory, but could be used to teach C, assembly, and other languages. The author would argue that inexperienced programmers *should* learn to program on low-performance machines, because it reinforces a life-long tendency towards efficiency.

General Approach

The approach of this book is to present concepts in different ways throughout the book, slowly building from simple examples towards complex programming on bare-metal embedded systems. Students who don't understand a concept when it is explained in a certain way may easily grasp the concept when it is presented later from a different viewpoint.

The main objective of this book is to provide an improved course in assembly language by replacing the x86 platform with one that is less costly, more ubiquitous, well-designed, powerful, and easier to learn. Since students are able to master the basics of assembly language quickly, it is possible to teach a wider range of topics, such as fixed and floating point mathematics, ethical considerations, performance tuning, and interrupt processing. The author hopes that courses using this book will better prepare students for the junior and senior level courses in operating systems, computer architecture, and compilers.

Companion Website

Please visit the companion web site to access additional resources. Instructors may download the author's lecture slides and solution manual for the exercises. Students and instructors may also access the laboratory manual and additional code examples. The author welcomes suggestions for additional lecture slides, laboratory assignments, or other materials.

http://booksite.elsevier.com/9780128036983

Acknowledgments

I would like to thank Randy Warner for reading the manuscript, catching errors, and making helpful suggestions. I would also like to thank the following students for suggesting exercises with answers and catching numerous errors in the drafts: Zach Buechler, Preston Cook, Joshua Daybrest, Matthew DeYoung, Josh Dodd, Matt Dyke, Hafiza Farzami, Jeremy Goens, Lawrence Hoffman, Colby Johnson, Benjamin Kaiser, Lauren Keene, Jayson Kjenstad, Murray LaHood-Burns, Derek Lane, Yanlin Li, Luke Meyer, Matthew Mielke, Forrest Miller, Christopher Navarro, Girik Ranchhod, Josh Schweigert, Christian Sieh, Weston Silbaugh, Jacob St. Amand, Njaal Tengesdal, Dylan Thoeny, Michael Vortherms, Dicheng Wu, and Kekoa (Peter) Yamaguchi. Finally, I am also very grateful for my assistants, Scott Logan, Ian Carlson, and Derek Stotz, who gave very valuable feedback during the writing of this book.

PART I

Assembly as a Language

Introduction

Chapter Outline

An executable computer program is, ultimately, just a series of numbers that have very little or no meaning to a human being. We have developed a variety of human-friendly languages in which to express computer programs, but in order for the program to execute, it must eventually be reduced to a stream of numbers. Assembly language is one step above writing the stream of numbers. The stream of numbers is called the *instruction stream*. Each number in the instruction stream instructs the computer to perform one (usually small) operation. Although each instruction does very little, the ability of the programmer to specify any *sequence* of instructions and the ability of the computer to perform billions of these small operations every second makes modern computers very powerful and flexible tools. In assembly language, one line of code usually gets translated into one machine instruction. In high-level languages, a single line of code may generate *many* machine instructions.

A simplified model of a computer system, as shown in Fig. 1.1, consists of memory, input/output devices, and a central processing unit (CPU), connected together by a system bus. The bus can be thought of as a roadway that allows data to travel between the components of the computer system. The CPU is the part of the system where most of the computation occurs, and the CPU controls the other devices in the system.

Memory can be thought of as a series of mailboxes. Each mailbox can hold a single postcard with a number written on it, and each mailbox has a unique numeric identifier. The identifier, x is called the memory address, and the number stored in the mailbox is called the contents of

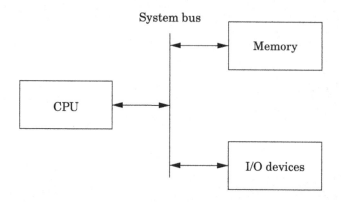

Figure 1.1
Simplified representation of a computer system.

address *x*. Some of the mailboxes contain data, and others contain instructions which control what actions are performed by the CPU.

The CPU also contains a much smaller set of mailboxes, which we call registers. Data can be copied from cards stored in memory to cards stored in the CPU, or vice-versa. Once data has been copied into one of the CPU registers, it can be used in computation. For example, in order to add two numbers in memory, they must first be copied into registers on the CPU. The CPU can then add the numbers together and store the result in one of the CPU registers. The result of the addition can then be copied back into one of the mailboxes in the memory.

Modern computers execute instructions sequentially. In other words, the next instruction to be executed is at the memory address immediately following the current instruction. One of the registers in the CPU, the program counter (PC), keeps track of the location from which the next instruction is to be fetched. The CPU follows a very simple sequence of actions. It fetches an instruction from memory, increments the PC, executes the instruction, and then repeats the process with the next instruction. However, some instructions may change the PC, so that the next instruction is fetched from a non-sequential address.

1.1 Reasons to Learn Assembly

There are many high-level *programming languages*, such as Java, Python, C, and C++ that have been designed to allow programmers to work at a high level of abstraction, so that they do not need to understand exactly what instructions are needed by a particular CPU. For *compiled* languages, such as C and C++, a compiler handles the task of translating the program, written in a high-level language, into *assembly language* for the particular CPU on the system. An assembler then converts the program from assembly language into the binary codes that the CPU reads as instructions.

High-level languages can greatly enhance programmer productivity. However, there are some situations where writing assembly code directly is desirable or necessary. For example, assembly language may be the best choice when writing

- the first steps in booting the computer,
- code to handle interrupts,
- low-level locking code for multi-threaded programs,
- code for machines where no compiler exists,
- code which needs to be optimized beyond the limits of the compiler,
- on computers with very limited memory, and
- code that requires low-level access to architectural and/or processor features.

Aside from sheer necessity, there are several other reasons why it is still important for computer scientists to learn assembly language.

One example where knowledge of assembly is indispensable is when designing and implementing compilers for high-level languages. As shown in Fig. 1.2, a typical compiler for a high-level language must generate assembly language as its output. Most compilers are designed to have multiple stages. In the input stage, the source language is read and converted into a graph representation. The graph may be optimized before being passed to the output, or code generation, stage where it is converted to assembly language. The assembly is then fed into the system's assembler to generate an object file. The object file is linked with other object files (which are often combined into *libraries*) to create an executable program.

The code generation stage of a compiler must traverse the graph and emit assembly code. The quality of the assembly code that is generated can have a profound influence on the performance of the executable program. Therefore, the programmer responsible for the code generation portion of the compiler must be well versed in assembly programming for the target CPU.

Some people believe that a good optimizing compiler will generate better assembly code than a human programmer. This belief is not justified. Highly optimizing compilers have lots of clever algorithms, but like all programs, they are not perfect. Outside of the cases that they were designed for, they do not optimize well. Many newer CPUs have instructions which operate on multiple items of data at once. However, compilers rarely make use of these powerful single instruction multiple data (SIMD) instructions. Instead, it is common for programmers to write functions in assembly language to take advantage of SIMD instructions. The assembly functions are assembled into object file(s), then linked with the object file(s) generated from the high-level language compiler.

Many modern processors also have some support for processing vectors (arrays). Compilers are usually not very good at making effective use of the vector instructions. In order to achieve excellent vector performance for audio or video codecs and other time-critical code, it is often necessary to resort to small pieces of assembly code in the performance-critical inner loops.

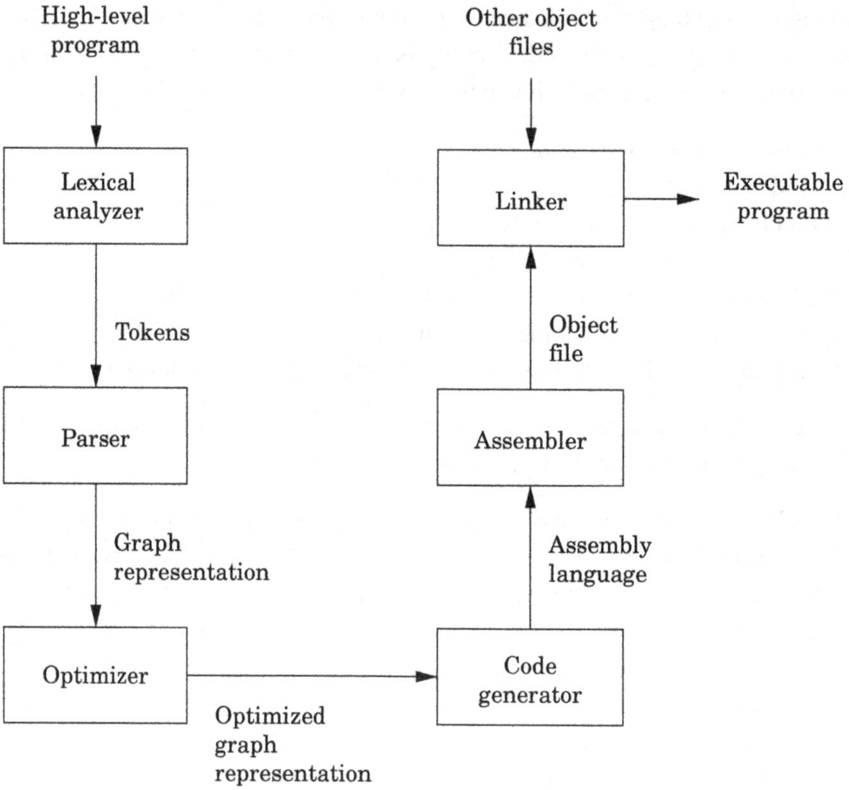

High-level
program

Other object
files

Lexical
analyzer

Linker → Executable
program

Tokens

Object
file

Parser

Assembler

Graph
representation

Assembly
language

Optimizer

Code
generator

Optimized
graph
representation

Figure 1.2
Stages of a typical compilation sequence.

A good example of this type of code is when performing vector and matrix multiplies. Such operations are commonly needed in processing images and in graphical applications. The ARM vector instructions are explained in Chapter 9.

Another reason for assembly is when writing certain parts of an operating system. Although modern operating systems are mostly written in high-level languages, there are some portions of the code that can only be done in assembly. Typical uses of assembly language are when writing device drivers, saving the state of a running program so that another program can use the CPU, restoring the saved state of a running program so that it can resume executing, and managing memory and memory protection hardware. There are many other tasks central to a modern operating system which can only be accomplished in assembly language. Careful design of the operating system can minimize the amount of assembly required, but cannot eliminate it completely.

Another good reason to learn assembly is for debugging. Simply understanding what is going on "behind the scenes" of compiled languages such as C and C++ can be very valuable when trying to debug programs. If there is a problem in a call to a third party library, sometimes the only way a developer can isolate and diagnose the problem is to run the program under a debugger and step through it one machine instruction at a time. This does not require a deep knowledge of assembly language coding but at least a passing familiarity with assembly is helpful in that particular case. Analysis of assembly code is an important skill for C and C++ programmers, who may occasionally have to diagnose a fault by looking at the contents of CPU registers and single-stepping through machine instructions.

Assembly language is an important part of the path to understanding how the machine works. Even though only a small percentage of computer scientists will be lucky enough to work on the code generator of a compiler, they all can benefit from the deeper level of understanding gained by learning assembly language. Many programmers do not really understand pointers until they have written assembly language.

Without first learning assembly language, it is impossible to learn advanced concepts such as microcode, pipelining, instruction scheduling, out-of-order execution, threading, branch prediction, and speculative execution. There are many other concepts, especially when dealing with operating systems and computer architecture, which require some understanding of assembly language. The best programmers understand why some language constructs perform better than others, how to reduce cache misses, and how to prevent buffer overruns that destroy security.

Every program is meant to run on a real machine. Even though there are many languages, compilers, virtual machines, and operating systems to enable the programmer to use the machine more conveniently, the strengths and weaknesses of that machine still determine what is easy and what is hard. Learning assembly is a fundamental part of understanding enough about the machine to make informed choices about how to write efficient programs, even when writing in a high-level language.

As an analogy, most people do not need to know a lot about how an internal combustion engine works in order to operate an automobile. A race car driver needs a much better understanding of exactly what happens when he or she steps on the accelerator pedal in order to be able to judge precisely when (and how hard) to do so. Also, who would trust their car to a mechanic who could not tell the difference between a spark plug and a brake caliper? Worse still, should we trust an engineer to *build* a car without that knowledge? Even in this day of computerized cars, someone needs to know the gritty details, and they are paid well for that knowledge. Knowledge of assembly language is one of the things that defines the computer scientist and engineer.

When learning assembly language, the specific instruction set is not critically important, because what is really being learned is the fine detail of how a typical stored-program machine uses different storage locations and logic operations to convert a string of bits into a meaningful calculation. However, when it comes to learning assembly languages, some processors make it more difficult than it needs to be. Because some processors have an instruction set that is extremely irregular, non-orthogonal, large, and poorly designed, they are not a good choice for learning assembly. The author feels that teaching students their first assembly language on one of those processors should be considered a crime, or at least a form of mental abuse. Luckily, there are processors that are readily available, low-cost, and relatively easy to learn assembly with. This book uses one of them as the model for assembly language.

1.2 The ARM Processor

In the late 1970s, the microcomputer industry was a fierce battleground, with several companies competing to sell computers to small business and home users. One of those companies, based in the United Kingdom, was Acorn Computers Ltd. Acorn's flagship product, the BBC Micro, was based on the same processor that Apple Computer had chosen for their Apple II^{TM} line of computers; the 8-bit 6502 made by MOS Technology. As the 1980s approached, microcomputer manufacturers were looking for more powerful 16-bit and 32-bit processors. The engineers at Acorn considered the processor chips that were available at the time, and concluded that there was nothing available that would meet their needs for the next generation of Acorn computers.

The only reasonably-priced processors that were available were the Motorola 68000 (a 32-bit processor used in the Apple Macintosh and most high-end Unix workstations) and the Intel 80286 (a 16-bit processor used in less powerful personal computers such as the IBM PC). During the previous decade, a great deal of research had been conducted on developing high-performance computer architectures. One of the outcomes of that research was the development of a new paradigm for processor design, known as Reduced Instruction Set Computing (RISC). One advantage of RISC processors was that they could deliver higher performance with a much smaller number of transistors than the older Complex Instruction Set Computing (CISC) processors such as the 68000 and 80286. The engineers at Acorn decided to design and produce their own processor. They used the BBC Micro to design and simulate their new processor, and in 1987, they introduced the Acorn Archimedes^{TM}. The Archimedes^{TM} was arguably the most powerful home computer in the world at that time, with graphics and audio capabilities that IBM PC^{TM} and Apple Macintosh^{TM} users could only dream about. Thus began the long and successful dynasty of the Acorn RISC Machine (ARM) processor.

Acorn never made a big impact on the global computer market. Although Acorn eventually went out of business, the processor that they created has lived on. It was re-named to the Advanced RISC Machine, and is now known simply as ARM. Stewardship of the ARM processor belongs to ARM Holdings, LLC which manages the design of new ARM architectures and licenses the manufacturing rights to other companies. ARM Holdings does not manufacture any processor chips, yet more ARM processors are produced annually than all other processor designs combined. Most ARM processors are used as components for embedded systems and portable devices. If you have a smart phone or similar device, then there is a very good chance that it has an ARM processor in it. Because of its enormous market presence, clean architecture, and small, orthogonal instruction set, the ARM is a very good choice for learning assembly language.

Although it dominates the portable device market, the ARM processor has almost no presence in the desktop or server market. However, that may change. In 2012, ARM Holdings announced the ARM64 architecture, which is the first major redesign of the ARM architecture in 30 years. The ARM64 is intended to compete for the desktop and server market with other high-end processors such as the Sun SPARC and Intel Xeon. Regardless of whether or not the ARM64 achieves much market penetration, the original ARM 32-bit processor architecture is so ubiquitous that it clearly will be around for a long time.

1.3 Computer Data

The basic unit of data in a digital computer is the binary digit, or *bit*. A bit can have a value of zero or one. In order to store numbers larger than 1, bits are combined into larger units. For instance, using two bits, it is possible to represent any number between zero and three. This is shown in Table 1.1. When stored in the computer, all data is simply a string of binary digits. There is more than one way that such a fixed-length string of binary digits can be interpreted.

Computers have been designed using many different bit group sizes, including 4, 8, 10, 12, and 14 bits. Today most computers recognize a basic grouping of 8 bits, which we call a *byte*. Some computers can work in units of 4 bits, which is commonly referred to as a *nibble* (sometimes spelled "nybble"). A nibble is a convenient size because it can exactly represent one hexadecimal digit. Additionally, most modern computers can also work with groupings of 16, 32 and 64 bits. The CPU is designed with a *default word size*. For most modern CPUs, the default word size is 32 bits. Many processors support 64-bit words, which is increasingly becoming the default size.

1.3.1 Representing Natural Numbers

A numeral system is a writing system for expressing numbers. The most common system is the Hindu-Arabic number system, which is now used throughout the world. Almost from the first day of formal education, children begin learning how to add, subtract, and perform other operations using the Hindu-Arabic system. After years of practice, performing basic mathematical operations using strings of digits between 0 and 9 seems natural. However, there are other ways to count and perform arithmetic, such as Roman numerals, unary systems, and Chinese numerals. With a little practice, it is possible to become as proficient at performing mathematics with other number systems as with the Hindu-Arabic system.

Table 1.1 Values represented by two bits

Bit 1	Bit 0	Value
0	0	0
0	1	1
1	0	2
1	1	3

The Hindu-Arabic system is a *base ten* or *radix ten* system, because it uses the ten digits 0, 1, 2, 3, 4, 5, 6, 7, 8, and 9. For our purposes, the words radix and base are equivalent, and refer to the number of individual digits available in the numbering system. The Hindu-Arabic system is also a *positional system*, or a place-value notation, because the value of each digit in a number depends on its position in the number. The radix ten Hindu-Arabic system is only one of an infinite family of closely related positional systems. The members of this family differ only in the radix used (and therefore, the number of characters used). For bases greater than base ten, characters are borrowed from the alphabet and used to represent digits. For example, the first column in Table 1.2 shows the character "A" being used as a single digit representation for the number 10.

In base ten, we think of numbers as strings of the 10 digits, "0"–"9". Each digit counts 10 times the amount of the digit to its right. If we restrict ourselves to integers, then the digit furthest to the right is always the ones digit. It is also referred to as the *least significant digit*. The digit immediately to the left of the ones digit is the tens digit. To the left of that is the hundreds digit, and so on. The leftmost digit is referred to as the *most significant digit*. The following equation shows how a number can be decomposed into its constituent digits:

$$57839_{10} = 5 \times 10^4 + 7 \times 10^3 + 8 \times 10^2 + 3 \times 10^1 + 9 \times 10^0.$$

Note that the subscript of "10" on 57839_{10} indicates that the number is given in base ten.

Table 1.2 The first 21 integers (starting with 0) in various bases

Base									
16	10	9	8	7	6	5	4	3	2
0	0	0	0	0	0	0	0	0	0
1	1	1	1	1	1	1	1	1	1
2	2	2	2	2	2	2	2	2	10
3	3	3	3	3	3	3	3	10	11
4	4	4	4	4	4	4	10	11	100
5	5	5	5	5	5	10	11	12	101
6	6	6	6	6	10	11	12	20	110
7	7	7	7	10	11	12	13	21	111
8	8	8	10	11	12	13	20	22	1000
9	9	10	11	12	13	14	21	100	1001
A	10	11	12	13	14	20	22	101	1010
B	11	12	13	14	15	21	23	102	1011
C	12	13	14	15	20	22	30	110	1100
D	13	14	15	16	21	23	31	111	1101
E	14	15	16	20	22	24	32	112	1110
F	15	16	17	21	23	30	33	120	1111
10	16	17	20	22	24	31	100	121	10000
11	17	18	21	23	25	32	101	122	10001
12	18	20	22	24	30	33	102	200	10010
13	19	21	23	25	31	34	103	201	10011
14	20	22	24	26	32	40	110	201	10100

Imagine that we only had 7 digits: 0, 1, 2, 3, 4, 5, and 6. We need 10 digits for base ten, so with only 7 digits we are limited to base seven. In base seven, each digit in the string represents a power of seven rather than a power of ten. We can represent any integer in base seven, but it may take more digits than in base ten. Other than using a different base for the power of each digit, the math works exactly the same as for base ten. For example, suppose we have the following number in base seven: 330425_7. We can convert this number to base ten as follows:

$$330425_7 = 3 \times 7^5 + 3 \times 7^4 + 0 \times 7^3 + 4 \times 7^2 + 2 \times 7^1 + 5 \times 7^0$$
$$= 50421_{10} + 7203_{10} + 0_{10} + 196_{10} + 14_{10} + 5_{10}$$
$$= 57839_{10}$$

Base two, or binary is the "native" number system for modern digital systems. The reason for this is mainly because it is relatively easy to build circuits with two stable states: on and off (or 1 and 0). Building circuits with more than two stable states is much more difficult and expensive, and any computation that can be performed in a higher base can also be performed in binary. The least significant (rightmost) digit in binary is referred to as the *least significant bit*, or LSB, while the leftmost binary digit is referred to as the *most significant bit*, or MSB.

1.3.2 Base Conversion

The most common bases used by programmers are base two (binary), base eight (octal), base ten (decimal) and base sixteen (hexadecimal). Octal and hexadecimal are common because, as we shall see later, they can be translated quickly and easily to and from base two, and are often easier for humans to work with than base two. Note that for base sixteen, we need 16 characters. We use the digits 0 through 9 plus the letters A through F. Table 1.2 shows the equivalents for all numbers between 0 and 20 in base two through base ten, and base sixteen.

Before learning assembly language it is essential to know how to convert from any base to any other base. Since we are already comfortable working in base ten, we will use that as an intermediary when converting between two arbitrary bases. For instance, if we want to convert a number in base three to base five, we will do it by first converting the base three number to base ten, then from base ten to base five. By using this two-stage process, we will only need to learn to convert between base ten and any arbitrary base b.

Base b to decimal
Converting from an arbitrary base b to base ten simply involves multiplying each base b digit d by b^n, where n is the significance of digit d, and summing all of the results. For example, converting the base five number 3421_5 to base ten is performed as follows:

$$3421_5 = 3 \times 5^3 + 4 \times 5^2 + 2 \times 5^1 + 1 \times 5^0$$
$$= 375_{10} + 100_{10} + 10_{10} + 1_{10}$$
$$= 486_{10}$$

This conversion procedure works for converting any integer from any arbitrary base b to its equivalent representation in base ten. Example 1.1 gives another specific example of how to convert from base b to base ten.

Example 1.1   Converting From an Arbitrary Base to Base Ten

Converting 7362_5 to base ten is accomplished by expanding and summing the terms:

$$7362_5 = 7 \times 5^3 + 3 \times 5^2 + 6 \times 5^1 + 2 \times 5^0$$
$$= 7 \times 125 + 3 \times 25 + 6 \times 5 + 2 \times 1$$
$$= 875 + 75 + 30 + 2$$
$$= 982_{10}$$

Decimal to base b

Converting from base ten to an arbitrary base b involves repeated division by the base, b. After each division, the remainder is used as the next *more significant* digit in the base b number, and the quotient is used as the dividend for the next iteration. The process is repeated until the quotient is zero. For example, converting 56_{10} to base four is accomplished as follows:

$$
\begin{array}{ccc}
14\searrow & 3\searrow & 0 \\
4\overline{)56} & 4\overline{)14} & 4\overline{)3} \\
\underline{40} & \underline{12} & \\
16 & 2 & \\
\underline{16} & & \\
0 & &
\end{array}
$$

Reading the remainders from right to left yields: 320_4. This result can be double-checked by converting it back to base ten as follows:

$$320_4 = 3 \times 4^2 + 2 \times 4^1 + 0 \times 4^0$$
$$= 48 + 8 + 0$$
$$= 56_{10}.$$

Since we arrived at the same number we started with, we have verified that $56_{10} = 320_4$. This conversion procedure works for converting any integer from base ten to any arbitrary base b. Example 1.2 gives another example of converting from base ten to another base b.

Example 1.2   Converting from Base Ten to an Arbitrary Base

Converting 8341_{10} to base seven is accomplished as follows:

$$
\begin{array}{ccccc}
1191 \searrow & 170 \searrow & 24 \searrow & 3 \searrow & 0 \\
7\,)\overline{8341} & 7\,)\overline{1191} & 7\,)\overline{170} & 7\,)\overline{24} & 7\,)\overline{3} \\
7000 & 700 & 140 & 21 & \\
\overline{1341} & \overline{491} & \overline{30} & \overline{3} & \\
700 & 490 & 28 & & \\
\overline{641} & \overline{1} & \overline{2} & & \\
630 & & & & \\
\overline{11} & & & & \\
7 & & & & \\
\overline{4} & & & &
\end{array}
$$

$$8341_{10} = 33214_7$$

Conversion between arbitrary bases

Although it is possible to perform the division and multiplication steps in any base, most people are much better at working in base ten. For that reason, the easiest way to convert from any base a to any other base b is to use a two step process. First step is to convert from base a to decimal. The second step is to convert from decimal to base b. Example 1.3 shows how to convert from any base to any other base.

Example 1.3   Converting from an Arbitrary Base to Another Arbitrary Base

Converting 84834_3 to base 11 is accomplished with two steps.
    The number is first converted to base ten as follows:

$$
\begin{aligned}
84835_3 &= 8 \times 3^4 + 4 \times 3^3 + 8 \times 3^2 + 3 \times 3^1 + 4 \times 3^0 \\
&= 8 \times 81 + 4 \times 27 + 8 \times 9 + 3 \times 3 + 4 \times 1 \\
&= 648 + 108 + 72 + 9 + 4 \\
&= 841_{10}
\end{aligned}
$$

Then the result is converted to base 11:

$$
\begin{array}{ccc}
76 \searrow & 6 \searrow & 0 \\
11\,)\overline{841} & 11\,)\overline{76} & 11\,)\overline{6} \\
770 & 66 & \\
\overline{71} & \overline{10} & \\
66 & & \\
\overline{5} & &
\end{array}
\qquad 84834_3 = 841_{10} = 6A5_{11}
$$

Bases that are powers-of-two

In addition to the methods above, there is a simple method for quickly converting between base two, base eight, and base sixteen. These shortcuts rely on the fact that 2, 8, and 16 are all powers of two. Because of this, it takes exactly four binary digits (bits) to represent exactly one hexadecimal digit. Likewise, it takes exactly three bits to represent an octal digit. Conversely, each hexadecimal digit can be converted to exactly four binary digits, and each octal digit can be converted to exactly three binary digits. This relationship makes it possible to do very fast conversions using the tables shown in Fig. 1.3.

When converting from hexadecimal to binary, all that is necessary is to replace each hex digit with the corresponding binary digits from the table. For example, to convert $5AC4_{16}$ to binary, we just replace "5" with "0101," replace "A" with "1010," replace "C" with "1100," and replace "4" with "0100." So, just by referring to the table, we can immediately see that $5AC4_{16} = 0101101011000100_2$. This method works exactly the same for converting from octal to binary, except that it uses the table on the right side of Fig. 1.3.

Base 2	Base 16
0000	0
0001	1
0010	2
0011	3
0100	4
0101	5
0110	6
0111	7
1000	8
1001	9
1010	A
1011	B
1100	C
1101	D
1110	E
1111	F

Base 2	Base 8
000	0
001	1
010	2
011	3
100	4
101	5
110	6
111	7

Figure 1.3
Tables used for converting between binary, octal, and hex.

Converting from binary to hexadecimal is also very easy using the table. Given a binary number, n, take the four least significant digits of n and find them in the table on the left side of Fig. 1.3. The hexadecimal digit on the matching line of the table is the least significant hex digit. Repeat the process with the next set of four bits and continue until there are no bits remaining in the binary number. For example, to convert 0011100101010111_2 to hexadecimal, just divide the number into groups of four bits, starting on the right, to get: $0011|1001|0101|0111_2$. Now replace each group of four bits by looking up the corresponding hex digit in the table on the left side of Fig. 1.3, to convert the binary number to 3957_{16}. In the case where the binary number does not have enough bits, simply pad with zeros in the high-order bits. For example, dividing the number 1001100010011_2 into groups of four yields $1|0011|0001|0011_2$ and padding with zeros in the high-order bits results in $0001|0011|0001|0011_2$. Looking up the four groups in the table reveals that $0001|0011|0001|0011_2 = 1313_{16}$.

1.3.3 Representing Integers

The computer stores groups of bits, but the bits by themselves have no meaning. The programmer gives them meaning by deciding what the bits represent, and how they are interpreted. Interpreting a group of bits as unsigned integer data is relatively simple. Each bit is weighted by a power-of-two, and the value of the group of bits is the sum of the non-zero bits multiplied by their respective weights. However, programmers often need to represent negative as well as non-negative numbers, and there are many possibilities for storing and interpreting integers whose value can be both positive and negative. Programmers and hardware designers have developed several standard schemes for encoding such numbers. The three main methods for storing and interpreting signed integer data are two's complement, sign-magnitude, and excess-N, Fig. 1.4 shows how the same binary pattern of bits can be interpreted as a number in four different ways.

Sign-magnitude representation
The sign-magnitude representation simply reserves the most significant bit to represent the sign of the number, and the remaining bits are used to store the magnitude of the number. This method has the advantage that it is easy for humans to interpret, with a little practice. However, addition and subtraction are slightly complicated. The addition/subtraction logic must compare the sign bits, complement one of the inputs if they are different, implement an end-around carry, and complement the result if there was no carry from the most significant bit. Complements are explained in Section 1.3.3. Because of the complexity, most integer CPUs do not directly support addition and subtraction of integers in sign-magnitude form. However, this method is commonly used for mantissa in floating-point numbers, as will be explained in Chapter 8. Another drawback to sign-magnitude is that it has two representations for zero, which can cause problems if the programmer is not careful.

Binary	Unsigned	Sign Magnitude	Excess-127	Two's Complement
00000000	0	0	-127	0
00000001	1	1	-126	1
⋮	⋮	⋮	⋮	⋮
01111110	126	126	-1	126
01111111	127	127	0	127
10000000	128	-0	1	-128
10000001	129	-1	2	-127
⋮	⋮	⋮	⋮	⋮
11111110	254	-126	127	-2
11111111	255	-127	128	-1

Figure 1.4
Four different representations for binary integers.

Excess-$(2^{n-1} - 1)$ representation
Another method for representing both positive and negative numbers is by using an *excess-N* representation. With this representation, the number that is stored is N greater than the actual value. This representation is relatively easy for humans to interpret. Addition and subtraction are easily performed using the complement method, which is explained in Section 1.3.3. This representation is just the same as unsigned math, with the addition of a *bias* which is usually $(2^{n-1} - 1)$. So, zero is represented as zero plus the bias. In $n = 12$ bits, the bias is $2^{12-1} - 1 = 2047_{10}$, or 011111111111_2. This method is commonly used to store the exponent in floating-point numbers, as will be explained in Chapter 8.

Complement representation
A very efficient method for dealing with signed numbers involves representing negative numbers as the *radix complements* of their positive counterparts. The complement is the amount that must be added to something to make it "whole." For instance, in geometry, two angles are complementary if they add to $90°$. In radix mathematics, the complement of a digit x in base b is simply $b - x$. For example, in base ten, the complement of 4 is $10 - 4 = 6$.

In complement representation, the most significant digit of a number is reserved to indicate whether or not the number is negative. If the first digit is less than $\frac{b}{2}$ (where b is the radix), then the number is positive. If the first digit is greater than or equal to $\frac{b}{2}$, then the number is negative. The first digit is not part of the magnitude of the number, but only indicates the sign of the number. For example, numbers in ten's complement notation are positive if the first digit is less than 5, and negative if the first digit is greater than 4. This works especially well in binary, since the number is considered positive if the first bit is zero and negative if the first bit is one. The magnitude of a negative number can be obtained by taking the radix complement. Because of the nice properties of the complement representation, it is the most common method for representing signed numbers in digital computers.

Finding the complement: The *radix complement* of an n digit number y in radix (base) b is defined as

$$C(y_b) = b^n - y_b. \tag{1.1}$$

For example, the ten's complement of the four digit number 8734_{10} is $10^4 - 8734 = 1266$. In this example, we directly applied the definition of the radix complement from Eq. (1.1). That is easy in base ten, but not so easy in an arbitrary base, because it involves performing a subtraction. However, there is a very simple method for calculating the complement which does not require subtraction. This method involves finding the *diminished radix complement*, which is $(b^n - 1) - y$ by substituting each digit with its complement from a *complement table*. The radix complement is found by adding one to the diminished radix complement. Fig. 1.5 shows the complement tables for bases ten and two. Examples 1.4 and 1.5 show how the complement is obtained in bases ten and two respectively. Examples 1.6 and 1.7 show additional conversions between binary and decimal.

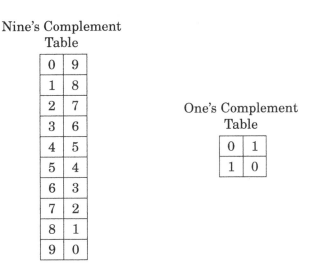

Nine's Complement
Table

0	9
1	8
2	7
3	6
4	5
5	4
6	3
7	2
8	1
9	0

One's Complement
Table

0	1
1	0

Figure 1.5
Complement tables for bases ten and two.

Example 1.4   The Complement in Base Ten

The nine's complement of the base ten number 593 is found by finding the digit '5' in the complement table, and replacing it with its complement, which is the digit '4.' The digit '9' is replaced with '0,' and '3' is replaced with '6.' Therefore the nine's complement of 593_{10} is 406. Likewise, the nine's complement of 1000_{10} is 8999_{10} and the nine's complement of 0999_{10} is 9000_{10}.

The ten's complement of 7266_{10} is $273_{10} + 1 = 274_{10}$.

Example 1.5   The One's and Two's Complement

The one's complement of a binary number is found in the same way as the nine's complement of a decimal number, but using the one's complement table instead of the nine's complement table. The one's complement of 01001101_2 is 10110010_2 and the one's complement of 0000000010110110 is 1111111101001001_2. Note that the one's complement of a base two number is equivalent to the bitwise logical "not" (Boolean complement) operator. This operator is very easy to implement in digital hardware.

The two's complement is the one's complement plus one. The two's complement of 1010100_2 is $0101011_2 + 1 = 0101100_2$.

Example 1.6   Conversion from Binary to Decimal

Suppose we want to convert a signed binary number to decimal.

1. If the most significant bit is '1', then
   a. Find the two's complement
   b. Convert the result to base 10
   c. Add a negative sign
2. else
   a. Convert the result to base 10

Number	One's Complement	Two's Complement	Base 10	Negative
11010010	00101101	00101110	46	−46
1111111100010110	0000000011101001	0000000011101010	234	−234
01110100	Not negative		116	
1000001101010110	0111110010101001	0111110010101010	31914	−31914
0101001111011011	Not negative		21467	

Example 1.7   Conversion from Decimal to Binary

Suppose we want to convert a negative number from decimal to binary.

1. Remove the negative sign
2. Convert the number to binary
3. Take the two's complement

Base 10	Positive Binary	One's Complement	Two's Complement
-46	00101110	11010001	11010010
-234	0000000011101010	1111111100010101	1111111100010110
-116	01110100	10001011	10001100
-31914	0111110010101010	1000001101010110	1000001101010111
-21467	0101001111011011	1010110000100100	1010110000100101

Subtraction using complements   One very useful feature of complement notation is that it can be used to perform subtraction by using addition. Given two numbers in base b, x_b, and y_b, the difference can be computed as:

$$z_b = x_b - y_b \tag{1.2}$$
$$= x_b + (b^n - y_b) - b^n \tag{1.3}$$
$$= x_b + C(y_b) - b^n, \tag{1.4}$$

where $C(y_b)$ is the radix complement of y_b. Assume that x_b and y_b are both positive where $y_b \leq x_b$ and both numbers have the same number of digits n (y_b may have leading zeros). In this case, the result of $x_b + C(y_b)$ will always be greater than or equal to b^n, but less than $2 \times b^n$. This means that the result of $x_b + C(y_b)$ will always begin with a '1' in the $n + 1$ digit position. Dropping the initial '1' is equivalent to subtracting b^n, making the result $x - y + b^n - b^n$ or just $x - y$, which is the desired result. This can be reduced to a simple procedure. When y and x are both positive and $y \leq x$, the following four steps are to be performed:

1. pad the subtrahend (y) with leading zeros, as necessary, so that both numbers have the same number of digits (n),
2. find the b's complement of the subtrahend,
3. add the complement to the minuend,
4. discard the leading '1'.

The complement notation provides a very easy way to represent both positive and negative integers using a fixed number of digits, and to perform subtraction by using addition. Since modern computers typically use a fixed number of bits, complement notation provides a very convenient and efficient way to store signed integers and perform mathematical operations on them. Hardware is simplified because there is no need to build a specialized subtractor circuit. Instead, a very simple complement circuit is built and the adder is reused to perform subtraction as well as addition.

1.3.4 Representing Characters

In the previous section, we discussed how the computer stores information as groups of bits, and how we can interpret those bits as numbers in base two. Given that the computer can only store information using groups of bits, how can we store textual information? The answer is that we create a table, which assigns a numerical value to each character in our language.

Early in the development of computers, several computer manufacturers developed such tables, or character coding schemes. These schemes were incompatible and computers from different manufacturers could not easily exchange textual data without the use of translation software to convert the character codes from one coding scheme to another.

Eventually, a standard coding scheme, known as the American Standard Code for Information Interchange (ASCII) was developed. Work on the ASCII standard began on October 6, 1960, with the first meeting of the American Standards Association's (ASA) X3.2 subcommittee. The first edition of the standard was published in 1963. The standard was updated in 1967 and again in 1986, and has been stable since then. Within a few years of its development, ASCII was accepted by all major computer manufacturers, although some continue to support their own coding schemes as well.

ASCII was designed for American English, and does not support some of the characters that are used by non-English languages. For this reason, ASCII has been extended to create more comprehensive coding schemes. Most modern multilingual coding schemes are based on ASCII, though they support a wider range of characters.

At the time that it was developed, transmission of digital data over long distances was very slow, and usually involved converting each bit into an audio signal which was transmitted over a telephone line using an acoustic modem. In order to maximize performance, the standards committee chose to define ASCII as a 7-bit code. Because of this decision, all textual data could be sent using seven bits rather than eight, resulting in approximately 10% better overall performance when transmitting data over a telephone modem. A possibly unforeseen benefit was that this also provided a way for the code to be extended in the future. Since there are 128 possible values for a 7-bit number, the ASCII standard provides 128 characters. However, 33 of the ASCII characters are *non-printing control characters*. These characters, shown in Table 1.3, are mainly used to send information about how the text is to be displayed and/or printed. The remaining 95 *printable characters* are shown in Table 1.4.

Non-printing characters

The non-printing characters are used to provide hints or commands to the device that is receiving, displaying, or printing the data. The FF character, when sent to a printer, will cause the printer to eject the current page and begin a new one. The LF character causes the printer or terminal to end the current line and begin a new one. The CR character causes the terminal

Table 1.3 The ASCII control characters

Binary	Oct	Dec	Hex	Abbr	Glyph	Name
000 0000	000	0	00	NUL	^@	Null character
000 0001	001	1	01	SOH	^A	Start of header
000 0010	002	2	02	STX	^B	Start of text
000 0011	003	3	03	ETX	^C	End of text
000 0100	004	4	04	EOT	^D	End of transmission
000 0101	005	5	05	ENQ	^E	Enquiry
000 0110	006	6	06	ACK	^F	Acknowledgment
000 0111	007	7	07	BEL	^G	Bell
000 1000	010	8	08	BS	^H	Backspace
000 1001	011	9	09	HT	^I	Horizontal tab
000 1010	012	10	0A	LF	^J	Line feed
000 1011	013	11	0B	VT	^K	Vertical tab
000 1100	014	12	0C	FF	^L	Form feed
000 1101	015	13	0D	CR	^M	Carriage return[g]
000 1110	016	14	0E	SO	^N	Shift out
000 1111	017	15	0F	SI	^O	Shift in
001 0000	020	16	10	DLE	^P	Data link escape
001 0001	021	17	11	DC1	^Q	Device control 1 (oft. XON)
001 0010	022	18	12	DC2	^R	Device control 2
001 0011	023	19	13	DC3	^S	Device control 3 (oft. XOFF)
001 0100	024	20	14	DC4	^T	Device control 4
001 0101	025	21	15	NAK	^U	Negative acknowledgement
001 0110	026	22	16	SYN	^V	Synchronous idle
001 0111	027	23	17	ETB	^W	End of transmission Block
001 1000	030	24	18	CAN	^X	Cancel
001 1001	031	25	19	EM	^Y	End of medium
001 1010	032	26	1A	SUB	^Z	Substitute
001 1011	033	27	1B	ESC	^[	Escape
001 1100	034	28	1C	FS	^\	File separator
001 1101	035	29	1D	GS	^]	Group separator
001 1110	036	30	1E	RS	^^	Record separator
001 1111	037	31	1F	US	^_	Unit separator
111 1111	177	127	7F	DEL	^?	Delete

or printer to move to the beginning of the current line. Many text editing programs allow the user to enter these non-printing characters by using the control key on the keyboard. For instance, to enter the BEL character, the user would hold the control key down and press the G key. This character, when sent to a character display terminal, will cause it to emit a beep. Many of the other control characters can be used to control specific features of the printer, display, or other device that the data is being sent to.

Converting character strings to ASCII codes

Suppose we wish to covert a string of characters, such as "Hello World" to an ASCII representation. We can use an 8-bit byte to store each character. Also, it is common practice to include an additional byte at the end of the string. This additional byte holds the ASCII NUL

Table 1.4 The ASCII printable characters

Binary	Oct	Dec	Hex	Glyph	Binary	Oct	Dec	Hex	Glyph
010 0000	040	32	20	␣	101 0000	120	80	50	P
010 0001	041	33	21	!	101 0001	121	81	51	Q
010 0010	042	34	22	"	101 0010	122	82	52	R
010 0011	043	35	23	#	101 0011	123	83	53	S
010 0100	044	36	24	$	101 0100	124	84	54	T
010 0101	045	37	25	%	101 0101	125	85	55	U
010 0110	046	38	26	&	101 0110	126	86	56	V
010 0111	047	39	27	'	101 0111	127	87	57	W
010 1000	050	40	28	(	101 1000	130	88	58	X
010 1001	051	41	29	)	101 1001	131	89	59	Y
010 1010	052	42	2A	*	101 1010	132	90	5A	Z
010 1011	053	43	2B	+	101 1011	133	91	5B	[
010 1100	054	44	2C	,	101 1100	134	92	5C	\
010 1101	055	45	2D	−	101 1101	135	93	5D	]
010 1110	056	46	2E	.	101 1110	136	94	5E	^
010 1111	057	47	2F	/	101 1111	137	95	5F	_
011 0000	060	48	30	0	110 0000	140	96	60	`
011 0001	061	49	31	1	110 0001	141	97	61	a
011 0010	062	50	32	2	110 0010	142	98	62	b
011 0011	063	51	33	3	110 0011	143	99	63	c
011 0100	064	52	34	4	110 0100	144	100	64	d
011 0101	065	53	35	5	110 0101	145	101	65	e
011 0110	066	54	36	6	110 0110	146	102	66	f
011 0111	067	55	37	7	110 0111	147	103	67	g
011 1000	070	56	38	8	110 1000	150	104	68	h
011 1001	071	57	39	9	110 1001	151	105	69	i
011 1010	072	58	3A	:	110 1010	152	106	6A	j
011 1011	073	59	3B	;	110 1011	153	107	6B	k
011 1100	074	60	3C	<	110 1100	154	108	6C	l
011 1101	075	61	3D	=	110 1101	155	109	6D	m
011 1110	076	62	3E	>	110 1110	156	110	6E	n
011 1111	077	63	3F	?	110 1111	157	111	6F	o
100 0000	100	64	40	@	111 0000	160	112	70	p
100 0001	101	65	41	A	111 0001	161	113	71	q
100 0010	102	66	42	B	111 0010	162	114	72	r
100 0011	103	67	43	C	111 0011	163	115	73	s
100 0100	104	68	44	D	111 0100	164	116	74	t
100 0101	105	69	45	E	111 0101	165	117	75	u
100 0110	106	70	46	F	111 0110	166	118	76	v
100 0111	107	71	47	G	111 0111	167	119	77	w
100 1000	110	72	48	H	111 1000	170	120	78	x
100 1001	111	73	49	I	111 1001	171	121	79	y
100 1010	112	74	4A	J	111 1010	172	122	7A	z
100 1011	113	75	4B	K	111 1011	173	123	7B	{
100 1100	114	76	4C	L	111 1100	174	124	7C	|
100 1101	115	77	4D	M	111 1101	175	125	7D	}
100 1110	116	78	4E	N	111 1110	176	126	7E	~
100 1111	117	79	4F	O					

Table 1.5 Binary equivalents for each character in "Hello World"

Character	Binary
H	01001000
e	01100101
l	01101100
l	01101100
o	01101111
	00100000
W	01010111
o	01101111
r	01110010
l	01101100
d	01100100
NUL	00000000

character, which indicates the end of the string. Such an arrangement is referred to as a null-terminated string.

To convert the string "Hello World" into a null-terminated string, we can build a table with each character on the left and its equivalent binary, octal, hexadecimal, or decimal value (as defined in the ASCII table) on the right. Table 1.5 shows the characters in "Hello World" and their equivalent binary representations, found by looking in Table 1.4. Since most modern computers use 8-bit bytes (or multiples thereof) as the basic storage unit, an extra zero bit is shown in the most significant bit position.

Reading the Binary column from top to bottom results in the following sequence of bytes: 01001000 01100101 01101100 01101100 01101111 00100000 01010111 01101111 01110010 01101100 01100100 0000000. To convert the same string to a hexadecimal representation, we can use the shortcut method that was introduced previously to convert each 4-bit nibble into its hexadecimal equivalent, or read the hexadecimal value from the ASCII table. Table 1.6 shows the result of extending Table 1.5 to include hexadecimal and decimal equivalents for each character. The string can now be converted to hexadecimal or decimal simply by reading the correct column in the table. So "Hello World" expressed as a null-terminated string in hexadecimal is "48 65 6C 6C 6F 20 57 6F 62 6C 64 00" and in decimal it is "72 101 108 108 111 32 87 111 98 108 100 0".

Interpreting data as ASCII strings
It is sometimes necessary to convert a string of bytes in hexadecimal into ASCII characters. This is accomplished simply by building a table with the hexadecimal value of each byte in the left column, then looking in the ASCII table for each value and entering the equivalent

**Table 1.6 Binary, hexadecimal, and decimal
equivalents for each character in "Hello World"**

Character	Binary	Hexadecimal	Decimal
H	01001000	48	72
e	01100101	65	101
l	01101100	6C	108
l	01101100	6C	108
o	01101111	6F	111
	00100000	20	32
W	01010111	57	87
o	01101111	6F	111
r	01110010	62	98
l	01101100	6C	108
d	01100100	64	100
NUL	00000000	00	0

**Table 1.7 Interpreting a
hexadecimal string as ASCII**

Hexadecimal	ASCII
46	F
61	a
62	b
75	u
6C	l
6F	o
75	u
73	s
21	!
00	NUL

character representation in the right column. Table 1.7 shows how the ASCII table is used to interpret the hexadecimal string "466162756C6F75732100" as an ASCII string.

ISO extensions to ASCII

ASCII was developed to encode all of the most commonly used characters in North American English text. The encoding uses only 128 of the 256 codes that are available in a 8-bit byte. ASCII does not include symbols frequently used in other countries, such as the British pound symbol (£) or accented characters (ü). However, the International Standards Organization

(ISO) has created several extensions to ASCII to enable the representation of characters from a wider variety of languages.

The ISO has defined a set of related standards known collectively as ISO 8859. ISO 8859 is an 8-bit extension to ASCII which includes the 128 ASCII characters along with an additional 128 characters, such as the British Pound symbol and the American cent symbol. Several variations of the ISO 8859 standard exist for different language families. Table 1.8 provides a brief description of the various ISO standards.

Unicode and UTF-8

Although the ISO extensions helped to standardize text encodings for several languages that were not covered by ASCII, there were still some issues. The first issue is that the display and input devices must be configured for the correct encoding, and displaying or printing documents with multiple encodings requires some mechanism for changing the encoding on-the-fly. Another issue has to do with the lexicographical ordering of characters. Although two languages may share a character, that character may appear in a different place in the alphabets of the two languages. This leads to issues when programmers need to sort strings into lexicographical order. The ISO extensions help to unify character encodings across

Table 1.8 Variations of the ISO 8859 standard

Name	Alias	Languages
ISO8859-1	Latin-1	Western European languages
ISO8859-2	Latin-2	Non-Cyrillic Central and Eastern European languages
ISO8859-3	Latin-3	Southern European languages and Esperanto
ISO8859-4	Latin-4	Northern European and Baltic languages
ISO8859-5	Latin/Cyrillic	Slavic languages that use a Cyrillic alphabet
ISO8859-6	Latin/Arabic	Common Arabic language characters
ISO8859-7	Latin/Greek	Modern Greek language
ISO8859-8	Latin/Hebrew	Modern Hebrew languages
ISO8859-9	Latin-5	Turkish
ISO8859-10	Latin-6	Nordic languages
ISO8859-11	Latin/Thai	Thai language
ISO8859-12	Latin/Devanagari	Never completed. Abandoned in 1997
ISO8859-13	Latin-7	Some Baltic languages not covered by Latin-4 or Latin-6
ISO8859-14	Latin-8	Celtic languages
ISO8859-15	Latin-9	Update to Latin-1 that replaces some characters. Most notably, it includes the euro symbol (€), which did not exist when Latin-1 was created
ISO8859-16	Latin-10	Covers several languages not covered by Latin-9 and includes the euro symbol (€)

multiple languages, but do not solve all of the issues involved in defining a universal character set.

In the late 1980s, there was growing interest in developing a universal character encoding for all languages. People from several computer companies worked together and, by 1990, had developed a draft standard for Unicode. In 1991, the Unicode Consortium was formed and charged with guiding and controlling the development of Unicode. The Unicode Consortium has worked closely with the ISO to define, extend, and maintain the international standard for a Universal Character Set (UCS). This standard is known as the ISO/IEC 10646 standard. The ISO/IEC 10646 standard defines the mapping of code points (numbers) to glyphs (characters). but does not specify character collation or other language-dependent properties. UCS code points are commonly written in the form U+XXXX, where XXXX in the numerical code point in hexadecimal. For example, the code point for the ASCII DEL character would be written as U+007F. Unicode extends the ISO/IEC standard and specifies language-specific features.

Originally, Unicode was designed as a 16-bit encoding. It was not fully backward-compatible with ASCII, and could encode only 65,536 code points. Eventually, the Unicode character set grew to encompass 1,112,064 code points, which requires 21 bits per character for a straightforward binary encoding. By early 1992, it was clear that some clever and efficient method for encoding character data was needed.

UTF-8 (UCS Transformation Format-8-bit) was proposed and accepted as a standard in 1993. UTF-8 is a variable-width encoding that can represent every character in the Unicode character set using between one and four bytes. It was designed to be backward compatible with ASCII and to avoid the major issues of previous encodings. Code points in the Unicode character set with lower numerical values tend to occur more frequently than code points with higher numerical values. UTF-8 encodes frequently occurring code points with fewer bytes than those which occur less frequently. For example, the first 128 characters of the UTF-8 encoding are exactly the same as the ASCII characters, requiring only 7 bits to encode each ASCII character. Thus any valid ASCII text is also valid UTF-8 text. UTF-8 is now the most common character encoding for the World Wide Web, and is the recommended encoding for email messages.

In November 2003, UTF-8 was restricted by RFC 3629 to end at code point $10FFFF_{16}$. This allows UTF-8 to encode 1,114,111 code points, which is slightly more than the 1,112,064 code points defined in the ISO/IEC 10646 standard. Table 1.9 shows how ISO/IEC 10646 code points are mapped to a variable-length encoding in UTF-8. Note that the encoding allows each byte in a stream of bytes to be placed in one of the following three distinct categories:

1. If the most significant bit of a byte is zero, then it is a single-byte character, and is completely ASCII-compatible.

Table 1.9 UTF-8 encoding of the ISO/IEC 10646 code points

UCS Bits	First Code Point	Last Code Point	Bytes	Byte 1	Byte 2	Byte 3	Byte 4
7	U+0000	U+007F	1	0xxxxxxx			
11	U+0080	U+07FF	2	110xxxxx	10xxxxxx		
16	U+0800	U+FFFF	3	1110xxxx	10xxxxxx	10xxxxxx	
21	U+10000	U+10FFFF	4	11110xxx	10xxxxxx	10xxxxxx	10xxxxxx

2. If the two most significant bits in a byte are set to one, then the byte is the beginning of a multi-byte character.

3. If the most significant bit is set to one, and the second most significant bit is set to zero, then the byte is part of a multi-byte character, but is not the first byte in that sequence.

The UTF-8 encoding of the UCS characters has several important features:

Backwards compatible with ASCII: This allows the vast number of existing ASCII documents to be interpreted as UTF-8 documents without any conversion.

Self-synchronization: Because of the way code points are assigned, it is possible to find the beginning of each character by looking only at the top two bits of each byte. This can have important performance implications when performing searches in text.

Encoding of code sequence length: The number of bytes in the sequence is indicated by the pattern of bits in the first byte of the sequence. Thus, the beginning of the next character can be found quickly. This feature can also have important performance implications when performing searches in text.

Efficient code structure: UTF-8 efficiently encodes the UCS code points. The high-order bits of the code point go in the lead byte. Lower-order bits are placed in continuation bytes. The number of bytes in the encoding is the minimum required to hold all the significant bits of the code point.

Easily extended to include new languages: This feature will be greatly appreciated when we contact intelligent species from other star systems.

With UTF-8 encoding, the first 128 characters of the UCS are each encoded in a single byte. The next 1,920 characters require two bytes to encode. The two-byte encoding covers almost all Latin alphabets, and also Arabic, Armenian, Cyrillic, Coptic, Greek, Hebrew, Syriac and Tāna alphabets. It also includes combining diacritical marks, which are used in combination with another character, such as á, ñ, and ö. Most of the Chinese, Japanese, and Korean (CJK) characters are encoded using three bytes. Four bytes are needed for the less common CJK characters, various historic scripts, mathematical symbols, and emoji (pictographic symbols).

Consider the UTF-8 encoding for the British Pound symbol (£), which is UCS code point U+00A3. Since the code point is greater than $7F_{16}$, but less than 800_{16}, it will require two bytes to encode. The encoding will be 110xxxxx 10xxxxxx, where the x characters are replaced with the 11 least-significant bits of the code point, which are 00010100011. Thus, the character £ is encoded in UTF-8 as 11000010 10100011 in binary, or C2 A3 in hexadecimal.

The UCS code point for the Euro symbol (€) is U+20AC. Since the code point is between 800_{16} and $FFFF_{16}$, it will require three bytes to encode in UTF-8. The three-byte encoding is 1110xxxx 10xxxxxx 10xxxxxx where the x characters are replaced with the 16 least-significant bits of the code point. In this case the code point, in binary is 0010000010101100. Therefore, the UTF-8 encoding for € is 11100010 10000010 10101100 in binary, or E2 82 AC in hexadecimal.

In summary, there are three components to modern language support. The ISO/IEC 10646 defines a mapping from code points (numbers) to glyphs (characters). UTF-8 defines an efficient variable-length encoding for code points (text data) in the ISO/IEC 10646 standard. Unicode adds language specific properties to the ISO/IEC 10646 character set. Together, these three elements currently provide support for textual data in almost every human written language, and they continue to be extended and refined.

1.4 Memory Layout of an Executing Program

Computer memory consists of number of storage locations, or cells, each of which has a unique numeric *address*. Addresses are usually written in hexadecimal. Each storage location can contain a fixed number of binary digits. The most common size is one byte. Most computers group bytes together into words. A computer CPU that is capable of accessing a single byte of memory is said to have *byte addressable* memory. Some CPUs are only capable of accessing memory in word-sized groups. They are said to have *word addressable* memory.

Fig. 1.6A shows a section of memory containing some data. Each byte has a unique address that is used when data is transferred to or from that memory cell. Most processors can also move data in word-sized chunks. On a 32-bit system, four bytes are grouped together to form a word. There are two ways that this grouping can be done. Systems that store the most significant byte of a word in the smallest address, and the least significant byte in the largest address, are said to be *big-endian*. The big-endian interpretation of a region of memory is shown in Fig. 1.6B. As shown in Fig. 1.6C, *little-endian* systems store the least significant byte in the lowest address and the most significant byte in the highest address. Some processors, such as the ARM, can be configured as either little-endian or big-endian. The Linux operating system, by default, configures the ARM processor to run in little-endian mode.

The memory layout for a typical program is shown in Fig. 1.7. The program is divided into four major memory regions, or *sections*. The programmer specifies the contents of the Text

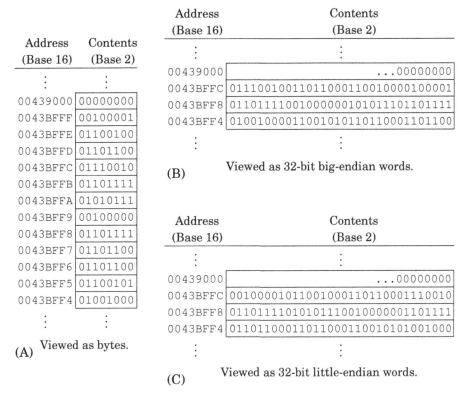

Address (Base 16)	Contents (Base 2)
⋮	⋮
00439000	00000000
0043BFFF	00100001
0043BFFE	01100100
0043BFFD	01101100
0043BFFC	01110010
0043BFFB	01101111
0043BFFA	01010111
0043BFF9	00100000
0043BFF8	01101111
0043BFF7	01101100
0043BFF6	01101100
0043BFF5	01100101
0043BFF4	01001000
⋮	⋮

(A) Viewed as bytes.

Address (Base 16)	Contents (Base 2)
⋮	⋮
00439000	...00000000
0043BFFC	01110010011011000110010000100001
0043BFF8	01101111001000000101011101101111
0043BFF4	01001000011001010110110001101100
⋮	⋮

(B) Viewed as 32-bit big-endian words.

Address (Base 16)	Contents (Base 2)
⋮	⋮
00439000	...00000000
0043BFFC	00100001011001000110110001110010
0043BFF8	01101111010101110010000001101111
0043BFF4	01101100011011000110010101001000
⋮	⋮

(C) Viewed as 32-bit little-endian words.

Figure 1.6
A section of memory.

and Data sections. The Stack and Heap segments are defined when the program is loaded for execution. The Stack and Heap may grow and shrink as the program executes, while the Text and Data segments are set to fixed sizes by the compiler, linker, and loader. The Text section contains the executable instructions, while the Data section contains constants and statically allocated variables. The sizes of the Text and Data segments depend on how large the program is, and how much static data storage has been declared by the programmer. The heap contains variables that are allocated dynamically, and the stack is used to store parameters for function calls, return addresses, and local (automatic) variables.

In a high-level language, storage space for a variable can be allocated in one of three ways: statically, dynamically, and automatically. Statically allocated variables are allocated from the .data section. The storage space is reserved, and usually initialized, when the program is loaded and begins execution. The address of a statically allocated variable is fixed at the time

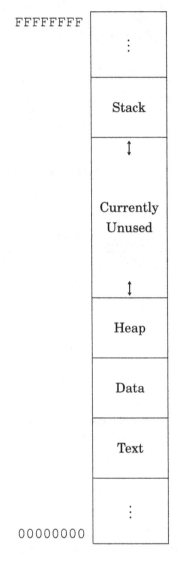

Figure 1.7
Typical memory layout for a program with a 32-bit address space.

the program begins running, and cannot be changed. Automatically allocated variables, often referred to as local variables, are stored on the stack. The stack pointer is adjusted down to make space for the newly allocated variable. The address of an automatic variable is always computed as an offset from the stack pointer. Dynamic variables are allocated from the heap, using malloc, new, or a language-dependent equivalent. The address of a dynamic variable is always stored in another variable, known as a pointer, which may be an automatic or static variable, or even another dynamic variable. The four major sections of program memory correspond to executable code, statically allocated variables, dynamically allocated variables, and automatically allocated variables.

1.5 Chapter Summary

There are several reasons for Computer Scientists and Computer Engineers to learn at least one assembly language. There are programming tasks that can only be performed using assembly language, and some tasks can be written to run much more efficiently and/or quickly if written in assembly language. Programmers with assembly language experience tend to write better code even when using a high-level language, and are usually better at finding and fixing bugs.

Although it is possible to construct a computer capable of performing arithmetic in any base, it is much cheaper to build one that works in base two. It is relatively easy to build an electrical circuit with two states, using two discrete voltage levels, but much more difficult to build a stable circuit with 10 discrete voltage levels. Therefore, modern computers work in base two.

Computer data can be viewed as simple bit strings. The programmer is responsible for supplying interpretations to give meaning to those bit strings. A set of bits can be interpreted as a number, a character, or anything that the programmer chooses. There are standard methods for encoding and interpreting characters and numbers. Fig. 1.4 shows some common methods for encoding integers. The most common encodings for characters are UTF-8 and ASCII.

Computer memory can be viewed as a sequence of bytes. Each byte has a unique address. A running program has four regions of memory. One region holds the executable code. The other three regions hold different types of variables.

Exercises

1.1 What is the two's complement of 11011101?

1.2 Perform the base conversions to fill in the blank spaces in the following table:

Base 10	Base 2	Base 16	Base 21
23			
	010011		
		ABB	
			2HE

1.3 What is the 8-bit ASCII binary representation for the following characters?

    (a) "A"

    (b) "a"

    (c) "!"

1.4   What is \ minus ! given that \ and ! are ASCII characters? Give your answer in binary.

1.5   Representing characters:

   (a)   Convert the string "Super!" to its ASCII representation. Show your result as a sequence of hexadecimal values.

   (b)   Convert the hexadecimal sequence into a sequence of values in base four.

1.6   Suppose that the string "This is a nice day" is stored beginning at address $4B3269AC_{16}$. What are the contents of the byte at address $4B3269B1_{16}$ in hexadecimal?

1.7   Perform the following:

   (a)   Convert 101101_2 to base ten.

   (b)   Convert 1023_{10} to base nine.

   (c)   Convert 1023_{10} to base two.

   (d)   Convert 301_{10} to base 16.

   (e)   Convert 301_{10} to base 2.

   (f)   Represent 301_{10} as a null-terminated ASCII string (write your answer in hexadecimal).

   (g)   Convert 3420_5 to base ten.

   (h)   Convert 2314_5 to base nine.

   (i)   Convert 116_7 to base three.

   (j)   Convert 1294_{11} to base 5.

1.8   Given the following binary string:

   01001001 01110011 01101110 00100111 01110100 00100000 01000001 01110011
   01110011 01100101 01101101 01100010 01101100 01111001 00100000 01000110
   01110101 01101110 00111111 00000000

   (a)   Convert it to a hexadecimal string.

   (b)   Convert the first four bytes to a string of base ten numbers.

   (c)   Convert the first (little-endian) halfword to base ten.

   (d)   Convert the first (big-endian) halfword to base ten.

   (e)   If this string of bytes were sent to an ASCII printer or terminal, what would be printed?

1.9   The number 1,234,567 is stored as a 32-bit word starting at address $F0439000_{16}$. Show the address and contents of each byte of the 32-bit word on a

   (a)   little-endian system,

   (b)   big-endian system.

1.10   The ISO/IEC 10646 standard defines 1,112,064 code points (glyphs). Each code point could be encoded using 24 bits, or three bytes. The UTF-8 encoding uses up to four bytes to encode a code point. Give three reasons why UTF-8 is preferred over a simple 3-byte per code point encoding.

1.11   UTF-8 is often referred to as Unicode. Why is this not correct?

1.12 Skilled assembly programmers can convert small numbers between binary, hexadecimal, and decimal in their heads. Without referring to any tables or using a calculator or pencil, fill in the blanks in the following table:

Binary	Decimal	Hexadecimal
	5	
1010		
		C
	23	
0101101		
		4B

1.13 What are the differences between a CPU register and a memory location?

GNU Assembly Syntax

All modern computers consist of three main components: the central processing unit (CPU), memory, and devices. It can be argued that the major factor that distinguishes one computer from another is the CPU architecture. The architecture determines the set of instructions that can be performed by the CPU. The human-readable language which is closest to the CPU architecture is assembly language.

When a new processor architecture is developed, its creators also define an assembly language for the new architecture. In most cases, a precise assembly language syntax is defined and an assembler is created by the processor developers. Because of this, there is no single syntax for assembly language, although most assembly languages are similar in many ways and have certain elements in common.

The GNU assembler (GAS) is a highly portable re-configurable assembler. GAS uses a simple, general syntax that works for a wide variety of architectures. Although the syntax used by GAS for the ARM processor is slightly different from the syntax defined by the developers of the ARM processor, it provides the same capabilities.

Modern Assembly Language Programming with the ARM Processor. http://dx.doi.org/10.1016/B978-0-12-803698-3.00002-4

2.1 Structure of an Assembly Program

An assembly program consists of four basic elements: assembler directives, labels, assembly instructions, and comments. Assembler directives allow the programmer to reserve memory for the storage of variables, control which program section is being used, define macros, include other files, and perform other operations that control the conversion of assembly instructions into machine code. The assembly instructions are given as *mnemonics,* or short character strings that are easier for human brains to remember than sequences of binary, octal, or hexadecimal digits. Each assembly instruction may have an optional label, and most assembly instructions require the programmer to specify one or more operands.

Most assembly language programs are written in lines of 80 characters organized into four columns. The first column is for optional labels. The second column is for assembly instructions or assembler directives. The third column is for specifying operands, and the fourth column is for comments. Traditionally, the first two columns are 8 characters wide, the third column is 16 characters wide, and the last column is 48 characters wide. However, most modern assemblers (including GAS) do not require a fixed column widths. Listing 2.1 shows a basic "Hello World" program written in GNU ARM Assembly to run under Linux. For comparison, Listing 2.2 shows an equivalent program written in C. The assembly language version of the program is significantly longer than the C version, and will only work on an ARM processor. The C version is at a higher level of abstraction, and can be compiled to run on any system that has a C compiler. Thus, C is referred to as a high-level language, and assembly is a low-level language.

```
 1        .data
 2  str:  .asciz  "Hello World\n" @ Define a null-terminated string
 3
 4        .text
 5        .globl  main
 6        /* This is the beginning of the main() function.
 7           It will print "Hello World" and then return.
 8        */
 9  main: stmfd   sp!,{lr}       @ push return address onto stack
10        ldr     r0, =str       @ load pointer to format string
11        bl      printf         @ printf("Hello World\n");
12        mov     r0, #0         @ move return code into r0
13        ldmfd   sp!,{lr}       @ pop return address from stack
14        mov     pc, lr         @ return from main
```

Listing 2.1
"Hello World" program in ARM assembly.

```
1   #include <stdio.h>
2   static char str[] = "Hello World\n";
3   int main()
4   {
5     printf(str);
6     return 0;
7   }
```

Listing 2.2
"Hello World" program in C.

2.1.1 Labels

Most modern assemblers are called two-pass assemblers because they read the input file twice. On the first pass, the assembler keeps track of the location of each piece of data and each instruction, and assigns an address or numerical value to each label and symbol in the input file. The main goal of the first pass is to build a symbol table, which maps each label or symbol to a numerical value.

On the second pass, the assembler converts the assembly instructions and data declarations into binary, using the symbol table to supply numerical values whenever they are needed. In Listing 2.1, there are two labels: main and str. During assembly, those labels are assigned the value of the *address counter* at the point where they appear. Labels can be used anywhere in the program to refer to the address of data, functions, or blocks of code. In GNU assembly syntax, labels always end with a colon (:) character.

2.1.2 Comments

There are two basic comment styles: multi-line and single-line. Multi-line comments start with /* and everything is ignored until a matching sequence of */ is found. These comments are exactly the same as multi-line comments in C and C++. In ARM assembly, single line comments begin with an @ character, and all remaining characters on the line are ignored. Listing 2.1 shows both types of comment. In addition, if the name of the file ends in .S, then single line comments can begin with //. If the file name does not end with a *capital* .S, then the // syntax is not allowed.

2.1.3 Directives

Directives are used mainly to define symbols, allocate storage, and control the behavior of the assembler, allowing the programmer to control how the assembler does its job. The GNU assembler has many directives, but assembly programmers typically need to know only a few of them. All assembler directives begin with a period "." which is followed by a sequence of letters, usually in lower case. Listing 2.1 uses the .data, .asciz, .text, and .globl directives. The most commonly used directives are discussed later in this chapter. There are many other directives available in the GNU Assembler which are not covered here. Complete documentation is available online as part of the GNU Binutils package.

2.1.4 Assembly Instructions

Assembly instructions are the program statements that will be executed on the CPU. Most instructions cause the CPU to perform one low-level operation, In most assembly languages, operations can be divided into a few major types. Some instructions move data from one location to another. Others perform addition, subtraction, and other computational operations. Another class of instructions is used to perform comparisons and control which part of the program is to be executed next. Chapters 3 and 4 explain most of the assembly instructions that are available on the ARM processor.

2.2 What the Assembler Does

Listing 2.3 shows how the GNU assembler will assemble the "Hello World" program from Listing 2.1. The assembler converts the string on input line 2 into the binary representation of the string. The results are shown in hexadecimal in the Code column of the listing. The first byte of the string is stored at address zero in the .data section of the program, as shown by the 0000 in the Addr column on line 2.

On line 4, the assembler switches to the .text section of the program and begins converting instructions into binary. The first instruction, on line 9, is converted into its 4-byte machine code, $00402DE9_{16}$, and stored at location 0000 in the .text section of the program, as shown in the Code and Addr columns on line 6.

Next, the assembler converts the ldr instruction on line 10 into the four-byte machine instruction $0C009FE5_{16}$ and stores it at address 0004. It repeats this process with each remaining instruction until the end of the program. The assembler writes the resulting data into a specially formatted file, called an object file. Note that the assembler was unable to locate the printf function. The linker will take care of that. The object file created by the assembler, hello.o, contains the data in the Code column of Listing 2.3, along with information to help the linker to link (or "patch") the instruction on line 11 so that printf is called correctly.

After creating the object file, the next step in creating an executable program would be to invoke the linker and request that it link hello.o with the C standard library. The linker will

```
ARM GAS  hello.S                    page 1

  Addr Code               Source
1                         .data
2 0000 48656C6C  str:     .asciz  "Hello World\n"
2      6F20576F
2      726C640A
2      00
3
4                         .text
5                         .globl  main
6                         /* This is the beginning of the main() function.
7                            It will print Hello World'' and then return.
8                         */
9 0000 00402DE9  main:    stmfd   sp!,{lr} @ push return address onto stack
10 0004 0C009FE5          ldr     r0, =str @ load pointer to format string
11 0008 FEFFFFEB          bl      printf   @ printf("Hello World - %d\n",i);
12 000c 0000A0E3          mov     r0, #0   @ move return code into r0
13 0010 0040BDE8          ldmfd   sp!,{lr} @ pop return address from stack
14 0014 0EF0A0E1          mov     pc, lr   @ return from main
14      00000000


ARM GAS  hello.S                    page 2



DEFINED SYMBOLS
          hello.S:2    .data:0000000000000000 str
          hello.S:9    .text:0000000000000000 main
          hello.S:9    .text:0000000000000000 $a
          hello.S:14   .text:0000000000000018 $d

UNDEFINED SYMBOLS
printf
```

Listing 2.3
"Hello World" assembly listing.

generate the final executable file, containing the code assembled from hello.S, along with the printf function and other start-up code from the C standard library. The GNU C compiler is capable of automatically invoking the assembler for files that end in .s or .S, and can also be used to invoke the linker. For example, if Listing 2.1 is stored in a file named hello.S in the current directory, then the command

```
gcc -o hello hello.S
```

will run the GNU C compiler, telling it to create an executable program file named `hello`, and to use `hello.S` as the source file for the program. The C compiler will notice the `.S` extension, and invoke the assembler to create an object file which is stored in a temporary file, possibly named `hello.o`. Then the C compiler will invoke the linker to link `hello.o` with the C standard library, which provides the `printf` function and some start-up code which calls the `main` function. The linker will create an executable file named `hello`. When the linker has finished, the C compiler will remove the temporary object file.

2.3 GNU Assembly Directives

Each processor architecture has its own assembly language, created by the designers of the architecture. Although there are many similarities between assembly languages, the designers may choose different names for various directives. The GNU assembler supports a relatively large set of directives, some of which have more than one name. This is because it is designed to handle assembling code for many different processors without drastically changing the assembly language designed by the processor manufacturers. We will now cover some of the most commonly used directives for the GNU assembler.

2.3.1 Selecting the Current Section

The instructions and data that make up a program are stored in different *sections* of the program file. There are several standard sections that the programmer can choose to put code and data in. Sections can also be further divided into numbered subsections. Each section has its own address counter, which is used to keep track of the location of bytes within that section. When a label is encountered, it is assigned the value of the current address counter for the currently active section.

Selecting a section and subsection is done by using the appropriate assembly directive. Once a section has been selected, all of the instructions and/or data will go into that section until another section is selected. The most important directives for selecting a section are:

`.data subsection`

    Instructs the assembler to append the following instructions or data to the data subsection numbered `subsection`. If the subsection number is omitted, it defaults to zero. This section is normally used for global variables and constants which have labels.

`.text subsection`

Tells the assembler to append the following statements to the end of the text subsection numbered `subsection`. If the subsection number is omitted, subsection number zero is used. This section is normally used for executable instructions, but may also contain constant data.

`.bss subsection`

The bss (short for Block Started by Symbol) section is used for defining data storage areas that should be initialized to zero at the beginning of program execution. The `.bss` directive tells the assembler to append the following statements to the end of the bss subsection numbered `subsection`. If the subsection number is omitted, subsection number zero is used. This section is normally used for global variables which need to be initialized to zero. Regardless of what is placed into the section at compile-time, all bytes will be set to zero when the program begins executing. This section does not actually consume any space in the object or executable file. It is really just a request for the loader to reserve some space when the program is loaded into memory.

`.section name`

In addition to the three common sections, the programmer can create other sections using this directive. However in order for custom sections to be linked into a program, the linker must be made aware of them. Controlling the linker is covered in Section 14.4.3.

2.3.2 Allocating Space for Variables and Constants

There are several directives that allow the programmer to allocate and initialize static storage space for variables and constants. The assembler supports bytes, integer types, floating point types, and strings. These directives are used to allocate a fixed amount of space in memory and optionally initialize the memory. Some of these directives allow the memory to be initialized using an expression. An expression can be a simple integer, or a C-style expression. The directives for allocating storage are as follows:

`.byte expressions`

`.byte` expects zero or more expressions, separated by commas. Each expression is assembled into the next byte. If no expressions are given, then the address counter is not advanced and no bytes are reserved.

`.hword expressions`

`.short expressions`

For the ARM processor, these two directives do exactly the same thing. They expect zero or more expressions, separated by commas, and emit a 16-bit number for each expression. If no expressions are given, then the address counter is not advanced and no bytes are reserved.

```
.word expressions
.long expressions
```
For the ARM processor, these two directives do exactly the same thing. They expect zero or more expressions, separated by commas. They will emit four bytes for each expression given. If no expressions are given, then the address counter is not advanced and no bytes are reserved.

```
.ascii "string"
```
The .ascii directive expects zero or more string literals, each enclosed in quotation marks and separated by commas. It assembles each string (with no trailing ASCII NULL character) into consecutive addresses.

```
.asciz "string"
.string "string"
```
The .asciz directive is similar to the .ascii directive, but each string is followed by an ASCII NULL character (zero). The "z" in .asciz stands for zero. .string is just another name for .asciz.

```
.float flonums
.single flonums
```
This directive assembles zero or more floating point numbers, separated by commas. On the ARM, they are 4-byte IEEE standard single precision numbers. .float and .single are synonymous.

```
.double flonums
```
The .double directive expects zero or more floating point numbers, separated by commas. On the ARM, they are stored as 8-byte IEEE standard double precision numbers.

Fig. 2.1A shows how these directives are used to declare variables and constants. Fig. 2.1B shows the equivalent statements for creating global variables in C or C++. Note that in both cases, the variables created will be visible anywhere within the file that they are declared, but not visible in other files which are linked into the program.

```
        .data
i:      .word   0
j:      .word   1
fmt:    .asciz  "Hello\n"
ch:     .byte   'A','B',0
ary:    .word   0,1,2,3,4
```

(A)   Declarations in assembly

```
static int i = 0;
static int j = 1;
static char fmt[] = "Hello\n";
static char ch[] = {'A','B',0};
static int ary[] = {0,1,2,3,4};
```

(B)   Declarations in C

Figure 2.1
Equivalent static variable declarations in assembly and C.

```
ARM GAS   variable1.S                    page 1

line addr value         code
  1                             .data
  2 0000 00000000      i:      .word   0
  3 0004 01000000      j:      .word   1
  4 0008 48656C6C      fmt:    .asciz  "Hello\n"
  4      6F0A00
  5 000f 414200        ch:     .byte   'A','B',0
  6 0012 00000000      ary:    .word   0,1,2,3,4
  6      01000000
  6      02000000
  6      03000000
  6      04000000
```

Listing 2.4
A listing with mis-aligned data.

In C, the declaration of an array can be performed by leaving out the number of elements and specifying an initializer, as shown in the last three lines of Fig. 2.1B. In assembly, the equivalent is accomplished by providing a label, a type, and a list of values, as shown in the last three lines of Fig. 2.1A. The syntax is different, but the result is precisely the same.

Listing 2.4 shows how the assembler assigns addresses to these labels. The second column of the listing shows the address (in hexadecimal) that is assigned to each label. The variable i is assigned the first address. Since it is a word variable, the address counter is incremented by four bytes and the next address is assigned to the variable j. The address counter is incremented again, and fmt is assigned the address 0008. The fmt variable consumes seven bytes, so the ch variable gets address 000f. Finally, the array of words named ary begins at address 0012. Note that $12_{16} = 18_{10}$ is not evenly divisible by four, which means that the word variables in ary are not aligned on word boundaries.

2.3.3 Filling and Aligning

On the ARM CPU, data can be moved to and from memory one byte at a time, two bytes at a time (half-word), or four bytes at a time (word). Moving a word between the CPU and memory takes significantly more time if the address of the word is not aligned on a four-byte boundary (one where the least significant two bits are zero). Similarly, moving a half-word between the CPU and memory takes significantly more time if the address of the half-word is not aligned on a two-byte boundary (one where the least significant bit is zero). Therefore, when declaring storage, it is important that words and half-words are stored on appropriate boundaries. The following directives allow the programmer to insert as much space as necessary to align the next item on any boundary desired.

`.align abs-expr, abs-expr, abs-expr`

Pad the location counter (in the current subsection) to a particular storage boundary. For the ARM processor, the first expression specifies the number of low-order zero bits the location counter must have after advancement. The second expression gives the fill value to be stored in the padding bytes. It (and the comma) may be omitted. If it is omitted, then the fill value is assumed to be zero. The third expression is also optional. If it is present, it is the maximum number of bytes that should be skipped by this alignment directive.

`.balign[lw] abs-expr, abs-expr, abs-expr`

These directives adjust the location counter to a particular storage boundary. The first expression is the byte-multiple for the alignment request. For example, `.balign 16` will insert fill bytes until the location counter is an even multiple of 16. If the location counter is already a multiple of 16, then no fill bytes will be created. The second expression gives the fill value to be stored in the fill bytes. It (and the comma) may be omitted. If it is omitted, then the fill value is assumed to be zero. The third expression is also optional. If it is present, it is the maximum number of bytes that should be skipped by this alignment directive.

The `.balignw` and `.balignl` directives are variants of the `.balign` directive. The `.balignw` directive treats the fill pattern as a 2-byte word value, and `.balignl` treats the fill pattern as a 4-byte long word value. For example, "`.balignw 4,0x368d`" will align to a multiple of four bytes. If it skips two bytes, they will be filled in with the value `0x368d` (the exact placement of the bytes depends upon the endianness of the processor).

`.skip size, fill`

`.space size, fill`

Sometimes it is desirable to allocate a large area of memory and initialize it all to the same value. This can be accomplished by using these directives. These directives emit `size` bytes, each of value `fill`. Both `size` and `fill` are absolute expressions. If the comma and `fill` are omitted, `fill` is assumed to be zero. For the ARM processor, the `.space` and `.skip` directives are equivalent. This directive is very useful for declaring large arrays in the `.bss` section.

Listing 2.5 shows how the code in Listing 2.4 can be improved by adding an alignment directive at line 6. The directive causes the assembler to emit two zero bytes between the end of the `ch` variable and the beginning of the `ary` variable. These extra "padding" bytes cause the following word data to be word aligned, thereby improving performance when the word data is accessed. It is a good practice to always put an alignment directive after declaring character or half-word data.

```
ARM GAS  variable2.S                    page 1

line addr value         code
  1                             .data
  2 0000 00000000      i:      .word   0
  3 0004 01000000      j:      .word   1
  4 0008 48656C6C      fmt:    .asciz  "Hello\n"
  4      6F0A00
  5 000f 414200        ch:     .byte   'A','B',0
  6 0012 0000                  .align  2
  7 0014 00000000      ary:    .word   0,1,2,3,4
  7      01000000
  7      02000000
  7      03000000
  7      04000000
```

Listing 2.5
A listing with properly aligned data.

2.3.4 Setting and Manipulating Symbols

The assembler provides support for setting and manipulating symbols that can then be used in other places within the program. The labels that can be assigned to assembly statements and directives are one type of symbol. The programmer can also declare other symbols and use them throughout the program. Such symbols may not have an actual storage location in memory, but they are included in the assembler's symbol table, and can be used anywhere that their value is required. The most common use for defined symbols is to allow numerical constants to be declared in one place and easily changed. The .equ directive allows the programmer to use a label instead of a number throughout the program. This contributes to readability, and has the benefit that the constant value can then be easily changed every place that it is used, just by changing the definition of the symbol. The most important directives related to symbols are:

.equ symbol, expression
.set symbol, expression
>   This directive sets the value of symbol to expression. It is similar to the C language
>   #define directive.

.equiv symbol, expression
>   The .equiv directive is like .equ and .set, except that the assembler will signal an error if
>   the symbol is already defined.

```
.global symbol
.globl symbol
```
> This directive makes the symbol visible to the linker. If `symbol` is defined within a file, and this directive is used to make it global, then it will be available to any file that is linked with the one containing the symbol. Without this directive, symbols are visible only within the file where they are defined.

```
.comm symbol, length
```
> This directive declares `symbol` to be a common symbol, meaning that if it is defined in more than one file, then all instances should be merged into a single symbol. If the symbol is not defined anywhere, then the linker will allocate `length` bytes of uninitialized memory. If there are multiple definitions for `symbol`, and they have different sizes, the linker will merge them into a single instance using the largest size defined.

Listing 2.6 shows how the `.equ` directive can be used to create a symbol holding the number of elements in an array. The symbol `arysize` is defined as the value of the current address counter (denoted by the `.`) minus the value of the `ary` symbol, divided by four (each word in the array is four bytes). The listing shows all of the symbols defined in this program segment. Note that the four variables are shown to be in the data segment, and the `arysize` symbol is marked as an "absolute" symbol, which simply means that it is a number and not an address. The programmer can now use the symbol `arysize` to control looping when accessing the array data. If the size of the array is changed by adding or removing constant values, the value of `arysize` will change automatically, and the programmer will not have to search through the code to change the original value, 5, to some other value in every place it is used.

2.3.5 Conditional Assembly

Sometimes it is desirable to skip assembly of portions of a file. The assembler provides some directives to allow conditional assembly. One use for these directives is to optionally assemble code to aid in debugging.

```
.if expression
```
> `.if` marks the beginning of a section of code which is only considered part of the source program being assembled if the argument (which must be an absolute expression) is non-zero. The end of the conditional section of code must be marked by the `.endif` directive. Optionally, code may be included for the alternative condition by using the `.else` directive.

```
.ifdef symbol
```
> Assembles the following section of code if the specified symbol has been defined.

```
.ifndef symbol
```
> Assembles the following section of code if the specified symbol has not been defined.

```
ARM GAS   variable3.S                    page 1

line addr value         code
  1                             .data
  2 0000 00000000      i:       .word    0
  3 0004 01000000      j:       .word    1
  4 0008 48656C6C      fmt:     .asciz   "Hello\n"
  4      6F0A00
  5 000f 414200        ch:      .byte    'A','B',0
  6 0012 0000                   .align   2
  7 0014 00000000      ary:     .word    0,1,2,3,4
  7      01000000
  7      02000000
  7      03000000
  7      04000000
  8                             .equ     arysize,(. - ary)/4
  9


DEFINED SYMBOLS
          variable3.S:2     .data:0000000000000000 i
          variable3.S:3     .data:0000000000000004 j
          variable3.S:4     .data:0000000000000008 fmt
          variable3.S:5     .data:000000000000000f ch
          variable3.S:6     .data:0000000000000012 $d
          variable3.S:7     .data:0000000000000014 ary
          variable3.S:8     *ABS*:0000000000000005 arysize


NO UNDEFINED SYMBOLS
```

Listing 2.6
Defining a symbol for the number of elements in an array.

.else

Assembles the following section of code only if the condition for the preceding .if or .ifdef was false.

.endif

Marks the end of a block of code that is only assembled conditionally.

2.3.6 Including Other Source Files

.include "file"

This directive provides a way to include supporting files at specified points in the source program. The code from the included file is assembled as if it followed the point of the .include directive. When the end of the included file is reached, assembly of the original file continues. The search paths used can be controlled with the '-I' command line parameter. Quotation marks are required around file. This assembler directive is similar to including header files in C and C++ using the #include compiler directive.

2.3.7 Macros

The directives .macro and .endm allow the programmer to define *macros* that the assembler expands to generate assembly code. The GNU assembler supports simple macros. Some other assemblers have much more powerful macro capabilities.

```
.macro macname
.macro macname macargs ...
```
Begin the definition of a macro called macname. If the macro definition requires arguments, their names are specified after the macro name, separated by commas or spaces. The programmer can supply a default value for any macro argument by following the name with '=deflt'.

The following begins the definition of a macro called reserve_str, with two arguments. The first argument has a default value, but the second does not:

```
1        .macro reserve_str p1=0 p2
```

When a macro is called, the argument values can be specified either by position, or by keyword. For example, reserve_str 9,17 is equivalent to reserve_str p2=17,p1=9. After the definition is complete, the macro can be called either as

```
reserve_str x,y
```

(with \p1 evaluating to x and \p2 evaluating to y), or as

```
reserve_str ,y
```

(with \p1 evaluating as the default, in this case 0, and \p2 evaluating to y). Other examples of valid .macro statements are:

```
1        @ Begin the definition of a macro called comm,
2        @ which takes no arguments:
3        .macro comm
```

```
1        @ Begin the definition of a macro called plus1,
2        @ which takes two arguments:
3        .macro plus1 p, p1
4        @ Write \p or \p1 to use the arguments.
```

`.endm`

End the current macro definition.

`.exitm`

Exit early from the current macro definition. This is usually used only within a `.if` or `.ifdef` directive.

`\@`

This is a pseudo-variable used by the assembler to maintain a count of how many macros it has executed. That number can be accessed with '`\@`', but only within a macro definition.

Macro example

The following definition specifies a macro `SHIFT` that will emit the instruction to shift a given register left by a specified number of bits. If the number of bits specified is negative, then it will emit the instruction to perform a right shift instead of a left shift.

```
1        .macro SHIFT a,b
2        .if \b < 0
3        mov \a, \a, asr #-\b
4        .else
5        mov \a, \a, lsl #\b
6        .endm
```

After that definition, the following code:

```
1        SHIFT   r1, 3
2        SHIFT   r4, -6
```

will generate these instructions:

```
1        mov     r1, r1, asr #3
2        mov     r4, r4, lsl #6
```

The meaning of these instructions will be covered in Chapters 3 and 4.

Recursive macro example

The following definition specifies a macro `enum` that puts a sequence of numbers into memory by using a recursive macro call to itself:

```
1        .macro  enum first=0, last=5
2        .long   \first
3        .if     \last-\first
4        enum    "(\first+1)",\last
5        .endif
6        .endm
```

With that definition, 'enum 0,5' is equivalent to this assembly input:

```
1          .long   0
2          .long   1
3          .long   2
4          .long   3
5          .long   4
6          .long   5
```

2.4 Chapter Summary

There are four elements to assembly syntax: labels, directives, instructions, and comments. Directives are used mainly to define symbols, allocate storage, and control the behavior of the assembler. The most common assembler directives were introduced in this chapter, but there are many other directives available in the GNU assembler. Complete documentation is available online as part of the GNU Binutils package.

Directives are used to declare statically allocated storage, which is equivalent to declaring global static variables in C. In assembly, labels and other symbols are visible only within the file that they are declared, unless they are explicitly made visible to other files with the .global directive. In C, variables that are declared outside of any function are visible to all files in the program, unless the static keyword is used to make them visible only within the file where they are declared. Thus, both C and assembly support file and global scope for static variables, but with the opposite defaults and different syntax.

Directives can also be used to declare macros. Macros are expanded by the assembler and may generate multiple statements. Careful use of macros can automate some simple tasks, allowing several lines of assembly code to be replaced with a single macro invocation.

Exercises

2.1   What is the difference between
    (a)   the .data section and .bss section?
    (b)   the .ascii and .asciz directives?
    (c)   the .word and the .long directives?

2.2   What is the purpose of the .align assembler directive? What does ".align 2" do in GNU ARM assembly?

2.3   Assembly language has four main elements. What are they?

2.4   Using the directives presented in this chapter, show three different ways to create a null-terminated string containing the phrase "segmentation fault".

2.5   What is the total memory, in bytes, allocated for the following variables?

```
1   var1:   .word    23
2   var2:   .long    0xC
3   expr:   .ascii   ">>"
```

2.6   Identify the directive(s), label(s), comment(s), and instruction(s) in the following code:

```
1           .global main
2   main:
3           mov     r0,#1   @ the program return code is 1
4           mov     pc,lr   @ return and exit the program
```

2.7   Write assembly code to declare variables equivalent to the following C code:

```
1   /* these variables are declared outside of any function */
2   static int foo[3];   /* visible anywhere in the current file */
3   static char bar[4];  /* visible anywhere in the current file */
4   char barfoo;         /* visible anywhere in the program */
5   int foobar;          /* visible anywhere in the program */
```

2.8   Show how to store the following text as a single string in assembly language, while making it readable and keeping each line shorter than 80 characters:

```
The three goals of the mission are:
1) Keep each line of code under 80 characters,
2) Write readable comments,
3) Learn a valuable skill for readability.
```

2.9   Insert the minimum number of .align directives necessary in the following code so that all word variables are aligned on word boundaries and all halfword variables are aligned on halfword boundaries, while minimizing the amount of wasted space.

```
1           .data
2           .align  2
3   a:      .byte   0
4   b:      .word   32
5   c:      .byte   3
6   d:      .hword  45
7   e:      .hword  0
8   f:      .byte   0
9   g:      .word   128
```

2.10   Re-order the directives in the previous problem so that no .align directives are necessary to ensure proper alignment. How many bytes of storage were wasted by the original ordering of directives, compared to the new one?

2.11   What are the most important directives for selecting a section?

2.12   Why are .ascii and .asciz directives usually followed by an .align directive, but .word directives are not?

2.13   Using the "Hello World" program shown in Listing 2.1 as a template, write a program that will print your name.

2.14   Listing 2.3 shows that the assembler will assign the location 00000000_{16} to the main symbol and also to the str symbol. Why does this not cause problems?

Load/Store and Branch Instructions

The part of the computer architecture related to programming is referred to as the *instruction set architecture* (ISA). The ISA includes the *set of registers* that the user program can access, and the *set of instructions* that the processor supports, as well as *data paths* and *processing elements* within the processor. The first step in learning a new assembly language is to become familiar with the ISA. For most modern computer systems, data must be loaded in a register before it can be used for any data processing instruction, but there are a limited number of registers. Memory provides a place to store data that is not currently needed. Program instructions are also stored in memory and fetched into the CPU as they are needed. This chapter introduces the ISA for the ARM processor.

Modern Assembly Language Programming with the ARM Processor. http://dx.doi.org/10.1016/B978-0-12-803698-3.00003-6

3.1 CPU Components and Data Paths

The CPU is composed of data storage and computational components connected together by a set of buses. The most important components of the CPU are the *registers*, where data is stored, and the *arithmetic and logic unit* (ALU), where arithmetic and logical operations are performed on the data. Some CPUs also have dedicated hardware units for multiplication and/or division. Fig. 3.1 shows the major components of the ARM CPU and the buses that connect the components together. These buses provide pathways for the data to move between

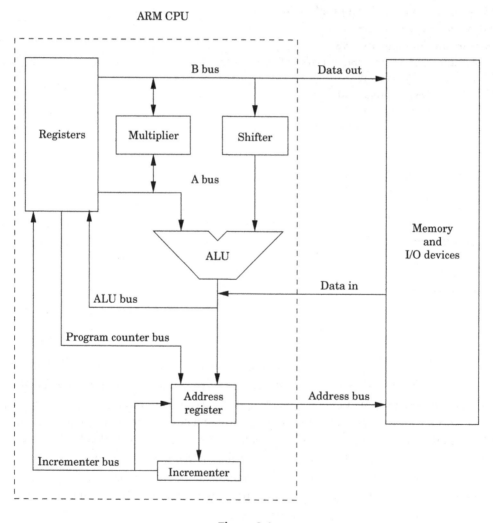

Figure 3.1
The ARM processor architecture.

the computational and storage components. The organization of the components and buses in a CPU govern what types of operations can be performed.

The set of instructions and addressing modes available on the ARM processor is closely related to the architecture shown in Fig. 3.1. The architecture provides for certain operations to be performed efficiently, and this has a direct relationship to the types of instructions that are supported.

Note that on the ARM, two source registers can be selected for an instruction, using the A and B buses. The data on the B bus is routed through a shifter, and then to the ALU. This allows the second operand of most instructions to be shifted an arbitrary amount before it reaches the ALU. The data on the A bus goes directly to the ALU. Additionally, the A and B buses can provide operands for the multiplier, and the multiplier can provide data for the A and B buses.

Data coming in from memory or an input/output device is fed directly onto the ALU bus. From there, it can be stored in one of the general-purpose registers. Data being written to memory or an input/output device is taken directly from the B bus, which means that store operations can move data from a register, but cannot modify the data on the way to memory or input/output devices.

The address register is a temporary register that is used by the CPU whenever it needs to read or write to memory or I/O devices. It is used every time an instruction is fetched from memory, and is used for all load and store operations. The address register can be loaded from the program counter, for fetching the next instruction. Also the address register can be loaded from the ALU, which allows the processor to support *addressing modes* where a register is used as a base pointer and an offset is calculated on-the-fly. After its contents are used to access memory or I/O devices, the base address can be incremented and the incremented value can be stored back into a register. This allows the processor to increment the program counter after each instruction, and to implement certain addressing modes where a pointer is automatically incremented after each memory access.

3.2 ARM User Registers

As shown in Fig. 3.2, the ARM processor provides 13 general-purpose registers, named r0 through r12. These registers can each store 32 bits of data. In addition to the 13 general-purpose registers, the ARM has three other special-purpose registers.

The program counter, r15, always contains the *address* of the next instruction that will be executed. The processor increments this register by four, automatically, after each instruction is fetched from memory. By moving an address into this register, the programmer can cause the processor to fetch the next instruction from the new address. This gives the programmer the ability to jump to any address and begin executing code there.

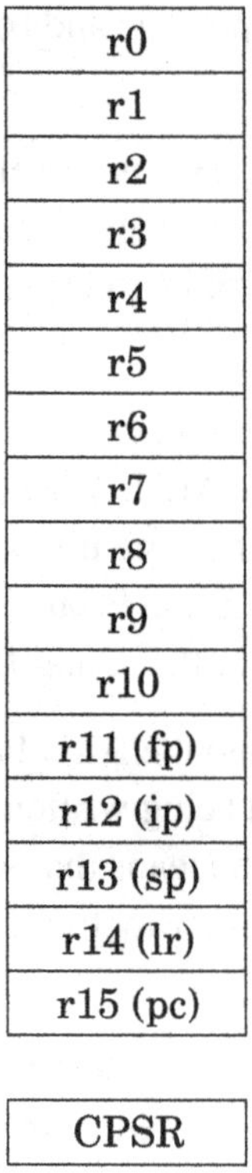

r0
r1
r2
r3
r4
r5
r6
r7
r8
r9
r10
r11 (fp)
r12 (ip)
r13 (sp)
r14 (lr)
r15 (pc)

CPSR

- Thirteen general-purpose registers (r0–r12)
- The stack pointer (r13 or sp)
- The link register (r14 or lr)
- The program counter (r15 or pc)
- Current Program Status Register (CPSR)

Figure 3.2
The ARM user program registers.

The link register, r14, is used to hold the *return address* for subroutines. Certain instructions cause the program counter to be copied to the link register, then the program counter is loaded with a new address. These *branch-and-link* instructions are briefly covered in Section 3.5 and in more detail in Section 5.4.

The program stack was introduced in Section 1.4. The stack pointer, r13, is used to hold the address where the stack ends. This is commonly referred to as the *top* of the stack, although on most systems the stack grows downwards and the stack pointer really refers to the *bottom* of the stack. The address where the stack ends may change when registers are pushed onto the stack, or when temporary local variables (*automatic variables*) are allocated or deleted. The use of the stack for storing automatic variables is described in Chapter 5. The use of r13 as the stack pointer is a *programming convention*. Some instructions (eg, branches) implicitly modify the program counter and link registers, but there are no special instructions involving the stack pointer. As far as the hardware is concerned, r13 is exactly the same as registers r0–r12, but all ARM programmers use it for the stack pointer.

Although register r13 is normally used as the stack pointer, it can be used as a general-purpose register if the stack is not used. However the high-level language compilers always use it as the stack pointer, so using it as a general-purpose register will result in code that cannot inter-operate with code generated using high-level languages. The link register, r14, can also be used as a general-purpose register, but its contents are modified by hardware when a subroutine is called. Using r13 and r14 as general-purpose registers is dangerous and strongly discouraged.

There are also two other registers which may have special purposes. As with the stack pointer, these are programming conventions. There are no special instructions involving these registers. The frame pointer (r11) is used by high-level language compilers to track the current *stack frame*. This is sometimes useful when running your program under a debugger, and can sometimes help the compiler to generate more efficient code for returning from a subroutine. The GNU C compiler can be instructed to use r11 as a general-purpose register by using the --omit-frame-pointer command line option. The inter-procedure scratch register r12 is used by the C library when calling functions in *dynamically linked libraries*. The contents may change, seemingly at random, when certain functions (such as printf) are called.

The final register in the ARM user programming model is the Current Program Status Register (CPSR). This register contains bits that indicate the status of the current program, including information about the results of previous operations. Fig. 3.3 shows the bits in the CPSR. The first four bits, N, Z, C, and V are the *condition flags*. Most instructions can modify these flags, and later instructions can use the flags to modify their operation. Their meaning is as follows:

Negative: This bit is set to one if the signed result of an operation is negative, and set to zero if the result is positive or zero.

Zero: This bit is set to one if the result of an operation is zero, and set to zero if the result is non-zero.

Carry: This bit is set to one if an add operation results in a carry out of the most significant bit, or if a subtract operation results in a borrow. For shift operations, this flag is set to the last bit shifted out by the shifter.

oVerflow: For addition and subtraction, this flag is set if a signed overflow occurred.

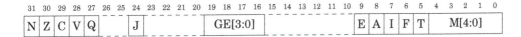

Figure 3.3
The ARM process status register.

The remaining bits are used by the operating system or for bare-metal programs, and are described in Section 14.1.

3.3 Instruction Components

The ARM processor supports a relatively small set of instructions grouped into four basic instruction types. Most instructions have optional *modifiers* which can be used to change their behavior. For example, many instructions can have modifiers which set or check condition codes in the CPSR. The combination of basic instructions with optional modifiers results in an extremely rich assembly language. There are four general instruction types, or categories. The following sections give a brief overview of the features which are common to instructions in each category. The individual instructions are explained later in this chapter, and in the following chapter.

3.3.1 Setting and Using Condition Flags

As mentioned previously, the CPSR contains four flag bits (bits 28–31), which can be used to control whether or not certain instructions are executed. Most of the data processing instructions have an optional modifier to control whether or not the flag bits are affected when the instruction is executed. For example, the basic instruction for addition is add. When the add instruction is executed, the result is stored in a register, but the flag bits in the CPSR are not affected.

However, the programmer can add the *s modifier* to the add instruction to create the adds instruction. When it is executed, this instruction *will* affect the CPSR flag bits. The flag bits can be used by subsequent instructions to control execution and branching. The meaning of the flags depends on the type of instruction that last set the flags. Table 3.1 shows the names and

Table 3.1 Flag bits in the CPSR register

Name	Logical Instruction	Arithmetic Instruction
N (Negative)	No meaning	Bit 31 of the result is set. Indicates a negative number in signed operations
Z (Zero)	Result is all zeroes	Result of operation was zero
C (Carry)	After Shift operation, '1' was left in carry flag	Result was greater than 32 bits
V (oVerflow)	No meaning	The signed two's complement result requires more than 32 bits. Indicates a possible corruption of the result

Table 3.2 ARM condition modifiers

<cond>	English Meaning
al	always (this is the default <cond>
eq	Z set (=)
ne	Z clear (≠)
ge	N set and V set, or N clear and V clear (≥)
lt	N set and V clear, or N clear and V set (<)
gt	Z clear, and either N set and V set, or N clear and V set (>)
le	Z set, or N set and V clear, or N clear and V set (≤)
hi	C set and Z clear (unsigned >)
ls	C clear or Z (unsigned ≤)
hs	C set (unsigned ≥)
cs	Alternate name for HS
lo	C clear (unsigned <)
cc	Alternate name for LO
mi	N set (result < 0)
pl	N clear (result ≥ 0)
vs	V set (overflow)
vc	V clear (no overflow)

meanings of the four bits depending on the type of instruction that set or cleared them. Most instructions support the s modifier to control setting the flags.

Most ARM instructions can have a *condition modifier* attached. If present, the modifier controls, at run-time, whether or not the instruction is actually executed. These condition modifiers are added to basic instructions to create *conditional instructions*. Table 3.2 shows the condition modifiers that can be attached to base instructions. For example, to create an instruction that adds only if the CPSR Z flag is set, the programmer would add the eq condition modifier to the basic add instruction to create the addeq instruction.

Setting and using condition flags are *orthogonal* operations. This means that they can be used in combination. Using the previous example, the programmer could add the s modifier to create the addeqs instruction, which executes only if the Z bit is set, and updates the CPSR flags only if it executes.

3.3.2 Immediate Values

An immediate value in assembly language is a constant value that is specified by the programmer. Some assembly languages encode the immediate value as part of the instruction. Other assembly languages create a table of immediate values in a *literal pool* and insert appropriate instructions to access them. ARM assembly language provides both methods.

Immediate values can be specified in decimal, octal, hexadecimal, or binary. Octal values must begin with a zero, and hexadecimal values must begin with "0x". Likewise immediate values

Table 3.3 Legal and illegal values for #<immediate|symbol>

#32	Ok because it can be stored as an 8-bit value
#1021	Illegal because the number cannot be created from an 8-bit value using shift or rotate and complement
#1024	Ok because it is 1 shifted left 10 bits
#0b1011	Ok because it fits in 8 bits
#-1	Ok because it is the one's complement of 0
#0xFFFFFFFE	Ok because it is the one's complement of 1
#0xEFFFFFFF	Ok because it is the one's complement of 1 shifted left 31 bits
#strsize	Ok if the value of strsize can be created from an 8-bit value using shift or rotate and complement

that start with "0b" are interpreted as binary numbers. Any value that does not begin with zero, 0x, or 0b will be interpreted as a decimal value.

There are two ways that immediate values can be specified in GNU ARM assembly. The =<immediate|symbol> syntax can be used to specify any immediate 32-bit number, or to specify the 32-bit value of any symbol in the program. Symbols include program labels (such as main) and symbols that are defined using .equ and similar assembler directives. However, this syntax can only be used with load instructions, and not with data processing instructions. This restriction is necessary because of the way the ARM machine instructions are encoded. For data processing instructions, there are a limited number of bits that can be devoted to storing immediate data as part of the instruction.

The #<immediate|symbol> syntax is used to specify immediate data values for data processing instructions. The #<immediate|symbol> syntax has some restrictions. Basically, the assembler must be able to construct the specified value using only eight bits of data, a shift or rotate, and/or a complement. For immediate values that can cannot be constructed by shifting or rotating and complementing an 8-bit value, the programmer must use an ldr instruction with the =<immediate|symbol> to specify the value. That method is covered in Section 3.4. Some examples of immediate values are shown in Table 3.3.

3.4 Load/Store Instructions

The ARM processor has a strict separation between instructions that perform computation and those that move data between the CPU and memory. Because of this separation between load/store operations and computational operations, it is a classic example of a *load-store architecture*. The programmer can transfer bytes (8 bits), half-words (16 bits), and words (32 bits), from memory into a register, or from a register into memory. The programmer can also perform computational operations (such as adding) using two source operands and one register as the destination for the result. All *computational* instructions assume that the registers already contain the data. Load instructions are used to move data into the registers, while store instructions are used to move data from the registers to memory.

3.4.1 Addressing Modes

Most of the load/store instructions use an `<address>` which is one of the six options shown in Table 3.4. The `<shift_op>` can be any of shift operations from Table 3.5, and `shift` should be a number between 0 and 31. Although there are really only six addressing modes, there are eleven variations of the assembly language syntax. Four of the variations are simply shorthand notations. One of the variations allows an immediate data value or the address of a label to be loaded into a register, and may result in the assembler generating more than one instruction. The following section describes each addressing mode in detail.

Immediate offset: `[Rn, #±<offset_12>]`

The immediate offset (which may be positive or negative) is added to the contents of `Rn`. The result is used as the address of the item to be loaded or stored. For example, the following line of code:

```
ldr     r0, [r1, #12]
```

calculates a memory address by adding 12 to the contents of register `r1`. It then loads four bytes of data, starting at the calculated memory address, into register `r0`. Similarly, the line:

```
str     r9, [r6, #-8]
```

Table 3.4 ARM addressing modes

Syntax	Name
`[Rn, #±<offset_12>]`	Immediate offset
`[Rn, ±Rm, <shift_op> #<shift>]`	Scaled register offset
`[Rn, #±<offset_12>]!`	Immediate pre-indexed
`[Rn, ±Rm, <shift_op> #<shift>]!`	Scaled register pre-indexed
`[Rn], #±<offset_12>`	Immediate post-indexed
`[Rn], ±Rm, <shift_op> #<shift>`	Scaled register post-indexed

Table 3.5 ARM shift and rotate operations

`<shift>`	Meaning
`lsl`	Logical **S**hift **L**eft by specified amount
`lsr`	Logical **S**hift **R**ight by specified amount
`asr`	Arithmetic **S**hift **R**ight by specified amount

subtracts 8 from the contents of r6 and uses that as the address where it stores the contents of r9 in memory.

Register immediate: [Rn]

When using immediate offset mode with an offset of zero, the comma and offset can be omitted. That is, [Rn] is just shorthand notation for [Rn, #0]. This shorthand is referred to as *register immediate* mode. For example, the following line of code:

```
ldr     r3, [r2]
```

uses the contents of register r2 as a memory address and loads four bytes of data, starting at that address, into register r3. Likewise,

```
str     r8, [r0]
```

copies the contents of r8 to the four bytes of memory starting at the address that is in r0.

Scaled register offset: [Rn, ±Rm, <shift_op> #<shift>]

Rm is shifted as specified, then added to or subtracted from Rn. The result is used as the address of the item to be loaded or stored. For example,

```
ldr r3, [r2, r1, lsl #2]
```

shifts the contents of r1 left two bits, adds the result to the contents of r2 and uses the sum as an address in memory from which it loads four bytes into r3. Recall that shifting a binary number left by two bits is equivalent to multiplying that number by four. This addressing mode is typically used to access an array, where r2 contains the address of the beginning of the array, and r1 is an integer index. The integer shift amount depends on the size of the objects in the array. To store an item from register r0 into an array of half-words, the following instruction could be used:

```
strh r0, [r4, r5, lsl #1]
```

where r4 holds the address of the first byte of the array, and r5 holds the integer index for the desired array item.

Register offset: [Rn, ±Rm]

When using scaled register offset mode with a shift amount of zero, the comma and shift specification can be omitted. That is, [Rn, ±Rm] is just shorthand notation for [Rn, ±Rm, lsl #0]. This shorthand is referred to as *register offset* mode.

Immediate pre-indexed: [Rn, #±<offset_12>]!

The address is computed in the same way as immediate offset mode, but after the load or store, the address that was used is stored in Rn. This mode can be used to step through elements in an array, updating a pointer to the next array element before each element is accessed.

Scaled register pre-indexed: `[Rn, ±Rm, <shift_op> #<shift>]!`

The address is computed in the same way as scaled register offset mode, but after the load or store, the address that was used is stored in `Rn`. This mode can be used to step through elements in an array, updating a pointer to the current array element before each access.

Register pre-indexed: `[Rn, ±Rm]!`

When using scaled register pre-indexed mode with a shift amount of zero, the comma and shift specification can be omitted. That is, `[Rn, ±Rm]!` is shorthand notation for `[Rn, ±Rm, lsl #0]!`. This shorthand is referred to as *register pre-indexed* mode.

Immediate post-indexed: `[Rn], #±<offset_12>`

Register `Rn` is used as the address of the value to be loaded or stored. After the value is loaded or stored, the value in `Rn` is updated by adding the immediate offset, which may be negative or positive. This mode can be used to step through elements in an array, updating a pointer to point at the next array element after each one is accessed.

Scaled register post-indexed: `[Rn], ±Rm, <shift_op> #<shift>`

Register `Rn` is used as the address of the value to be loaded or stored. After the value is loaded or stored, the value in `Rn` is updated by adding or subtracting the contents of `Rm` shifted as specified. This mode can be used to step through elements in an array, updating a pointer to point at the next array element after each one is accessed.

Register post-indexed: `[Rn], ±Rm`

When using scaled register post-indexed mode with a shift amount of zero, the comma and shift specification can be omitted. That is, `[Rn], ±Rm` is shorthand notation for `[Rn], ±Rm, lsl #0`. This shorthand is referred to as *register post-indexed* mode.

Load Immediate: `[Rn], =<immediate|symbol>`

This is really a pseudo-instruction. The assembler will generate a `mov` instruction if possible. Otherwise it will store the value of `immediate` or the address of `symbol` in a "literal table" and generate a load instruction, using one of the previous addressing modes, to load the value into a register. This addressing mode can only be used with the `ldr` instruction.

The load and store instructions allow the programmer to move data from memory to registers or from registers to memory. The load/store instructions can be grouped into the following types:

- single register,
- multiple register, and
- atomic.

The following sections describe the seven load and store instructions that are available, and all of their variations.

3.4.2 Load/Store Single Register

These instructions transfer a single word, half-word, or byte from a register to memory or from memory to a register:

ldr	Load Register, and
str	Store Register.

Syntax

```
<op>{<cond>}{<size>}  Rd, <address>
```

- `<op>` is either `ldr` or `str`.
- The optional `<cond>` can be any of the codes from Table 3.2 on page 59 specifying conditional execution.
- The optional `<size>` is one of:
b	unsigned byte
h	unsigned half-word
sb	signed byte
sh	signed half-word
- The `<address>` is any valid address specifier described in Section 3.4.1.

Operations

Name	Effect	Description
ldr	$Rd \leftarrow Mem[address]$	Load register from memory at address
str	$Mem[address] \leftarrow Rd$	Store register in memory at address

Examples

```
 1      ldrsh   r5, [r2]        @ Load r5 with signed half-
 2                              @ word at the address in r2
 3      strb    r1, [r9, #4]    @ Store the byte in r1 at
 4                              @ the address (r9 + 4)
 5      ldr     r5, [r3, r2]!   @ Load r5 with word at the
 6                              @ address (r3 + r2), then
 7                              @ store the address in r3
 8      ldrh    r9, [r2, #2]!   @ Load r9 with half-word at
 9                              @ the address (r2 + 2), then
10                              @ store the address in r2
```

3.4.3 Load/Store Multiple Registers

ARM has two instructions for loading and storing multiple registers:

ldm   Load Multiple Registers, and
stm   Store Multiple Registers.

These instructions are used to store registers on the program stack, and for copying blocks of data. The ldm and stm instructions each have four variants, and each variant has two equivalent names. So, although there are only two basic instructions, there are sixteen mnemonics. These are the most complex instructions in the ARM assembly language.

Syntax

```
<op><variant>     Rd{!}, <register_list>{^}
```

- <op> is either ldm or stm.
- <variant> is chosen from the following tables:

Block Copy Method		Stack Type	
Variant	**Description**	**Variant**	**Description**
ia	Increment After	ea	Empty Ascending
ib	Increment Before	fa	Full Ascending
da	Decrement After	ed	Empty Descending
db	Decrement Before	fd	Full Descending

- The optional ! specifies that the address register Rd should be modified after the registers are stored.
- An optional trailing ^ can only be used by operating system code. It causes the transfer to affect user registers instead of operating system registers.

There are two equivalent mnemonics for each load/store multiple instruction. For example, ldmia is exactly the same instruction as ldmfd, and stmdb is exactly the same instruction as stmfd. There are two different names so that the programmer can indicate what the instruction is being used for.

The mnemonics in the Block Copy Method table are used when the programmer is using the instructions to move blocks of data. For instance, the programmer may want to copy eight words from one address in memory to another address. One very efficient way to do that is to:

1. load the address of the first byte of the source into a register,
2. load the address of the first byte of the destination into another register,
3. use ldmia (load multiple increment after) to load eight registers from the source address, then
4. use stmia (store multiple increment after) to store the registers to the destination address.

Assuming `source` and `dest` are labeled blocks of data declared elsewhere, the following listing shows the exact instructions needed to move eight words from `source` to `dest`:

```
1    ldr     r8,=source @ load address of source
2    ldr     r9,=dest   @ load address of destination
3    ldmia   r8,{r0-r7} @ load eight words from source
4    stmia   r9,{r0-r7} @ store them in destination
```

The mnemonics in the Stack Type table are used when the programmer is performing stack operations. The most common variants are `stmfd` and `ldmfd`, which are used for pushing registers onto the program stack and later popping them back off, respectively. In Linux, the C compiler always uses the `stmfd` and `ldmfd` versions for accessing the stack. The following code shows how the programmer could save the contents of registers `r0-r9` on the stack, use them to perform a block copy, then restore their contents:

```
1    stmfd   sp!{r0-r9} @ push r0...r9 to the stack
2    ldr     r8,=source @ load address of source
3    ldr     r9,=dest   @ load address of destination
4    ldmia   r8,{r0-r7} @ load eight words from source
5    stmia   r9,{r0-r7} @ store them in destination
6    ldmfd   sp!{r0-r9} @ restore contents of r0...r9
```

Note that in the previous example, after the `stmfd sp!,{r0-r9}` instruction, `sp` will contain the address of the last word on the stack, because the optional `!` was used to indicate that the register should be updated.

Operations

Name	Effect	Description

ldmia and ldmfd	*addr ← Rd*
	for all *i* ∈ *register_list* **do**
	    i ← Mem[addr]
	    addr ← addr + 4
	end for
	if ! is present **then**
	    Rd ← addr
	end if
		Load multiple registers from memory, starting at the address in Rd and increment the address by four bytes after each load.

stmia and stmea	*addr ← Rd*
	for all *i* ∈ *register_list* **do**
	    Mem[addr] ← i
	    addr ← addr + 4
	end for
	if ! is present **then**
	    Rd ← addr
	end if
		Store multiple registers in memory, starting at the address in Rd and increment the address by four bytes after each store.

ldmib and ldmed	*addr* ← *Rd*
	 for all *i* ∈ *register_list* **do**
	    addr ← *addr* + 4
	    i ← *Mem*[*addr*]
	 end for
	 if ! is present **then**
	    Rd ← *addr*
	 end if
		Load multiple registers from memory, starting at the address in Rd and increment the address by four bytes before each load.

stmib and stmfa	*addr* ← *Rd*
	 for all *i* ∈ *register_list* **do**
	    addr ← *addr* + 4
	    Mem[*addr*] ← *i*
	 end for
	 if ! is present **then**
	    Rd ← *addr*
	 end if
		Store multiple registers in memory, starting at the address in Rd and increment the address by four bytes before each store.

ldmda and ldmfa	*addr* ← *Rd*
	 for all *i* ∈ *register_list* **do**
	    i ← *Mem*[*addr*]
	    addr ← *addr* − 4
	 end for
	 if ! is present **then**
	    Rd ← *addr*
	 end if
		Load multiple registers from memory, starting at the address in Rd and decrement the address by four bytes after each load.

stmda and stmed	*addr* ← *Rd*
	 for all *i* ∈ *register_list* **do**
	    Mem[*addr*] ← *i*
	    addr ← *addr* − 4
	 end for
	 if ! is present **then**
	    Rd ← *addr*
	 end if
		Store multiple registers in memory, starting at the address in Rd and decrement the address by four bytes after each store.

ldmdb and ldmea	*addr* ← *Rd*
	 for all *i* ∈ *register_list* **do**
	    addr ← *addr* − 4
	    i ← *Mem*[*addr*]
	 end for
	 if ! is present **then**
	    Rd ← *addr*
	 end if
		Load multiple registers from memory, starting at the address in Rd and decrement the address by four bytes before each load.

stmdb and stmfd	*addr* ← *Rd*
	 for all *i* ∈ *register_list* **do**
	    addr ← *addr* − 4
	    Mem[*addr*] ← *i*
	 end for
	 if ! is present **then**
	    Rd ← *addr*
	 end if
		Store multiple registers in memory, starting at the address in Rd and decrement the address by four bytes before each store.

Examples

```
1       stmfd sp!,{r4-r7,fp,lr}  @ store r4, r5, r6, r7, r11
2                                @ and r14 (lr) on the stack,
3                                @ and store the new stack
4                                @ pointer in sp
5       ldmfd sp!,{r4-r7,fp,lr}  @ load r4, r5, r6, r7, r11,
6                                @ and r14 from the stack,
7                                @ and store the new stack
8                                @ pointer in sp
9       stmib r9!,{r0-r7}        @ Store 8 registers at the
10                               @ location pointed to by r9,
11                               @ and increment r9 BEFORE
12                               @ each store. After executing,
13                               @ r9 will point to the last
14                               @ item stored.
15      ldmia r4,{r0,r2,r3}      @ Load r0, r2, and r3 at the
16                               @ location pointed to by r4
17                               @ and increment the address
18                               @ AFTER each store. After
19                               @ executing, r4 will contain
20                               @ its original value.
```

3.4.4 Swap

Multiprogramming and threading require the ability to set and test values *atomically*. This instruction is used by the operating system or threading libraries to guarantee *mutual exclusion*:

swp   Atomic Load and Store

Note: swp and swpb are deprecated in favor of ldrex and strex, which work on multiprocessor systems as well as uni-processor systems.

Syntax

```
swp{<cond>}{s}      Rd, Rm, [Rn]
swp{<cond>}{s}b     Rd, Rm, [Rn]
```

- The optional s specifies whether or not the instruction should affect the bits in the CPSR.
- The optional <cond> can be any of the codes from Table 3.2 specifying conditional execution.

Operations

Name	Effect	Description
swp	$Rd \leftarrow Mem[Rn]$ $Mem[Rn] \leftarrow Rm$	Atomically load Rd and store Rm
swpb	$Rd \leftarrow Mem[Rn]$ $Mem[Rn] \leftarrow Rm$	Atomically load Rd and store Rm

Example

```
 1    swpeqb  r1, r4, [r3]  @ if (eq) then load r1 with byte
 2                          @ at address in r3 and store byte
 3                          @ in r4 at address in r3
```

3.4.5 Exclusive Load/Store

These instructions are used by the operating system or threading libraries to guarantee *mutual exclusion*, even on multiprocessor systems:

ldrex   Load Multiple Registers, and
strex   Store Multiple Registers.

Exclusive load (ldrex) reads data from memory, tagging the memory address at the same time. Exclusive store (strex) stores data to memory, but only if the tag is still valid. A strex to the same address as the previous ldrex will invalidate the tag. A str to the same address *may* invalidate the tag (implementation defined). The strex instruction sets a bit in the specified register which indicates whether or not the store succeeded. This allows the programmer to implement semaphores on uni-processor and multiprocessor systems.

Syntax

```
 ldrex{<cond>}   Rd, Rn
 strex{<cond>}   Rd, Rn, Rm
```

* The optional <cond> can be any of the codes from Table 3.2 specifying conditional execution.

Operations

Name	Effect	Description
ldrex	$Rd \leftarrow Mem[Rn]$ $tag_{Mem[Rn]} \leftarrow true$	Load register and tag memory address

strex	**if** $tag_{Mem[Rn]}$ **then**
	    $Mem[Rn] \leftarrow Rd$
	end if
		Store register in memory if tag is valid

Example

```
1           ldr     r12, =sem     @ preload semaphore address
2           ldr     r1, =LOCKED   @ preload "locked" value
3   splck:  ldrex   r0, [r12]     @ load semaphore value
4           cmp     r0, r1        @ if semaphore was not locked
5           strexne r0, r1, [r12] @    try to claim
6           cmpne   r0, #1        @    and check success
7           beq     splck         @ retry if claiming failed
```

3.5 Branch Instructions

Branch instructions allow the programmer to change the address of the next instruction to be executed. They are used to implement loops, if-then structures, subroutines, and other flow control structures. There are two basic branch instructions:

- Branch, and
- Branch and Link (subroutine call).

3.5.1 Branch

This instruction is used to perform conditional and unconditional branches in program execution:

b   Branch.

It is used for creating loops and if-then-else constructs.

Syntax

```
    b{<cond>}  <target_label>
```

- The optional <cond> can be any of the codes from Table 3.2 specifying conditional execution.
- The target_label can be any label in the current file, or any label that is defined as .global or .globl in any file that is linked in.

Operations

Name	Effect	Description
b	pc ← target_address	load pc with new address (branch)

Examples

```
1       blt    ll       @ Branch to label 'll' if flags
2                       @ indicate "less than"
3       b      there    @ Always branch to label 'there'
4       bcc    done     @ Branch to label 'done' if carry
5                       @ flag is clear (zero)
6       beq    145      @ Branch to '145' if Z flag is set (one)
```

3.5.2 Branch and Link

The following instruction is used to call subroutines:

bl   Branch and Link.

The branch and link instruction is identical to the branch instruction, except that it copies the current program counter to the link register before performing the branch. This allows the programmer to copy the link register back into the program counter at some later point. This is how subroutines are called, and how subroutines return and resume executing at the next instruction after the one that called them.

Syntax

```
bl{<cond>}  <target_address>
```

- The optional <cond> can be any of the codes from Table 3.2 specifying conditional execution.
- The target_address can be any label in the current file, or any label that is defined as .global or .globl in any file that is linked in.

Operations

Name	Effect	Description
bl	$lr \leftarrow pc$ $pc \leftarrow target\_address$	Save pc in lr, then load pc with new address

Examples

```
1     mov   r0, =fmt_string @ load pointer to format string
2     bl    printf          @ call printf
```

Example 3.1 shows how the bl instruction can be used to call a function from the C standard library to read a single character from standard input. By convention, when a function is called, it will leave its return value in r0. Example 3.2 shows how the bl instruction can be used to call another function from the C standard library to print a message to standard output. By convention, when a function is called, it will expect to find its first argument in r0. There are other rules, which all ARM programmers must follow, regarding which registers are used when passing arguments to functions and procedures. Those rules will be explained fully in Section 5.4.

Example 3.1   Using the bl Instruction to Read a Character

Suppose we want to read a single character from standard input. This can be accomplished in C by calling the getchar() function from the C standard library as follows:

```
1     :
2     c = getchar();
3     :
```

The above C code assumes that the variable c has been declared to hold the result of the function. In ARM assembly language, functions always return their results in r0. The assembly programmer may then move the result to any register or memory location they choose. In the following example, it is assumed that r9 was chosen to hold the value of the variable c:

```
1     :
2     bl     getchar  @ Call the getchar function
3     mov    r9,r0    @ Move the result to register 9
4     :
```

Example 3.2 Using the bl Instruction to Print a Message

To print a string to standard output in C, we can use the `printf()` function from the C standard library as follows:

```
1       ⋮
2       printf("This is a message\n");
3       ⋮
```

The C compiler will automatically create a constant array of characters and initialize it to hold the message. Then it will load the address of the first character in the array into register r0 before calling `printf()`. The `printf()` function will expect to see an address in r0, which it will assume is the address of the format string to be printed. The function call can be made as follows in ARM assembly:

```
1          ⋮
2          .data
3          ⋮
4    msg:  .asciz  "This is a message\n"
5          ⋮
6          .text
7          ⋮
8          mov    r0, =msg @ Load address of message string
9          bl     printf   @ Call the printf function
10         ⋮
```

3.6 Pseudo-Instructions

The assembler provides a small number of pseudo-instructions. From the perspective of the programmer, these instructions are indistinguishable from standard instructions. However, when the assembler encounters a pseudo-instruction, it may substitute a different instruction or generate a short sequence of machine instructions.

3.6.1 Load Immediate

This pseudo-instruction loads a register with any 32-bit value:

ldr   Load Immediate

When this pseudo-instruction is encountered, the assembler first determines whether or not it can substitute a `mov Rd,#<immediate>` or `mvn Rd,#<immediate>` instruction. If that is not possible, then it reserves four bytes in a "literal pool" and stores the immediate value there.

Then, the pseudo-instruction is translated into an `ldr` instruction using Immediate Offset addressing mode with the `pc` as the base register.

Syntax

```
ldr{<cond>}  Rd, =<immediate>
```

- The optional `<cond>` can be any of the codes from Table 3.2 specifying conditional execution.
- The `<immediate>` parameter is any valid 32-bit quantity.

Operations

Name	Effect	Description
ldr	Rd ← value	Load register with immediate value

Example

Example 3.3 shows how the assembler generates code from the load immediate pseudo-instruction. Line 2 of the example listing just declares two 32-bit words. They cause the next variable to be given a non-zero address for demonstration purposes, and are not used anywhere in the program, but line 3 declares a string of characters in the data section. The string is located at offset `0x00000008` from the beginning of the data section. The linker is responsible for calculating the actual address, when it assigns a location for the data section. Line 6 shows how a register can be loaded with an immediate value using the `mov` instruction. The next line shows the equivalent using the `ldr` pseudo-instruction. Note that the assembler generates the same machine instruction (`FD5FE0E3`) for both lines.

Example 3.3   Assembly of the Load Immediate Pseudo-Instruction

```
1                           .data
2 0000 0A000000 dummy:    .word   10,11
2      0B000000
3 0008 48656C6C str:      .asciz  "Hello World\n"
3      6F20576F
3      726C640A
3      00
4                         .text
5                         .global main
6 0000 FD5FE0E3 main:     mov     r5, #-1013  @ Load r5
7 0004 FD5FE0E3           ldr     r5, =-1013  @ Load r5
8 0008 B470DFE1           ldrh    r7, =0xFFF  @ Load r7
```

```
    9 000c 04409FE5      ldr    r4, =str    @ Load r4
   10                                        @ with addr
   11 0010 0EF0A0E1      mov    pc,lr       @ return...
   11      FF0F0000
   11      08000000


DEFINED SYMBOLS
        pseudoload.s:2      .data:00000000 dummy
        pseudoload.s:3      .data:00000008 str
        pseudoload.s:6      .text:00000000 main
        pseudoload.s:6      .text:00000000 $a
        pseudoload.s:11     .text:00000014 $d
```

Line 8 shows the ldr pseudo-instruction being used to load a value that cannot be loaded using the mov instruction. The assembler generated a load half-word instruction using the program counter as the base register, and an offset to the location where the value is stored. The value is actually stored in a literal pool at the end of the text segment. The listing has three lines labeled 11. The first line 11 is an instruction. The remaining lines are the literal pool.

On line 9, the programmer used the ldr pseudo-instruction to request that the address of str be loaded into r4. The assembler created a storage location to hold the address of str, and generated a load word instruction using the program counter as the base register and an offset to the location where the address is stored. The address of str is actually stored in the text segment, on the third line 11.

3.6.2 Load Address

These pseudo instructions are used to load the address associated with a label:

adr   Load Address
adrl   Load Address Long

They are more efficient than the ldr rx,=label instruction, because they are translated into one or two add or subtract operations, and do not require a load from memory. However, the address must be in the same section as the adr or adrl pseudo-instruction, so they cannot be used to load addresses of labels in the .data section.

Syntax

```
    <op>{<cond>}{s}  Rd, label
```

- <op> is either `adr` or `adrl`.
- The `adr` pseudo-instruction will be translated into one or two pc-relative `add` or `sub` instructions.
- The `adrl` pseudo-instruction will always be translated into two instructions. The second instruction may be a `nop` instruction.
- The label must be defined in the same file and section where these pseudo-instructions are used.

Operations

Name	Effect	Description
adr	*Rd* ← Address of Label	Load Address
adrl	*Rd* ← Address of Label	Load Address

Examples

```
1    adr    r0, str    @ load address of str into r0
```

3.7 Chapter Summary

The ARM Instruction Set Architecture includes 17 registers and a four basic instruction types. This chapter explained the instructions used for

- moving data between memory and registers, and
- branching and calling subroutines.

The load and store operations are used to move data between memory and registers. The basic load and store operations, `ldr` and `str`, have a very powerful set of addressing modes. To facilitate moving multiple registers to or from memory, the ARM ISA provides the `ldm` and `stm` instructions, which each have several variants. The assembler provides two pseudo-instructions for loading addresses and immediate values.

The ARM processor provides only two types of branch instruction. The `bl` instruction is used to call subroutines (functions). The `b` instruction can be used to create loops and to create if-then-else constructs. The ability to append a condition to almost any instruction results in a very rich instruction set.

Exercises

3.1   Which registers hold the stack pointer, return address, and program counter?

3.2   Which is more efficient for loading a constant value, the `ldr` pseudo-instruction, or the `mov` instruction? Explain.

3.3   Which two variants of the Store Multiple instruction are used most often, and why?

3.4   The `stm` and `ldm` instructions include an optional '!' after the address register. What does it do?

3.5   The following C statement declares an array of four integers, and initializes their values to 7, 3, 21, and 10, in that order.

```
int nums[]={7,3,21,10};
```

(a)   Write the equivalent in GNU ARM assembly.

(b)   Write the ARM assembly instructions to load all four numbers into registers `r3`, `r5`, `r6`, and `r9`, respectively, using:

i.   a single `ldm` instruction, and

ii.   four `ldr` instructions.

3.6   What is the difference between a memory location and a CPU register?

3.7   How many registers are provided by the ARM Instruction Set Architecture?

3.8   Use `ldm` and `stm` to write a short sequence of ARM assembly language to copy 16 words of data from a source address to a destination address. Assume that the source address is already loaded in `r0` and the destination address is already loaded in `r1`. You may use registers `r2` through `r5` to hold values as needed. Your code is allowed to modify `r0` and/or `r1`.

3.9   Assume that `x` is an array of integers. Convert the following C statements into ARM assembly language.

(a)   `x[8] = 100;`

(b)   `x[10] = x[0];`

(c)   `x[9] = x[3];`

3.10   Assume that `x` is an array of integers, and `i` and `j` are integers. Convert the following C statements into ARM assembly language.

(a)   `x[i] = j;`

(b)   `x[j] = x[i];`

(c)   `x[i] = x[j*2];`

3.11   What is the difference between the `b` instruction and the `bl` instruction? What is each used for?

3.12   What are the meanings of the following instructions?

(a)   `ldreq`

(b)   `ldrlt`

(c)   `bgt`

(d)   `bne`

(e)   `bge`

CHAPTER 4

Data Processing and Other Instructions

Chapter Outline

The ARM processor has approximately 25 data processing instructions. The exact number depends on the processor version. For example, older versions of the architecture did not have the six multiply instructions, and the Cortex M3 and newer processors have two division instructions. There are also a few special instructions that are used infrequently to perform operations that are not classified as load/store, branch, or data processing.

4.1 Data Processing Instructions

The data processing instructions operate only on CPU registers, so data must first be moved from memory into a register before processing can be performed. Most of these instructions use two source operands and one destination register. Each instruction performs one basic arithmetical or logical operation. The operations are grouped in the following categories:

- Arithmetic Operations,
- Logical Operations,
- Comparison Operations,
- Data Movement Operations,
- Status Register Operations,
- Multiplication Operations, and
- Division Operations.

4.1.1 Operand2

Most of the data processing instructions require the programmer to specify two *source operands* and one *destination register* for the result. Because three items must be specified for these instructions, they are known as *three address instructions*. The use of the word *address* in this case has nothing to do with memory addresses. The term *three address instruction* comes from earlier processor architectures that allow arithmetic operations to be performed with data that is stored in memory rather than registers. The first source operand specifies a register whose contents will be on the A bus in Fig. 3.1. The second source operand will be on the B bus and is referred to as Operand2. Operand2 can be any one of the following three things:

- a register (r0-r15),
- a register (r0-r15) and a *shift operation* to modify it, or
- a 32-bit *immediate value* that can be constructed by shifting, rotating, and/or complementing an 8-bit value.

The options for Operand2 allow a great deal of flexibility. Many operations that would require two instructions on most processors can be performed using a single ARM instruction. Table 4.1 shows the mnemonics used for specifying shift operations, which we refer to as <shift_op>.

Table 4.1 Shift and rotate operations in Operand2

<shift_op>	Meaning	
lsl	Logical **S**hift **L**eft by specified amount	
lsr	Logical **S**hift **R**ight by specified amount	Can be used as
asr	**A**rithmetic **S**hift **R**ight by specified amount	<shift_op>
ror	**RO**tate **R**ight by specified amount	in instructions.
rrx	**R**otate **R**ight by one bit with e**X**tend	

Table 4.2 Formats for Operand2

`#<immediate|symbol>`	A 32-bit immediate value that can be constructed from an 8 bit value
`Rm`	Any of the 16 registers `r0-r15`
`Rm, <shift_op> #<shift_imm>`	The contents of a register shifted or rotated by an immediate amount between 0 and 31
`Rm, <shift_op> Rs`	The contents of a register shifted or rotated by an amount specified by the contents of another register
`Rm, rrx`	The contents of a register rotated right by one bit through the carry flag

The `lsl` operation shifts each bit left by a specified amount *n*. Zero is shifted into the *n* least significant bits, and the most significant *n* bits are lost. The `lsr` operation shifts each bit right by a specified amount *n*. Zero is shifted into the *n* most significant bits, and the least significant *n* bits are lost. The `asr` operation shifts each bit right by a specified amount *n*. The *n* most significant bits become copies of the sign bit (bit 31), and the least significant *n* bits are lost. The `ror` operation rotates each bit right by a specified amount *n*. The *n* most significant bits become the least significant *n* bits. The `RRX` operation rotates one place to the right but the `CPSR` carry flag, C, is included. The carry flag and the register together create a 33 bit quantity to be rotated. The carry flag is rotated into the most significant bit of the register, and the least significant bit of the register is rotated into the carry flag. Table 4.2 shows all of the possible forms for Operand2.

4.1.2 Comparison Operations

These four comparison operations update the `CPSR` flags, but have no other effect:

`cmp`   Compare,
`cmn`   Compare Negative,
`tst`   Test Bits, and
`teq`   Test Equivalence.

They each perform an arithmetic operation, but the result of the operation is discarded. Only the CPSR carry flags are affected.

Syntax

```
<op>{<cond>}    Rn, Operand2
```

- `<op>` is either `cmp`, `cmn`, `tst`, or `teq`.
- The optional `<cond>` can be any of the codes from Table 3.2 specifying conditional execution.

Operations

Name	Effect	Description
cmp	$Rn - operand2$	Compare and set CPSR flags
cmn	$Rn + operand2$	Compare negative and set CPSR flags
tst	$Rn \wedge operand2$	Test bits and set CPSR flags
teq	$Rn \oplus operand2$	Test equivalence and set CPSR flags

Examples

```
1      cmp    r0, r1    @ Compare r0 to r1 and set CPSR flags
2      tsteq  r2, #5    @ Compare r2 to 5 and set CPSR flags
```

Example 4.1 shows how conditional execution and the test instruction can be used together to create an if-then-else structure. Note that in this case, the assembly code is more concise than the C code. That is not generally true.

Example 4.1   Making an If-Then-Else Construct

The following C code adds three to a if a is odd, and adds seven to a if a is even.

```
1      :
2    if( a & 1 )
3      a += 3;
4    else
5      a += 7;
6      :
```

Assuming that the value of a is currently being stored in register r4, the following ARM assembly code performs the same function:

```
1      :
2      tst    r4,#1    @ Compare bit zero of a to 1
3      addne  r4,r4,#3 @ if bit 0 is set, add 3 to a
4      addeq  r4,r4,#7 @ else add 7 to a
5      :
```

4.1.3 Arithmetic Operations

There are six basic arithmetic operations:

add Add,
adc Add with Carry,
sub Subtract,
sbc Subtract with Carry,
rsb Reverse Subtract, and
rsc Reverse Subtract with Carry.

All of them involve two 32-bit source operands and a destination register.

Syntax

```
<op>{<cond>}{s}  Rd, Rn, Operand2
```

- `<op>` is one of add, adc, sub, sbc, or rsb, or rsc.
- The optional s specifies whether or not the instruction should affect the bits in the CPSR.
- The optional `<cond>` can be any of the codes from Table 3.2 on page 59 specifying conditional execution.

Operations

Name	Effect	Description
add	$Rd \leftarrow Rn + operand2$	Add
adc	$Rd \leftarrow Rn + operand2 + carry$	Add with carry
sub	$Rd \leftarrow Rn - operand2$	Subtract
sbc	$Rd \leftarrow Rn - operand2 + carry - 1$	Subtract with carry
rsb	$Rd \leftarrow operand2 - Rn$	Reverse subtract
rsc	$Rd \leftarrow operand2 - Rn + carry - 1$	Reverse subtract with carry

Examples

```
1        add    r0, r1, r2  @ r0=r1+r2 and don't set
2                           @ the CPSR flags
3        subgt  r3, r3, #1  @ if (gt) then r3=r3-1 and
4                           @ don't set the CPSR flags
5        rsbles r4, r5, #5  @ if (le) then r4=5-r5 and
6                           @ set the CPSR flags
7        orn    r6,r6,r6    @ Complement the value in r6
```

Example 4.2 shows a complete program for adding the contents of two statically allocated variables and printing the result. The `printf()` function expects to find the address of a string in `r0`. As it prints the string, it finds the `%d` formatting command, which indicates that the value of an integer variable should be printed. It expects the variable to be stored in `r1`. Note that the variable `sum` does not need to be stored in memory. It is stored in `r1`, where `printf()` expects to find it.

Example 4.2   Adding the Contents of Two Variables

The following C program will add together two numbers stored in memory and print the result.

```c
1   #include <stdio.h>
2   static int x = 5;
3   static int y = 8;
4   int main()
5   {
6     int sum;
7     sum = x + y;
8     printf("The sum is %d\n",sum);
9     return 0;
10  }
```

The equivalent ARM assembly program is as follows:

```
1            .data
2   fmt:     .asciz  "The sum is %d\n"
3            .align
4   x:       .word   5
5   y:       .word   8
6            .text
7            .global main
8            @ The bl instruction to call printf() will overwrite
9            @ the link register, so we save it to the stack.
10  main:    stmfd   sp!,{lr}  @ push link register to stack
11           ldr     r1,=x     @ Load address of x
12           ldr     r1,[r1]   @ Load value of x
13           ldr     r2,=y     @ Load address of y
14           ldr     r2,[r2]   @ Load value of y
15           add     r1,r1,r2  @ add x and y
16           ldr     r0,=fmt   @ Load address of format string
17           bl      printf    @ Call the printf function
18           ldmfd   sp!,{lr}  @ Pop link register from the stack
19           mov     r0,#0     @ Load zero as return value
20           mov     pc,lr     @ Return from main
```

Example 4.3 shows how the compare, branch, and add instructions can be used to create a loop. There are basically three steps for creating a loop: allocating and initializing the loop variable, testing the loop variable, and modifying the loop variable. In general, any of the registers r0-r12 can be used to hold the loop variable. Section 5.4 introduces some considerations for choosing an appropriate register. For now, it is assumed that r0 is available for use as the loop variable for this example.

Example 4.3   Making a Loop

Suppose we want to implement a loop that is equivalent to the following C code:

```
1      ⋮
2    for(i=1;i<=10;i++)
3      {
4        ⋮
5        /* insert loop body statements here */
6        ⋮
7      }
8      ⋮
```

The loop can be written with the following ARM assembly code:

```
1          ⋮
2          mov    r0,#1     @ Use r0 as the loop counter (i)
3  loop:                    @ Provide a label
4          cmp    r0,#10    @ Loop from one to ten
5          bgt    endloop   @ Exit loop if r0 > 10
6          ⋮
7          @ Insert loop body instructions here
8          ⋮
9          add    r0,r0,#1  @ Increment the loop counter
10         b      loop      @ Go back to top of loop
11 endloop:                 @ Provide a label
12         ⋮
```

4.1.4 Logical Operations

There are five basic logical operations:

and   Bitwise AND,
orr   Bitwise OR,
eor   Bitwise Exclusive OR,

orn   Bitwise OR NOT, and

bic   Bit Clear.

All of them involve two source operands and a destination register.

Syntax

```
<op>{<cond>}{s}  Rd, Rn, Operand2
```

- <op> is either and, eor, orr, orn, or bic.
- The optional s specifies whether or not the instruction should affect the bits in the CPSR.
- The optional <cond> can be any of the codes from Table 3.2 specifying conditional execution.

Operations

Name	Effect	Description
and	$Rd \leftarrow Rn \wedge operand2$	Bitwise AND
orr	$Rd \leftarrow Rn \vee operand2$	Bitwise OR
eor	$Rd \leftarrow Rn \oplus operand2$	Bitwise Exclusive OR
orn	$Rd \leftarrow \neg(Rn \vee operand2)$	Complement of Bitwise OR
bic	$Rd \leftarrow Rn \wedge \neg operand2$	Bit Clear

Examples

```
1        and    r0, r1, r2    @ r0=r1&r2 and don't set
2                             @ the CPSR flags
3        biceq  r3, r3, #1    @ if (eq) then r3=r3&!0x00000001
4                             @ and don't set CPSR flags
5        eorles r4, r5, #5    @ if (le) then r4=r5^0x00000005
6                             @ and set CPSR flags
```

4.1.5 Data Movement Operations

The data movement operations copy data from one register to another:

mov   Move,

mvn   Move Not, and

movt   Move Top.

The movt instruction copies 16 bits of data into the upper 16 bits of the destination register, without affecting the lower 16 bits. It is available on ARMv6T2 and newer processors.

Syntax

```
<op>{<cond>}{s}        Rd, Operand2
movt{<cond>}           Rd, #immed16
```

- `<op>` is one of `mov` or `mvn`.
- The optional `s` specifies whether or not the instruction should affect the bits in the CPSR.
- The optional `<cond>` can be any of the codes from Table 3.2 specifying conditional execution.

Operations

Name	Effect	Description
mov	$Rd \leftarrow operand2$	Copy operand2 to Rd
mvn	$Rn \leftarrow \neg operand2$	Copy 1's complement of operand2
movt	$Rn \leftarrow (immed16 \ll 16) \vee (Rd \wedge 0xFFFF)$	Copy immed16 into upper 16 bits of Rd

Examples

```
1    mov    r0, r1           @ r0 = r1
2    movs   r2, #10          @ r2 = 10
3    mvneq  r1, #0           @ if (eq) then r1 = 0
4    movles r2, r2, asr #1   @ if (le) then r2 = r2 / 2
```

4.1.6 Multiply Operations with 32-bit Results

These two instructions perform multiplication using two 32-bit registers to form a 32-bit result:

mul   Multiply, and
mla   Multiply and Accumulate.

The `mla` instruction adds a third register to the result of the multiplication.

Syntax

```
mul{<cond>}{s}        Rd, Rm, Rs
mla{<cond>}{s}        Rd, Rm, Rs, Rn
```

- The optional `s` specifies whether or not the instruction should affect the bits in the CPSR.
- The optional `<cond>` can be any of the codes from Table 3.2 specifying conditional execution.

Operations

Name	Effect	Description
mul	$Rd \leftarrow Rm \times Rs$	Multiply
mla	$Rd \leftarrow Rm \times Rs + Rn$	Multiply and accumulate

Examples

```
1      mul    r0, r1, r2     @ multiply r1 by r2
2      mla    r0, r1, r2, r3 @ r0 <= r1 * r2 + r3
3      muleq  r0, r1, r2     @ if (eq) then r0<=r1*r2
4      mlas   r0, r1, r2, r3 @ r0 <= r1 * r2 + r3 and
5                            @ set CPSR flags
6      mulnes r0, r1, r2     @ if (ne) then r0<=r1*r2
7                            @ and set CPSR flags
```

4.1.7 Multiply Operations with 64-bit Results

These instructions perform multiplication using two 32-bit registers to form a 64-bit result:

smull   Signed Multiply Long,
umull   Unsigned Multiply Long,
smlal   Signed Multiply and Accumulate Long, and
umlal   Unsigned Multiply and Accumulate Long.

The smlal and umlal instructions add a 64-bit quantity to the result of the multiplication.

Syntax

```
<type><op>l{<cond>}{s}      RdLo, RdHi, Rm, Rs
```

- <type> must be either s for signed or u for unsigned.
- <op> must be either mul, or mla.
- The optional s specifies whether or not the instruction should affect the bits in the CPSR.
- The optional <cond> can be any of the codes from Table 3.2 specifying conditional execution.

Operations

Name	Effect	Description
smull	$RdHi : RdLo \leftarrow Rm \times Rs$	Signed Multiply
umull	$RdHi : RdLo \leftarrow Rm \times Rs$	Unsigned Multiply
smlal	$RdHi : RdLo \leftarrow Rm \times Rs +$ $RdHi : RdLo$	Signed Multiply and Accumulate
umlal	$RdHi : RdLo \leftarrow Rm \times Rs +$ $RdHi : RdLo$	Unsigned Multiply and Accumulate

Examples

```
1      smull    r0, r1, r3, r4 @ r1:r0<-r3*r4
2      smulls   r3, r5, r3, r4 @ r5:r3<-r3*r4 and
3                              @ set CPSR flags
4      umlaleq  r0, r1, r3, r4 @ if (eq) then
5                              @ r1:r0<-r3*r4 + r1:r0 and
6                              @ set CPSR flags
```

4.1.8 Division Operations

Some ARM processors have the following instructions to perform division:

sdiv   Signed Divide, and
udiv   Unsigned Divide.

The divide operations are available on Cortex M3 and newer ARM processors. The processor used on the Raspberry Pi does not have these instructions. The Raspberry Pi 2 does have them.

Syntax

```
<type>div{<cond>}{s}    Rd, Rm, Rn
```

- <type> must be either s for signed or u for unsigned.
- The optional <cond> can be any of the codes from Table 3.2 specifying conditional execution.
- The optional s specifies whether or not the instruction should affect the bits in the CPSR.

Operations

Name	Effect	Description
sdiv	$Rd \leftarrow Rm \div Rn$	Signed Divide
udiv	$Rd \leftarrow Rm \div Rn$	Unsigned Divide

Examples

```
1        sdiv  r0, r1, r2   @ r0 <= r1 / r2 (signed)
2        udivs r0, r1, r2   @ r0 <= r1 / r2 (unsigned)
3                           @ and set CPSR flags
```

4.2 Special Instructions

There are a few instructions that do not fit into any of the previous categories. They are used to request operating system services and access advanced CPU features.

4.2.1 Count Leading Zeros

This instruction counts the number of leading zeros in the operand register and stores the result in the destination register:

clz   Count Leading Zeros.

Syntax

```
    clz{<cond>}     Rd, Rm
```

- The optional <cond> can be any of the codes from Table 3.2 specifying conditional execution.

Operations

Name	Effect	Description
clz	$Rd \leftarrow 31 - \lfloor log_2(Rm) \rfloor$	Count leading zeros in Rm

Example

```
1        clz   r8,r0
```

4.2.2 Accessing the CPSR and SPSR

These two instructions allow the programmer to access the status bits of the CPSR and SPSR:

mrs   Move Status to Register, and
msr   Move Register to Status.

The SPSR is covered in Section 14.1.

Syntax

```
mrs{<cond>}      Rd, <CPSR|SPSR>{_<fields>}
msr{<cond>}      <CPSR|SPSR>{_<fields>}, Rd
```

- The optional <fields> is any combination of:

  c   control field
  x   extension field
  s   status field
  f   flags field

- The optional <cond> can be any of the codes from Table 3.2 specifying conditional execution.

Operations

Name	Effect	Description
mrs	$Rd \leftarrow CPSR|SPSR$	Move from Status Register
msr	$CPSR|SPSR \leftarrow Rn$	Move to Status Register

Examples

```
1      mrs  R0, CPSR              @ Read the CPSR into r0
2      bic  R0,R0, #0xF0000000    @ Clear all of the flags
3      msr  CPSR_f, R0            @ Write the flags to CPSR
```

4.2.3 Software Interrupt

The following instruction allows a user program to perform a *system call* to request operating system services:

swi   Software Interrupt.

In Unix and Linux, the system calls are documented in the second section of the online manual. Each system call has a unique id number which is defined in the `/usr/include/syscall.h` file.

Syntax

```
swi     <syscall_number>
```

- The `<syscall_number>` is encoded in the instruction. The operating system may examine it to determine which operating system service is being requested.
- In Linux, `<syscall_number>` is ignored. The system call number is passed in `r7`, and up to seven parameters are passed in `r0-r6`. No Linux system call requires more than seven parameters.

Operations

Name	Effect	Description
swi	Request Operating System Service	Perform software interrupt

Example

```
1       @ the following code asks the operating system
2       @ to write some characters to standard output
3       mov  r0, #1    @ file descriptor 1 is stdout
4       ldr  r1, =msg  @ load address of data to write
5       ldr  r2, =len  @ load number of bytes to write
6       mov  r7, #4    @ syscall #4 is the write() function
7       swi  #0        @ invoke syscall
```

4.2.4 Thumb Mode

The ARM processor has an alternate mode where it executes a 16-bit instruction set known as Thumb. This instruction allows the programmer to change the processor mode and branch to Thumb code:

bx   Branch and Exchange.

The thumb instruction set is sometimes more efficient than the full ARM instruction set, and may offer advantages on small systems.

Syntax

```
bx{<cond>}  Rn
blx{<cond>} Rn
```

Operations

Name	Effect	Description
bx	*pc ← target_address*	Branch and change to ARM state. Bit 0 of Rn must be set to 1. Used to return from a Thumb subroutine
blx	*lr ← pc ∨ 1* *pc ← target_address*	Branch and link with change to Thumb state. Bit 0 of Rn must be set to 1. Bit 0 of lr will be set to 1

Example

```
blx     thumb_sub @ call thumb subroutine
  ⋮
bx      lr        @ return from subroutine
```

4.3 Pseudo-Instructions

The assembler provides a small number of pseudo-instructions. From the perspective of the programmer, these instructions are indistinguishable from standard instructions. However, when the assembler encounters a pseudo-instruction, it may substitute a different instruction or generate a short sequence of machine instructions.

4.3.1 No Operation

This pseudo instruction does nothing, but takes one clock cycle to execute.

nop   No Operation.

This is equivalent to a mov r0,r0 instruction.

Syntax

```
nop
```

Operations

Name	Effect	Description
nop	No effects	No Operation

Examples

```
1        ⋮
2        @@ a programmed delay loop
3        mov     r0,#100   @ load loop counter
4   loop: nop
5        nop
6        nop
7        nop
8        sub     r0,r0,#1  @ decrement counter
9        cmp     r0,#0
10       bgt     loop
11       ⋮
```

4.3.2 Shifts

These pseudo instructions are assembled into mov instructions with an appropriate shift of Operand2:

lsl   Logical Shift Left,

lsr   Logical Shift Right,

asr   Arithmetic Shift Right,

ror   Rotate Right, and

rrx   Rotate Right with eXtend.

Syntax

```
<op>{<cond>}{s}  Rd, Rn, Rs
<op>{<cond>}{s}  Rd, Rn, #shift
rrx{<cond>}{s}       Rd, Rn
```

- <op> must be either lsl, lsr, asr, or ror.
- Rs is a register holding the shift amount. Only the least significant byte is used.
- shift must be between 1 and 32.
- If the optional s is specified, then the N and Z flags are updated according to the result, and the C flag is updated to the last bit shifted out.
- The optional <cond> can be any of the codes from Table 3.2 on page 59 specifying conditional execution.

Operations

Name	Effect	Description
lsl	$Rd \leftarrow Rn \ll shift$	Shift Left
lsr	$Rd \leftarrow Rn \gg shift$	Shift Right
asr	$Rd \leftarrow Rn \gg shift$	Shift Right with sign extend
rrx	$Rd : Carry \leftarrow Carry : Rd$	Rotate Right with eXtend

The `rrx` operation rotates one place to the right but the `CPSR` carry flag, C, is included. The carry flag and the register together create a 33-bit quantity to be rotated. The carry flag is rotated into the most significant bit of the register, and the least significant bit of the register is rotated into the carry flag.

Examples

```
1    lsls   r0, r1, #1  @ r0<-r1<<1 and set CPSR flags
2    asr    r3, r3, r0  @ r3<-r3<<r0
3    lsr    r4, r5, #5  @ r4<-r5<<5
4    rrx    r0, r0      @ rotate r0 and carry one bit right
```

4.4 Alphabetized List of ARM Instructions

This chapter and the previous one introduced the core set of ARM instructions. Most of these instructions were introduced with the very first ARM processors. There are approximately 50 additional instructions and pseudo instructions that were introduced with the ARMv6 and later versions of the architecture, or that only appear in specific versions of the ARM. There are also additional instructions available on systems that have the Vector Floating Point (VFP) coprocessor and/or the NEON extensions. The instructions introduced so far are:

Name	Page	Operation
adc	83	Add with Carry
add	83	Add
adr	75	Load Address
adrl	75	Load Address Long
and	85	Bitwise AND
asr	94	Arithmetic Shift Right
b	70	Branch
bic	86	Bit Clear
bl	71	Branch and Link
bx	92	Branch and Exchange

clz	90	Count Leading Zeros
cmn	81	Compare Negative
cmp	81	Compare
eor	85	Bitwise Exclusive OR
ldm	65	Load Multiple Registers
ldr	73	Load Immediate
ldr	64	Load Register
ldrex	69	Load Multiple Registers
lsl	94	Logical Shift Left
lsr	94	Logical Shift Right
mla	87	Multiply and Accumulate
mov	86	Move
movt	86	Move Top
mrs	91	Move Status to Register
msr	91	Move Register to Status
mul	87	Multiply
mvn	86	Move Not
nop	93	No Operation
orn	86	Bitwise OR NOT
orr	85	Bitwise OR
ror	94	Rotate Right
rrx	94	Rotate Right with eXtend
rsb	83	Reverse Subtract
rsc	83	Reverse Subtract with Carry
sbc	83	Subtract with Carry
sdiv	89	Signed Divide
smlal	88	Signed Multiply and Accumulate Long
smull	88	Signed Multiply Long
stm	65	Store Multiple Registers
str	64	Store Register
strex	69	Store Multiple Registers
sub	83	Subtract
swi	91	Software Interrupt
swp	68	Load Multiple Registers
teq	81	Test Equivalence
tst	81	Test Bits
udiv	89	Unsigned Divide
umlal	88	Unsigned Multiply and Accumulate Long
umull	88	Unsigned Multiply Long

4.5 Chapter Summary

The ARM Instruction Set Architecture includes 17 registers and four basic instruction types. This chapter introduced the instructions used for

- moving data from one register to another,
- performing computational operations with two source operands and one destination register,

- multiplication and division,
- performing comparisons, and
- performing special operations.

Most of the data processing instructions are three address instructions, because they involve two source operands and produce one result. For most instructions, the second source operand can be a register, a rotated or shifted register, or an immediate value. This flexibility results in a relatively powerful assembly language. In addition, almost all instructions can be executed conditionally, which, if used properly, results in very efficient and compact code.

Exercises

4.1   If r0 initially contains 1, what will it contain after the third instruction in the sequence below?

```
1        add     r0,r0,#1
2        mov     r1,r0
3        add     r0,r1,r0 lsl #1
```

4.2   What will r0 and r1 contain after each of the following instructions? Give your answers in base 10.

```
1        mov     r0,#1
2        mov     r1,#0x20
3        orr     r1,r1,r0
4        lsl     r1,#0x2
5        orr     r1,r1,r0
6        eor     r0,r0,r1
7        lsr     r1,r0,#3
```

4.3   What is the difference between lsr and asr?

4.4   Write the ARM assembly code to load the numbers stored in num1 and num2, add them together, and store the result in numsum. Use only r0 and r1.

4.5   Given the following variable definitions:

```
1    num1:    .word    x
2    num2:    .word    y
```

where you do not know the values of *x* and *y*, write a short sequence of ARM assembly instructions to load the two numbers, compare them, and move the largest number into register r0.

4.6 Assuming that a is stored in register r0 and b is stored in register r1, show the ARM assembly code that is equivalent to the following C code.

```
1    if ( a & 1 )
2        a = -a;
3    else
4        b = b+7;
```

4.7 Without using the `mul` instruction, give the instructions to multiply r3 by the following constants, leaving the result in r0. You may also use r1 and r2 to hold temporary results, and you do not need to preserve the original contents of r3.
  (a) 10
  (b) 100
  (c) 575
  (d) 123

4.8 Assume that r0 holds the least significant 32 bits of a 64-bit integer *a*, and r1 holds the most significant 32 bits of *a*. Likewise, r2 holds the least significant 32 bits of a 64-bit integer *b*, and r3 holds the most significant 32 bits of *b*. Show the shortest instruction sequences necessary to:
  (a) compare *a* to *b*, setting the CPSR flags,
  (b) shift *a* left by one bit, storing the result in *b*,
  (c) add *b* to *a*, and
  (d) subtract *b* from *a*.

4.9 Write a loop to count the number of bits in r0 that are set to 1. Use any other registers that are necessary.

4.10 The C standard library provides the `open()` function, which is documented in the second section of the Linux manual pages. This function is a very small "wrapper" to allow C programmers to access the `open()` system call. Assembly programmers can access the system call directly. In ARM Linux, the system call number for `open()` is 5. The values for flag constants used with `open()` are defined in

`/usr/include/bits/fcntl-linux.h`.

Write the ARM assembly instructions and directives necessary to make a Linux system call to open a file named `input.txt` for reading, without using the C standard library. In other words, write the assembly equivalent to: `open("input.txt",O_RDONLY);` using the `swi` instruction.

Structured Programming

Chapter Outline

Before IBM released FORTRAN in 1957, almost all programming was done in assembly language. Part of the reason for this is that nobody knew how to design a good high-level language, nor did they know how to write a compiler to generate efficient code. Early attempts at high-level languages resulted in languages that were not well structured, difficult to read, and difficult to debug. The first release of FORTRAN was not a particularly elegant language by today's standards, but it *did* generate efficient code.

In the 1960s, a new paradigm for designing high-level languages emerged. This new paradigm emphasized grouping program statements into blocks of code that execute from beginning to end. These *basic blocks* have only one entry point and one exit point. Control of which basic blocks are executed, and in what order, is accomplished with highly structured *flow control* statements. The *structured program theorem* provides the theoretical basis of structured programming. It states that there are three ways of combining basic blocks: sequencing, selection, and iteration. These three mechanisms are sufficient to express any computable function. It has been proven that all programs can be written using only basic blocks, the *pre-test loop*, and *if-then-else* structure. Although most high-level languages provide additional statements for the convenience of the programmer, they are just "syntactical sugar." Other structured programming concepts include well-formed functions and procedures, pass-by-reference and pass-by-value, separate compilation, and information hiding.

These structured programming languages enabled programmers to become much more productive. Well-written programs that adhere to structured programming principles are much easier to write, understand, debug, and maintain. Most successful high-level languages are designed to enforce, or at least facilitate, good programming techniques. This is not generally true for assembly language. The burden of writing a well-structured code lies with the programmer, and not with the language.

The best assembly programmers rely heavily on structured programming concepts. Failure to do so results in code that contains unnecessary branch instructions and, in the worst cases, results in something called *spaghetti code*. Consider a code listing where a line has been drawn from each branch instruction to its destination. If the result looks like someone spilled a plate of spaghetti on the page, then the listing is spaghetti code. If a program is spaghetti code, then the flow of control is difficult to follow. Spaghetti code is much more likely to have bugs and is extremely difficult to debug. If the flow of control is too complex for the programmer to follow, then it cannot be adequately debugged. It is the responsibility of the assembly language programmer to write code that uses a block-structured approach.

Adherence to structured programming principles results in code that has a much higher probability of working correctly. Well-written code also has fewer branch statements, making the percentage of data processing statements versus branch statements is higher. High data processing density results in higher throughput of data. In other words, writing code in a structured manner leads to higher efficiency.

5.1 Sequencing

Sequencing simply means executing statements (or instructions) in a linear sequence. When statement n is completed, statement $n + 1$ will be executed next. Uninterrupted sequences of statements form basic blocks. Basic blocks have exactly one entry point and one exit point. Flow control is used to select which basic block should be executed next.

5.2 Selection

The first control structure that we will examine is the basic *selection* construct. It is called selection because it selects one of the two (or possibly more) blocks of code to execute, based on some condition. In its most general form, the condition could be computed in a variety of ways, but most commonly it is the result of some comparison operation or the result of evaluating a Boolean expression.

Most languages support selection in the form of an if-then-else statement. Selection can be implemented very easily in ARM assembly language with a two-stage process:

1. perform an operation that updates the CPSR flags, and
2. use conditional execution to select a block of instructions to execute.

Because the ARM architecture supports conditional execution on almost every instruction, there are two basic ways to implement this control structure: by using conditional execution on all instructions in a block, or by using branch instructions. The conditional execution can be applied directly to instructions following the flag update, or to branch instructions that transfer execution to another location. Listing 5.1 shows a typical if-then-else statement in C.

5.2.1 Using Conditional Execution

Listing 5.2 shows the ARM code equivalent to Listing 5.1, using conditional execution. The then and else are written with one instruction each on lines 7 and 8. The then section is written as a conditional instruction with the lt condition attached. The else section is a single instruction with the opposite (ge) condition. Therefore only one of the two instructions will actually execute, depending on the results of the cmp instruction. If there are *three or fewer*

```
1          :
2   static int a,b,x;
3          :
4        if ( a < b )
5          x = 1;
6        else
7          x = 0;
8          :
```

Listing 5.1
Selection in C.

```
 1              ⋮
 2         ldr    r0, =a        @ load pointer to 'a'
 3         ldr    r1, =b        @ load pointer to 'b'
 4         ldr    r0, [r0]      @ load 'a'
 5         ldr    r1, [r1]      @ load 'b'
 6         cmp    r0, r1        @ compare them
 7         movlt  r0, #1        @ THEN section - load 1 into r0
 8         movge  r0, #0        @ ELSE section - load 0 into r0
 9         ldr    r1, =x        @ load pointer to 'x'
10         str    r0, [r1]      @ store r0 in 'x'
11              ⋮
```

Listing 5.2
Selection in ARM assembly using conditional execution.

instructions in each block that can be selected, then this is the preferred and most efficient method of writing the bodies of the then and else selections.

5.2.2 Using Branch Instructions

Listing 5.3 shows the ARM code equivalent to Listing 5.1, using branch instructions. Note that this method requires a conditional branch, an unconditional branch, and two labels. If there are *more than three* instructions in either basic block, then this is the preferred and most efficient method of writing the bodies of the then and else selections.

```
 1                    ⋮
 2            ldr     r0, =a        @ load pointer to 'a'
 3            ldr     r1, =b        @ load pointer to 'b'
 4            ldr     r0, [r0]      @ load 'a'
 5            ldr     r1, [r1]      @ load 'b'
 6            cmp     r0, r1        @ compare them
 7            bge     else          @ if a >= b then skip forward
 8            mov     r0, #1        @ THEN section - load 1 into r0
 9            b       after         @ skip the else section
10  else:     mov     r0, #0        @ ELSE section - load 0 into r0
11  after:    ldr     r1, =x        @ load pointer to 'x'
12            str     r0, [r1]      @ store r0 in 'x'
13                    ⋮
```

Listing 5.3
Selection in ARM assembly using branch instructions.

5.2.3 Complex Selection

More complex selection structures should be written with care. Listing 5.4 shows a fragment of C code which compares the variables *a*, *b*, and *c*, and sets the variable *x* to the least of the three values. In C, Boolean expressions use *short-circuit evaluation*. For example, consider the Boolean AND operator in the expression `((a<b)&&(a<c))`. If the first sub-expression evaluates to false, then the truth value of the complete expression can be immediately determined to be false, so the second sub-expression is not evaluated. This usually results in the compiler generating very efficient assembly code. Good programmers can take advantage of short-circuiting by checking array bounds early in a Boolean expression and accessing array elements later in the expression. For example, the expression `((i<15)&&(array[i]<0))` makes sure that the index i is less than 15 before attempting to access the array. If the index is greater than 14, the array access will not take place. This prevents the program from attempting to access the 16th element on an array that has only 15 elements.

Listing 5.5 shows an ARM assembly code fragment which is equivalent to Listing 5.4. In this code fragment, r0 is used to store a temporary value for the variable x, and the value is only stored to memory once at the end of the fragment of code. The outer if-then-else statement is implemented using branch instructions. The first comparison is performed on line 8. If the comparison evaluates to false, then it immediately branches to the else block of the outer if-then-else statement. But if the first comparison evaluates to true, then it performs the second comparison. Again, if that comparison evaluates to false, then it branches to the else block of the outer if-then-else statement. If both comparisons evaluate to true, then it executes the then block of the outer if-then-else statement, and then branches to the statement following the else block.

The if-then-else statement on line 5 of Listing 5.4 is implemented using conditional execution. The comparison is performed on line 13 of Listing 5.5. Lines 14 and 15 contain

```
1        :
2        if (( a < b ) && (a < c))
3          x = a;
4        else
5          if ( b < c )
6            x = b;
7          else
8            x = c;
9        :
```

Listing 5.4
Complex selection in C.

```
 1        ⋮
 2        ldr     r0, =a        @ load pointer to 'a'
 3        ldr     r1, =b        @ load pointer to 'b'
 4        ldr     r2, =c        @ load pointer to 'c'
 5        ldr     r0, [r0]      @ load 'a'
 6        ldr     r1, [r1]      @ load 'b'
 7        ldr     r1, [r2]      @ load 'c'
 8        cmp     r0, r1        @ compare 'a' and 'b'
 9        bge     else          @ first test failed, go to outer else
10        cmp     r0, r2        @ compare 'a' and 'c'
11        bge     else          @ second test failed, go to outer else
12        b       finish        @ outer THEN section - 'a' is in r0
13 else:  cmp     r1, r2        @ compare 'b' and 'c'
14        movlt   r0, r1        @ inner THEN section: move 'b' into r0
15        movge   r0, r2        @ inner ELSE section: move 'c' into r0
16 finish: ldr    r1, =x        @ load pointer to 'x'
17        str     r0, [r1]      @ store r0 in 'x'
18        ⋮
```

Listing 5.5
Complex selection in ARM assembly.

instructions that are conditionally executed. Since they have complementary conditions, it is guaranteed that one of them will move a value into r0. The comparison on line 13 determines which statement executes.

Note that the number of comparisons performed will always be minimized, and the number of branches has also been minimized. The only way that line 13 can be reached is if one of the first two comparisons evaluates to false. If line 2 is executed, then no matter which sequence of events occurs, the program fragment will *always* reach line 16 and a value will be stored in x. Thus, the ARM assembly code fragment in Listing 5.5 can be considered to be a block of code with exactly one entry point and one exit point.

When writing nested selection structures, it is important to maintain a block structure, even if the bodies of the blocks consist of only a single instruction. It is often very helpful to write the algorithm in pseudo-code or a high-level language, such as C or Java, before converting it to assembly. Prolific commenting of the code is also strongly encouraged.

5.3 Iteration

Iteration involves the transfer of control from a statement in a sequence to a *previous* statement in the sequence. The simplest type of iteration is the *unconditional* loop, also known

```
1       :
2           ldr     r0,=hellostr @ load pointer to "Hello World\n\0"
3   loop:   bl      printf       @ print "Hello World\n"
4           b       loop         @ repeat loop unconditionally
5       :
```

Listing 5.6
Unconditional loop in ARM assembly.

as the *infinite* loop. This type of loop may be used in programs or tasks that should continue running indefinitely. Listing 5.6 shows an ARM assembly fragment containing an unconditional loop. Few high-level languages provide a true unconditional loop, but the high-level programmer can achieve a similar effect by using a conditional loop and specifying a condition that always evaluates to true.

5.3.1 Pre-Test Loop

A pre-test loop is a loop in which a test is performed *before* the block of instructions forming the loop body is executed. If the test evaluates to true, then the loop body is executed. The last instruction in the loop body is a branch back to the beginning of the test. If the test evaluates to false, then execution branches to the first instruction following the loop body. All structured programming languages have a pre-test loop construct. For example, in C, the pre-test loop is called a `while` loop. In assembly, a pre-test loop is constructed very similarly to an `if-then` statement. The only difference is that it includes an additional branch instruction at the end of the sequence of instructions that form the body. Listing 5.7 shows a pre-test loop in ARM assembly.

```
1       :
2   loop:   cmp     r0, r1       @ perform loop test
3           blt     done         @ exit loop if r0 < r1
4       :                        @ body of loop
5           b       loop         @ repeat loop
6   done:
7       :
```

Listing 5.7
Pre-test loop in ARM assembly.

5.3.2 Post-Test Loop

In a post-test loop, the test is performed *after* the loop body is executed. If the test evaluates to true, then execution branches to the first instruction in the loop body. Otherwise, execution continues sequentially. Most structured programming languages have a post-test loop construct. For example, in C, the post-test loop is called a do-while loop. Listing 5.8 shows a post-test loop in ARM assembly. The body of a post-test loop will always be executed at least once.

5.3.3 For Loop

Many structured programming languages have a for loop construct, which is a type of *counting* loop. The for loop is not essential, and is only included as a matter of syntactical convenience. In some cases, a for loop is easier to write and understand than an equivalent pre-test or post-test loop. However, with the addition of an if-then construct, *any* loop can be implemented as a pre-test loop. The following sections show how loops can be converted from one form to another.

Pre-test conversion
Listing 5.9 shows a simple C program with a for loop. The program prints "Hello World" 10 times, appending an integer to the end of each line.

```
1   loop:      :
2              :                  @ body of loop
3         cmp    r0, r1           @ perform loop test
4         blt    loop             @ repeat loop if r0 < r1
5              :
```

Listing 5.8
Post-test loop in ARM assembly.

```
1          :
2      for(i=0;i<10;i++)
3          printf("Hello World - %d\n",i);
4          :
```

Listing 5.9
for loop in C.

```
1        :
2        int i = 0;
3        while(i<10)
4        {
5          printf("Hello World - %d\n",i);
6          i++;
7        }
8        :
```

Listing 5.10
for loop rewritten as a pre-test loop in C.

```
1          :
2          mov    r4, #0      @ use r4 for i; i=0
3   loop:  cmp    r4, #10     @ perform comparison
4          bge    done        @ end loop if i >= 10
5          ldr    r0, =str     @ load pointer to format string
6          mov    r1, r4      @ copy i into r1
7          bl     printf      @ printf("Hello World - %d\n",i);
8          add    r4, r4, #1  @ i++
9          b      loop        @ repeat loop test
10  done:
11         :
```

Listing 5.11
Pre-test loop in ARM assembly.

In order to write an equivalent program in assembly, the programmer must first rewrite the for loop as a pre-test loop. Listing 5.10 shows the program rewritten so that it is easier to translate into assembly. Note that the initialization of the loop variable has been moved to its own line before the while statement. Also, the loop variable is modified on the last line of the loop body. This is a straightforward conversion from one type of loop to another type. Listing 5.11 shows a translation of the pre-test loop structure into ARM assembly.

Post-test conversion
If the programmer can guarantee that the body of a for loop will *always* execute at least once, then the for loop can be converted to an equivalent post-test loop. This form of loop is more efficient, because the loop control variable is tested one time less than for a pre-test loop. Also, a post-test loop requires only one label and one conditional branch instruction, whereas a pre-test loop requires two labels, a conditional branch, and an unconditional branch.

```
1        ⋮
2        int i = 0;
3        do {
4          printf("Hello World - %d\n",i);
5          i++;
6        } while(i<10);
7        ⋮
```

Listing 5.12

for loop rewritten as a post-test loop in C.

```
1              ⋮
2        ldr    r4, #0      @ use r4 for i; i=0
3  loop:  ldr    r0, =str    @ load pointer to format string
4         mov    r1, r4      @ copy i into r1
5         bl     printf      @ printf("Hello World - %d\n",i);
6         add    r4, r4, #1  @ i++
7         cmp    r4, #10     @ perform comparison
8         blt    loop        @ end loop if i >= 10
9              ⋮
```

Listing 5.13

Post-test loop in ARM assembly

Since the loop in Listing 5.9 always executes the body exactly 10 times, we know that the body will always execute at least once. Therefore, the loop can be converted to a post-test loop. Listing 5.12 shows the program rewritten as a post-test loop so that it is easier to translate into assembly. Note that, as in the previous example, the initialization of the loop variable has been moved to its own line before the do-while loop, and the loop variable is modified on the last line of the loop body. This post-test version will produce the same output as the pre-test version. This is a straightforward conversion from one type of loop to an equivalent type. Listing 5.13 shows a straightforward translation of the post-test loop structure into ARM assembly.

5.4 Subroutines

A subroutine is a sequence of instructions to perform a specific task, packaged as a single unit. Depending on the particular programming language, a subroutine may be called a procedure, a function, a routine, a method, a subprogram, or some other name. Some languages, such as Pascal, make a distinction between functions and procedures. A function must return a value and must not alter its input arguments or have any other *side effects* (such as producing output or changing static or global variables). A procedure returns no value, but may alter the value of its arguments or have other side effects.

Other languages, such as C, make no distinction between procedures and functions. In these languages, functions may be described as pure or impure. A function is pure if:

1. the function always evaluates the same result value when given the same argument value(s), and
2. evaluation of the result does not cause any semantically observable side effect or output.

The first condition implies that the result of the function cannot depend on any hidden information or state that may change as program execution proceeds, or between different executions of the program, nor can it depend on any external input from I/O devices. The result value of a pure function does not depend on anything other than the argument values. If the function returns multiple result values, then these two conditions must apply to all returned values. Otherwise the function is impure. Another way to state this is that impure functions have side effects while pure functions have no side effects.

Assembly language does not impose any distinction between procedures and functions, pure or impure. Although every assembly language will provide a way to call subroutines and return from them, it is up to the programmer to decide how to pass arguments to the subroutines and how to pass return values back to the section of code that called the subroutine. Once again, the expert assembly programmer will use structured programming concepts to write efficient, readable, debugable, and maintainable code.

5.4.1 Advantages of Subroutines

Subroutines help programmers to design reliable programs by decomposing a large problem into a set of smaller problems. It is much easier to write and debug a set of small code pieces than it is to work on one large piece of code. Careful use of subroutines will often substantially reduce the cost of developing and maintaining a large program, while increasing its quality and reliability. The advantages of breaking a program into subroutines include:

*   enabling reuse of code across multiple programs,
*   reducing duplicate code within a program,
*   enabling the programming task to be divided between several programmers or teams,
*   decomposing a complex programming task into simpler steps that are easier to write, understand, and maintain,
*   enabling the programming task to be divided into stages of development, to match various stages of a project, and
*   hiding implementation details from users of the subroutine (a programming principle known as *information hiding*).

5.4.2 Disadvantages of Subroutines

There are two minor disadvantages in using subroutines. First, invoking a subroutine (versus using in-line code) imposes overhead. The arguments to the subroutine must be put into some known location where the subroutine can find them. if the subroutine is a function, then the return value must be put into a known location where the caller can find it. Also, a subroutine typically requires some standard entry and exit code to manage the stack and save and restore the return address.

In most languages, the cost of using subroutines is hidden from the programmer. In assembly, however, the programmer is often painfully aware of the cost, since they have to explicitly write the entry and exit code for each subroutine, and must explicitly write the instructions to pass the data into the subroutine. However, the advantages usually outweigh the costs. Assembly programs can get very large and failure to modularize the code by using subroutines will result in code that cannot be understood or debugged, much less maintained and extended.

5.4.3 Standard C Library Functions

Subroutines may be defined within a program, or a set of subroutines may be packaged together in a library. Libraries of subroutines may be used by multiple programs, and most languages provide some built-in library functions. The C language has a very large set of functions in the C standard library. All of the functions in the C standard library are available to any program that has been *linked* with the C standard library. Even assembly programs can make use of this library. Linking is done automatically when gcc is used to assemble the program source. All that the programmer needs to know is the name of the function and how to pass arguments to it.

5.4.4 Passing Arguments

Listing 5.14 shows a very simple C program which reads an integer from standard input using scanf and prints the integer to standard output using printf. An equivalent program written in ARM assembly is shown in Listing 5.15. These examples show how arguments can be passed to subroutines in C and equivalently in assembly language.

All processor families have their own standard methods, or *function calling conventions*, which specify how arguments are passed to subroutines and how function values are returned. The function call standard allows programmers to write subroutines and libraries of subroutines that can be called by other programmers. In most cases, the function calling standards are not enforced by hardware, but assembly programmers and compiler writers conform to the standards in order to make their code accessible to other programmers. The basic subroutine calling rules for the ARM processor are simple:

```
 1   #include <stdio.h>
 2
 3   static char str1[] = "%d";
 4   static char str2[] = "You entered %d\n";
 5   static int n = 0;
 6
 7   int main()
 8   {
 9     scanf(str1,&n);
10     printf(str2,n);
11     return 0;
12   }
```

Listing 5.14
Calling scanf and printf in C.

```
 1          .data
 2   str1:  .asciz  "%d"
 3   str2:  .asciz  "You entered %d\n"
 4   n:     .word   0
 5          .text
 6          .globl  main
 7   main:  stmfd   sp!,{lr}    @ push link register onto stack
 8          ldr     r0, =str1   @ load pointer to format string
 9          ldr     r1, =n      @ load pointer to int variable
10          bl      scanf       @ call scanf("%d",&n)
11          ldr     r0, =str2   @ load pointer to format string
12          ldr     r1, =n      @ load pointer to int variable
13          ldr     r1, [r1]    @ load int variable
14          bl      printf      @ call printf("You entered %d\n",n)
15          mov     r0, #0      @ load return value
16          ldmfd   sp!,{lr}    @ pop link register from stack
17          mov     pc, lr      @ return from main
```

Listing 5.15
Calling scanf and printf in ARM assembly.

- The first four arguments go in registers r0-r3.
- Any remaining arguments are pushed to the stack.

If the subroutine returns a value, then it is stored in r0 before the function returns to its caller. Calling a subroutine in ARM assembly usually requires several lines of code. The number of lines required depends on how many arguments the subroutine requires and where the data for those arguments are stored. Some variables may already be in the correct register. Others may need to be moved from one register to another. Still others may need to be pushed onto the stack. Careful programming is required to minimize the amount of work that must be done just to move the subroutine arguments into their required locations.

The ARM register set was introduced in Chapter 3. Some registers have special purposes that are dictated by the hardware design. Others have special purposes that are dictated by *programming conventions*. Programmers follow these conventions so that their subroutines are compatible with each other. These conventions are simply a set of rules for how registers should be used. In ARM assembly, all registers have alternate names which can be used to help remember the rules for using them. Fig. 5.1 shows an expanded view of the ARM registers, including their alternate names and conventional use.

Registers r0-r3 are also known as a1-a4, because they are used for passing *arguments* to subroutines. Registers r4-r11 are also known as v1-v8, because they are used for holding

r0 (a1)	Used to pass argument values into a subroutine and to return a result value from a function. They may also be used to hold intermediate values within a routine. Caller assumes they will be modified.
r1 (a2)	
r2 (a3)	
r3 (a4)	
r4 (v1)	A subroutine must preserve (or save and restore) the contents of these registers. If they are used, they must be pushed to the stack at the beginning of the subroutine/function, and re-stored before returning.
r5 (v2)	
r6 (v3)	
r7 (v4)	
r8 (v5)	
r9 (v6)	
r10 (v7)	
r11 (fp) (v8)	
r12 (ip)	Intra-procedure scratch register.
r13 (sp)	Program stack pointer.
r14 (lr)	Link Register (return address). See bl instruction.
r15 (pc)	Program Counter. Changing this causes a branch.

CPSR

Figure 5.1
ARM user program registers

local *variables* in a subroutine. As mentioned in Section 3.2, register r11 can also be referred to as fp because it is used by the C compiler to track the *stack frame*, unless the code is compiled using the --omit-frame-pointer command line option.

The intra-procedure scratch register, r12, is used by the C library when calling dynamically linked functions. If a subroutine does not call any C library functions, then it can use r12 as another register to store local variables. If a C library function is called, it may change the contents of r12. Therefore, if r12 is being used to store a local variable, it should be saved to another register or to the stack before a C library function is called.

5.4.5 Calling Subroutines

The stack pointer (sp), link register (lr), and program counter (pc), along with the argument registers, are all involved in performing subroutine calls. The calling subroutine must place arguments in the argument registers, and possibly on the stack as well. Placing the arguments in their proper locations is known as *marshaling* the arguments. After marshaling the arguments, the calling subroutine executes the bl instruction, which will modify the program counter and link register. The bl instruction copies the contents of the program counter to the link register, then loads the program counter with the address of the first instruction in the subroutine that is being called. The CPU will then fetch and execute its next instruction from the address in the program counter, which is the first instruction of the subroutine that is being called.

Our first examples of calling a function will involve the printf function from the C standard library. The printf function can be a bit confusing at first, but it is an extremely useful and flexible function for printing formatted output. The printf function examines its first argument to determine how many other arguments have been passed to it. The first argument is a *format string*, which is a null-terminated ASCII string. The format string may include conversion specifiers, which start with the % character. For each conversion specifier, printf assumes that an argument has been passed in the correct register or location on the stack. The argument is retrieved, converted according to the specified format, and printed. Other specifiers include %X to print the matching argument as an integer in hexadecimal, %c to print the matching argument as an ASCII character, %s to print a zero-terminated string. The integer specifiers can include an optional width and zero-padding specification. For example %8X will print an integer in hexadecimal, using 8 characters. Any leading zeros will be printed as spaces. The format string %08X will print an integer in hexadecimal, using 8 characters. In this case, any leading zeros will be printed as zeros. Similarly, %15d can be used to print an integer in base 10 using spaces to pad the number up to 15 characters, while %015d will print an integer in base 10 using zeros to pad up to 15 characters.

```
1    printf("Hello World");
```

Listing 5.16
Simple function call in C.

```
1    @ load first argument (pointer to  format string) in r0
2    ldr   r0, =hellostr
3    @ call printf
4    bl    printf
```

Listing 5.17
Simple function call in ARM assembly.

```
1    printf("The results are: %d %d %d\n",i,j,k);
```

Listing 5.18
A larger function call in C.

Listing 5.16 shows a call to printf in C. The printf function requires one argument, and can accept more than one. In this case, there is only one argument, the format string. Listing 5.17 shows an equivalent call made in ARM assembly language. The single argument is loaded into r0 in conformance with the ARM subroutine calling convention.

Passing arguments in registers
Listing 5.18 shows a call to printf in C having four arguments. The format string is the first argument. The format string contains three conversion specifiers, and is followed by three more arguments. Arguments are matched to conversion specifiers according to their positions. The type of each argument matches the type indicated in the conversion specifier. The first conversion specifier is applied to the second argument, the second conversion specifier is applied to the third argument, and the third conversion specifier is applied to the fourth argument. The %d conversion specifiers indicate that the arguments are to be interpreted as integers and printed in base 10.

Listing 5.19 shows an equivalent call made in ARM assembly language. The arguments are loaded into r0-r3 in conformance with the ARM subroutine calling convention. Note that we assume that formatstr has previously been defined using a .asciz or .string assembler directive or equivalent method. As long as there are four or fewer arguments that must be passed, they can all fit in registers r0-r3 (a.k.a a1-a4), but when there are more arguments, things become a little more complicated. Any remaining arguments must be passed on the

```
 1   @ load first argument (pointer to format string) in r0
 2   ldr   r0, =formatstr
 3   ldr   r1, =i         @ load pointer to i in r1
 4   ldr   r1, [r1]       @ load value of i in r1
 5   mov   r2, r6         @ value of j was in r6. copy to r2
 6   ldr   r3, =k         @ load pointer to k in r3
 7   ldr   r3, [r3]       @ load value of k in r3
 8   @ call printf
 9   bl    printf
```

Listing 5.19
A larger function call in ARM assembly.

```
 1  printf("The results are: %d %d %d %d %d\n",i,j,k,l,m);
```

Listing 5.20
A function call using the stack in C.

program stack, using the stack pointer r13. Care must be taken to ensure that the arguments are pushed to the stack in the proper order. Also, after the function call, the arguments must be removed from the stack, so that the stack pointer is restored to its original value.

Passing arguments on the stack
Listing 5.20 shows a call to printf in C having more than four arguments. The format string is the first argument. The format string contains five conversion specifiers, which implies that the format string must be followed by five additional arguments. Arguments are matched to conversion specifiers according to their positions. The type of each argument matches the type indicated in the conversion specifier. The first conversion specifier is applied to the second argument, the second conversion specifier is applied to the third argument, the third conversion specifier is applied to the fourth argument, etc. The %d conversion specifiers indicate that the arguments are to be interpreted as integers and printed in base 10.

Listing 5.21 shows an equivalent call made in ARM assembly language. Since there are six arguments, the last two must be pushed to the program stack. The arguments are loaded into r0 one at a time and then the register pre-indexed addressing mode is used to subtract four bytes from the stack pointer and then store the argument at the top of the stack. Note that the sixth argument is pushed to the stack first, followed by the fifth argument. The remaining arguments are loaded in r0-r3. Note that we assume that formatstr has previously been defined to be "The results are: %d %d %d %d %d\n" using an .asciz or .string assembler directive.

```
1    ldr    r0,=m          @ load last argument ('m')
2    ldr    r0,[r0]
3    str    r0,[sp,#-4]! @ push it on the stack
4    ldr    r0,=1          @ load '1'
5    ldr    r0,[r0]
6    str    r0,[sp,#-4]! @ push it on the stack
7    @ load first argument (pointer to format string) in r0
8    ldr    r0, =resultstr
9    ldr    r1, =i         @ load pointer to i in r1
10   ldr    r1, [r1]       @ load value of i in r1
11   mov    r2, r6         @ value of j was in r6. copy to r2
12   mov    r3, r7         @ value of k was in r7. copy to r3
13   @ call printf
14   bl     printf
15   add    sp,sp,#8       @ pop 2 words from the stack
```

Listing 5.21
A function call using the stack in ARM assembly.

```
1    ldr    r3,=m          @ load last argument ('m')
2    ldr    r3,[r3]
3    ldr    r0,=1          @ load '1' in lower numbered reg
4    ldr    r0,[r0]
5    stmfd  sp!,{r0,r3}    @ push them on the stack
6    ldr    r0, =fmtstr    @ load pointer to format string
7    ldr    r1, =i         @ load pointer to i in r1
8    ldr    r1, [r1]       @ load value of i in r1
9    mov    r2, r6         @ copy value of j from r6 to r2
10   mov    r3, r7         @ copy value of k from r7 to r3
11   bl     printf         @ call printf
12   add    sp,sp,#8       @ pop 2 words from stack
```

Listing 5.22
A function call using stm to push arguments onto the stack.

Listing 5.22 shows how the fifth and sixth arguments can be pushed to the stack using a single stmfd instruction. The sixth argument is loaded into r3 and the fifth argument is loaded into r0, then the stmfd instruction is used to store them on the stack and adjust the stack pointer. A little care must be taken to ensure that the arguments are stored in the correct order on the stack. Remember that the stmfd instruction will always push the lowest-numbered register to the lowest address, and the stack grows downward. Therefore, r3, the sixth argument, will be pushed onto the stack first, making it grow downward by four bytes. Next, r0 is pushed, making the stack grow downward by four more bytes. As in the previous example, the remaining four arguments are loaded into `a1-a4`.

After the `printf` function is called, the fifth and sixth arguments must be popped from the stack. If those values are no longer needed, then there is no need to load them into registers. The quickest way to pop them from the stack is to simply adjust the stack pointer back to its original value. In this case, we pushed two arguments onto the stack, using a total of eight bytes. Therefore, all we need to do is add eight to the stack pointer, thereby restoring its original value.

5.4.6 Writing Subroutines

We have looked at the conventions that are followed for calling functions. Now we will examine these same conventions from the point of view of the function being called. Because of the calling conventions, the programmer writing a function can assume that

- the first four arguments are in `r0-r3`,
- any additional arguments can be accessed with `ldr rd,[sp,#offset]`,
- the *calling* function will remove arguments from the stack, if necessary,
- if the function return type is not `void`, then they must enusure that the return value is in `r0` (and possibly `r1, r2, r3`), and
- the return address will be in `lr`.

Also because of the conventions, there are certain registers that can be used freely while others must be preserved or restored so that the calling function can continue operating correctly. Registers which can be used freely are referred to as *volatile*, and registers which must be preserved or restored before returning are referred to as *non-volatile*. When writing a subroutine (function),

- registers `r0-r3` and `r12` are *volatile*,
- registers `r4-r11` and `r13` are *non-volatile* (they can be used, but their contents *must* be restored to their original value before the function returns),
- register `r14` can be used by the function, but its contents must be saved so that the return address can be loaded into `r15` when the function returns to its caller,
- if the function calls another function, then it must save register `r14` either on the stack or in a non-volatile register before making the call.

Listing 5.23 shows a small C function that simply returns the sum of its six arguments. The ARM assembly version of that function is shown in Listing 5.24. Note that on line 5, the fifth argument is loaded from the stack, and on line 7, the sixth argument is loaded in a similar way, using an offset from the stack pointer. If the calling function has followed the conventions,

```
1   int myfun(int a, int b, int c, int d, int e, int f)
2   {
3       return a+b+c+d+e+f;
4   }
```

Listing 5.23
A small function in C.

```
1
2   myfun:  add     r0,r0,r1    @ r0 = a + b
3           add     r0,r0,r2    @ r0 = a + b + c
4           add     r0,r0,r3    @ r0 = a + b + c + d
5           ldr     r1,[sp,#0]  @ load e from stack
6           add     r0,r0,r1    @ r0 = a + b + c + d + e
7           ldr     r1,[sp,#4]  @ load f from stack
8           add     r0,r0,r1    @ r0 = a + b + c + d + e + f
9           mov     pc,lr       @ return from function
```

Listing 5.24
A small function in ARM assembly.

then the fifth and sixth arguments will be where they are expected to be in relation to the stack pointer.

5.4.7 Automatic Variables

In block-structured high-level languages, an automatic variable is a variable that is local to a block of code and not declared with static duration. It has a lifetime that lasts only as long as its block is executing. Automatic variables can be stored in one of two ways:

1. the stack is temporarily adjusted to hold the variable, or
2. the variable is held in a register during its entire life.

When writing a subroutine in assembly, it is the responsibility of the programmer to decide what automatic variables are required and where they will be stored. In high-level languages this decision is usually made by the compiler. In some languages, including C, it is possible to request that an automatic variable be held in a register. The compiler will attempt to comply with the request, but it is not guaranteed. Listing 5.25 shows a small function which requests that one of its variables be kept in a register instead of on the stack.

Listing 5.26 shows how the function could be implemented in assembly. Note that the array of integers consumes 80 bytes of storage on the stack, and could not possibly fit into the registers

```
1   int doit()
2   { int x[20];
3     register int i;    /* try to keep i in a register */
4     for(i=0;i<20;i++) x[i] = i;
5     return i;
6   }
```

Listing 5.25
A small C function with a register variable.

```
1    doit: sub    sp,sp,#80 @ Allocate  'x' on stack
2          mov    r2,#0     @ use r2 as 'i'
3    loop: cmp    r2,#20    @ pre-test loop
4          bge    done      @ quit if i >= 20
5          str    r2,[sp,r2,asl#2] @  x[i] = i;
6          add    r2,r2,#1  @ i++
7          b      loop      @ go back to loop test
8    done: mov    r0,r2     @ return i
9          add    sp,sp,#80 @ destroy automatic variable
10         mov    pc,lr     @ return from function
```

Listing 5.26
Automatic variables in ARM assembly.

available on the ARM processor. However, the loop control variable can easily be stored in one of the registers for the duration of the function. Also notice that on line 1 the storage for the array is allocated simply by adjusting the stack pointer, and on line 9 the storage is released by restoring the stack pointer to its original contents. It is *critical* that the stack pointer be restored, no matter how the function returns. Otherwise, the calling function will probably mysteriously fail. For this reason, each function should have exactly one block of instructions for returning. If the function needs to return from some location other than the end, then it should branch to the return block rather than returning directly.

5.4.8 Recursive Functions

A function that calls itself is said to be *recursive*. Certain problems are easy to implement recursively, but are more difficult to solve iteratively. A problem exhibits recursive behavior when it can be defined by two properties:

1. a simple base case (or cases), and
2. a set of rules that reduce all other cases toward the base case.

For example, we can define person's ancestors recursively as follows:

1. one's parents are one's ancestors (base case),
2. the ancestors of one's ancestors are also one's ancestors (recursion step).

Recursion is a very powerful concept in programming. Many functions are naturally recursive, and can be expressed very concisely in a recursive way. Numerous mathematical axioms are based upon recursive rules. For example, the formal definition of the natural numbers by the Peano axioms can be formulated as:

1. 0 is a natural number, and
2. each natural number has a successor, which is also a natural number.

Using one base case and one recursive rule, it is possible to generate the set of all natural numbers. Other recursively defined mathematical objects include functions and sets.

Listing 5.27 shows the C code for a small program which uses recursion to reverse the order of characters in a string. The base case where recursion ends is when there are fewer than two characters remaining to be swapped. The recursive rule is that the reverse of a string can be created by swapping the first and last characters and then reversing the string between them. In short, a string is reversed if:

1. the string has a length of zero or one character, or
2. the first and last characters have been swapped and the remaining characters have been reversed.

```
1   void reverse(char *a,int left, int right)
2   { char tmp;
3     if(left<right)
4       {
5         tmp=a[left];
6         a[left]=a[right];
7         a[right]=tmp;
8         reverse(a,left+1,right-1);
9       }
10  }
11  int main()
12  { char *str="This is the string to reverse";
13    printf(str);
14    reverse(str,0,strlen(str)-1);
15    printf(str);
16    return 0;
17  }
```

Listing 5.27
A C program that uses recursion to reverse a string.

```
1   reverse:stmfd  sp!,{lr}    @ I may call myself:save lr
2          sub     sp,sp,#4    @ Allocate tmp on stack
3          cmp     r1,r2       @ if(left>=right)
4          bge     exit        @ then return
5          ldrb    r3,[r0,r1]  @ load character at a[left]
6          strb    r3,[sp,#0]  @ store in tmp
7          ldrb    r3,[r0,r2]  @ load character at a[right]
8          strb    r3,[r0,r1]  @ store in a[left]
9          ldrb    r3,[sp,#0]  @ load tmp
10         strb    r3,[r0,r2]  @ store in a[right]
11         add     r1,r1,#1    @ calculate left+1
12         sub     r2,r2,#1    @ calculate right-1
13         bl      reverse     @ make recursive call
14  exit:  ldr     lr,[sp,#4]  @ get lr from 4 bytes above sp
15         add     sp,sp,#8    @ restore sp to original value
16         mov     pc,lr       @ return from function
```

Listing 5.28

ARM assembly implementation of the reverse function.

In Listing 5.27, line 3 checks for the base case. If the string has not been reversed according to the first rule, then the second rule is applied. Lines 5–7 swap the first and last characters, and line 8 recursively reverses the characters between them.

Listing 5.28 shows how the reverse function can be implemented using recursion in ARM assembly. Line 1 saves the link register to the stack and decrements the stack pointer. Next, storage is allocated for an automatic variable. Lines 3 and 4 test for the base case. If the current case is the base case, then the function simply returns (restoring the stack as it goes). Otherwise, the first and last characters are swapped in lines 5 through 10 and a recursive call is made in lines 11 through 13.

The code in Listing 5.28 can be made a bit more efficient. First, the test for the base case can be performed before anything else is done, as shown in Listing 5.29. Also, the local variable tmp can be stored in a volatile register rather than stored on the stack, because it is only needed for lines 4 through 8. It is not needed after the recursive call, so there is really no need to preserve it on the stack. This means that our function can use half as much stack space and will run much faster. This further refined version is shown in Listing 5.30. This version uses ip (r12) as the tmp variable instead of using the stack.

The previous examples used the concept of an array of characters to access the string that is being reversed. Listing 5.31 shows how this problem can be solved in C using pointers to the first and last characters rather than array indices. This version only has two parameters in the reverse function, and uses pointer dereferencing rather than array indexing to access each character. Other than that difference, it works the same as the original version. Listing 5.32

```
1   reverse:cmp    r1,r2      @ if(left>=right)
2           bge    exit       @ then return
3           stmfd  sp!,{lr}   @ I WILL call myself-save lr
4           sub    sp,sp,#4   @ Allocate tmp on stack
5           ldrb   r3,[r0,r1] @ load character at a[left]
6           strb   r3,[sp,#0] @ store in tmp
7           ldrb   r3,[r0,r2] @ load character at a[right]
8           strb   r3,[r0,r1] @ store in a[left]
9           ldrb   r3,[sp,#0] @ load tmp
10          strb   r3,[r0,r2] @ store in a[right]
11          add    r1,r1,#1   @ calculate left+1
12          sub    r2,r2,#1   @ calculate right-1
13          bl     reverse    @ make recursive call
14          ldr    lr,[sp,#4] @ get lr from 4 bytes back
15          add    sp,sp,#8   @ pop tmp and lr
16  exit:   mov    pc,lr      @ return from function
```

Listing 5.29
Better implementation of the reverse function.

```
1   reverse:cmp    r1,r2      @ if(left>=right)
2           bge    exit       @ then return
3           stmfd  sp!,{lr}   @ I WILL call myself-save lr
4           ldrb   r3,[r0,r1] @ load character at a[left]
5           ldrb   ip,[r0,r2] @ load character at a[right]
6           strb   r3,[r0,r2] @ store in a[right]
7           strb   ip,[r0,r1] @ store in a[left]
8           add    r1,r1,#1   @ calculate left+1
9           sub    r2,r2,#1   @ calculate right-1
10          bl     reverse    @ make recursive call
11          ldmfd  sp!,{lr}   @ pop lr from the stack
12  exit:   mov    pc,lr      @ return from function
```

Listing 5.30
Even better implementation of the reverse function.

shows how the reverse function can be implemented efficiently in ARM assembly. This implementation has the same number of instructions as the previous version, but lines 4 through 7 use a different addressing mode. On the ARM processor, the pointer method and the array index method are equally efficient. However, many processors do not have the rich set of addressing modes available on the ARM. On those processors, the pointer method may be significantly more efficient.

```
1   void reverse(char *left, char *right)
2   {
3     char tmp;
4     if(left<=right)
5     {
6       tmp=*left;
7       *left=*right;
8       *right=tmp;
9       reverse(left+1,right-1);
10    }
11  }
12  int main()
13  { char *str="This is the string to reverse";
14    printf(str);
15    reverse(str,str+strlen(str)-1);
16    printf(str);
17    return 0;
18  }
```

Listing 5.31
String reversing in C using pointers.

```
1   reverse:cmp   r0,r1      @ if(left>=right)
2           bge   exit       @ then return
3           stmfd sp!,{lr}   @ I WILL call myself-save lr
4           ldrb  r3,[r0]    @ load character at *left
5           ldrb  ip,[r1]    @ load character at *right
6           strb  ip,[r0]    @ store in *left
7           strb  r3,[r1]    @ store in *right
8           add   r0,r0,#1   @ calculate left+1
9           sub   r1,r1,#1   @ calculate right-1
10          bl    reverse    @ make recursive call
11          ldmfd sp!,{lr}   @ pop lr from the stack
12  exit:   mov   pc,lr      @ return from function
```

Listing 5.32
String reversing in assembly using pointers.

5.5 Aggregate Data Types

An aggregate data item can be referenced as a single entity, and yet consists of more than one piece of data. Aggregate data types are used to keep related data together, so that the

```
1        ⋮
2       int x[100];
3       int i;
4
5       for(i=0;i<100;i++)
6         x[i] = 0;
7        ⋮
```

Listing 5.33
Initializing an array of integers in C.

programmer's job becomes easier. Some examples of aggregate data are arrays, structures or records, and objects, In most programming languages, aggregate data types can be defined to create higher-level structures. Most high-level languages allow aggregates to be composed of basic types as well as other aggregates. Proper use of structured data helps to make programs less complicated and easier to understand and maintain.

In high-level languages, there are several benefits to using aggregates. Aggregates make the relationships between data clear, and allow the programmer to perform operations on blocks of data. Aggregates also make passing parameters to functions simpler and easier to read.

5.5.1 Arrays

The most common aggregate data type is an array. An array contains zero or more values of the same data type, such as characters, integers, floating point numbers, or fixed point numbers. An array may also contain values of another aggregate data type. Every element in an array must have the same type. Each data item in an array can be accessed by its *array index*.

Listing 5.33 shows how an array can be allocated and initialized in C. Listing 5.34 shows the equivalent code in ARM assembly. Note that in this case, the scaled register offset addressing mode was used to access each element in the array. This mode is often convenient when the size of each element in the array is an integer power of 2. If that is not the case, then it may be necessary to use a different addressing mode. An example of this will be given in Section 5.5.3.

5.5.2 Structured Data

The second common aggregate data type is implemented as the struct in C or the record in Pascal. It is commonly referred to as a structured data type or a record. This data type can

```
1        ⋮
2        sub     sp, sp, #400      @ allocate 400 bytes in stack
3        mov     r0, #0            @ use r0 to hold the index
4        mov     r1, #0            @ value to initialize with
5  loop: str     r1, [sp, r0, lsl #2] @ set array element to zero
6        add     r0, r0, #1        @ increment index
7        cmp     r0, #100          @ loop test
8        blt     loop              @ loop while index < 100
9        ⋮
```

Listing 5.34
Initializing an array of integers in assembly.

```
1   struct student {
2       char first_name[30];
3       char last_name[30];
4       unsigned char class;
5       int grade;
6   };
7       ⋮
8   struct student newstudent;  /* allocate struct on the stack */
9   strcpy(newstudent.first_name,"Sam");
10  strcpy(newstudent.last_name,"Smith");
11  newstudent.class = 2;
12  newstudent.grade = 88;
13      ⋮
```

Listing 5.35
Initializing a structured data type in C.

contain multiple *fields*. The individual fields in the structured data may also be referred to as structured data elements, or simply elements. In most high-level languages, each element of a structured data type may be one of the base types, an array type, or another structured data type. Listing 5.35 shows how a struct can be declared, allocated, and initialized in C. Listing 5.36 shows the equivalent code in ARM assembly.

Care must be taken using assembly to access data structures that were declared in higher level languages such as C and C++. The compiler will typically pad a data structure to ensure that the data fields are aligned for efficiency. On most systems, it is more efficient for the processor to access word-sized data if the data is aligned to a word boundary. Some processors simply cannot load or store a word from an address that is not on a word boundary, and attempting to do so will result in an exception. The assembly programmer must somehow determine the

```
 1              .data
 2              .equ    s_first_name, 0
 3              .equ    s_last_name, 30
 4              .equ    s_class, 60
 5              .equ    s_grade, 64
 6              .equ    s_size, 68
 7  sam:        .asciz  "Sam"
 8  smith:      .asciz  "Smith"
 9              ⋮
10              sub     sp, sp,#s_size        @ allocate struct on the stack
11              mov     r0, sp                @ put pointer to struct in r0
12              add     r0, r0, #s_first_name @ offset to first name field
13              ldr     r1, =sam              @ load pointer to "Sam"
14              bl      strcpy                @ copy the string
15              mov     r0, sp                @ put pointer to struct in r0
16              add     r0, r0, #s_last_name  @ offset to last name field
17              ldr     r1, =smith            @ load pointer to "Smith"
18              bl      strcpy                @ copy the string
19              mov     r0, sp                @ put pointer to struct in r0
20              mov     r1, #2                @ load constant value of 2
21              strb    r1, [r0, #s_class]    @ store with offset
22              mov     r1, #88               @ load constant value of 88
23              str     r1, [r0, #s_grade]    @ store with offset
24              ⋮
```

Listing 5.36
Initializing a structured data type in ARM assembly.

relative address of each field within the higher-level language structure. One way that this can be accomplished in C is by writing a small function which prints out the offsets to each field in the C structure. The offsets can then be used to access the fields of the structure from assembly language. Another method for finding the offsets is to run the program under a debugger and examine the data structure.

5.5.3 Arrays of Structured Data

It is often useful to create arrays of structured data. For example, a color image may be represented as a two-dimensional array of pixels, where each pixel consists of three integers which specify the amount of red, green, and blue that are present in the pixel. Typically, each of the three values is represented using an unsigned eight bit integer. Image processing software often adds a fourth value, α, specifying the transparency of each pixel.

```
1    :
2    const int width = 100;
3    const int height = 100;
4    /* define structure for a pixel */
5    struct pixel {
6        unsigned char red;
7        unsigned char green;
8        unsigned char blue;
9    };
10   :
11   struct pixel *image; /* declare pointer to an image */
12   :
13   /* allocate storage for the image */
14   image = malloc(width * height * sizeof(struct pixel));
15   if(image != NULL)
16     {
17       /* initialize all pixels in the image to black */
18       for(j=0; j<height; j++)
19         for(i=0; i<width; i++)
20           {
21               image[j*width+i].red = 0;
22               image[j*width+i].blue = 0;
23               image[j*width+i].green = 0;
24           }
25     :
26     }
27   else
28     :
```

Listing 5.37
Initializing an array of structured data in C.

Listing 5.37 shows how an array of pixels can be allocated and initialized in C. The listing uses the malloc() function from the C standard library to allocate storage for the pixels from the heap (see Section 1.4). Note that the code uses the sizeof() function to determine how many bytes of memory are consumed by a single pixel, then multiplies that by the width and height of the image. Listing 5.38 shows the equivalent code in ARM assembly.

Note that the code in Listing 5.38 is far from optimal. It can be greatly improved by combining the two loops into one loop. This will remove the need for the multiply on line 28 and the addition on line 29, and will simplify the code structure. An additional improvement would be to increment the single loop counter by three on each loop iteration, making it very

```
 1              .data
 2              .equ     i_red, #0
 3              .equ     i_green, #1
 4              .equ     i_blue, #2
 5              .equ     i_size, #3
 6              .equ     NULL, #0
 7   width:   .word    100
 8   height:  .word    100
 9              :
10              @@ Calculate size of data to be allocated
11              ldr      r4, =width          @ load address of width
12              ldr      r4, [r4]            @ load value of width
13              ldr      r5, =height         @ load address of height
14              ldr      r5, [r5]            @ load value of height
15              mul      r6, r4, r5          @ calculate width x height
16              add      r0, r6, r6, lsl #1  @ multiply by 3
17              bl       malloc              @ allocate storage.
18              cmp      r0, NULL            @ pointer is returned in r0
19              beq      else
20              mov      r8, r0              @ copy array pointer to r8
21              mov      r2, #0              @ put constant 0 in r2
22              mov      r1, #0              @ use r1 for j; j=0
23   jtest:   cmp      r1, r5              @ is j < height ?
24              bge      jdone               @ if not, then end loop
25              mov      r0, #0              @ use r0 for i; i=0
26   itest:   cmp      r0, r4              @ is i < width ?
27              bge      idone               @ if not, then end loop
28              mul      r3, r1, r4          @ calculate row offset
29              add      r3, r3, r0          @ add column offset
30              add      r3, r8, r3          @ get pointer to image pixel
31              strb     r2, [r3,i_red]      @ set red value to 0
32              strb     r2, [r3,i_blue]     @ set blue value to 0
33              strb     r2, [r3,i_green]    @ set green value to 0
34              add      r0, r0, #1          @ i++
35              b        itest
36   idone:   add      r1, r1, #1          @ j++
37              b        jtest
38   jdone:   :
39              :
40   else:    :
```

Listing 5.38
Initializing an array of structured data in assembly.

```
1          .data
2          .equ     i_red, #0
3          .equ     i_green, #1
4          .equ     i_blue, #2
5          .equ     i_size, #4
6          .equ     NULL, #0
7   width:  .word    100
8   height: .word    100
9          :
10         @@ Calculate size of data to be allocated
11         ldr      r4, =width          @ load address of width
12         ldr      r4, [r4]            @ load value of width
13         ldr      r5, =height         @ load address of height
14         ldr      r5, [r5]            @ load value of height
15         mul      r6, r4, r5          @ calculate width x height
16         add      r0, r6, r6 lsl #1   @ multiply by 3
17         bl       malloc              @ allocate storage
18         cmp      r0, NULL            @ verify pointer
19         beq      else
20         mov      r8, r0              @ copy array pointer to r8
21         mov      r2, #0              @ put constant 0 in r2
22         mov      r0, #0              @ use r0 for i; i=0
23  itest:  cmp      r0, r6             @ is i < width x height x 3 ?
24         bge      idone               @ if not, then end loop
25         add      r3, r8, r0          @ get pointer to image[i]
26         strb     r2, [r3,i_red]      @ set red value to 0
27         strb     r2, [r3,i_blue]     @ set blue value to 0
28         strb     r2, [r3,i_green]    @ set green value to 0
29         add      r0, r0, #3          @ i+=3
30         b        itest
31  idone:  :
32         :
33  else:   :
```

Listing 5.39
Improved initialization in assembly.

easy to calculate the pointer for each pixel. Listing 5.39 shows the ARM assembly implementation with these optimizations.

Although the implementation shown in Listing 5.39 is more efficient than the previous version, there are several more improvements that can be made. If we consider that the goal of the code is to allocate some number of bytes and initialize them all to zero, then the code can be written more efficiently. Rather than using three separate store instructions to set 3 bytes to

```
1          .data
2          .equ    i_red, #0
3          .equ    i_green, #1
4          .equ    i_blue, #2
5          .equ    i_size, #3
6          .equ    NULL, #0
7  width:  .word   100
8  height: .word   100
9            ⋮
10         @@ Calculate size of data to be allocated
11         ldr     r4, =width           @ load address of width
12         ldr     r4, [r4]             @ load value of width
13         ldr     r5, =height          @ load address of height
14         ldr     r5, [r5]             @ load value of height
15         mul     r6, r4, r5           @ calculate width x height
16         add     r7, r6, r6 lsl #1    @ multiply by 3
17         mov     r0, r7               @ copy to r0 for malloc call
18         bl      malloc               @ allocate storage
19         cmp     r0, NULL             @ verify pointer
20         beq     else
21         mov     r8, r0               @ copy array pointer to r8
22         add     r7, r8, r7           @ get pointer to end of array
23         and     r0, r7, #0xFFFFFFFC  @ make it be word aligned
24         mov     r2, #0               @ put constant 0 in r2
25         mov     r3, r8               @ copy initial pointer to r3
26 testa:  str     r2, [r3], #4         @ clear word, increment ptr
27         cmp     r3, r0               @ done clearing full words?
28         blt     testa                @ if no, then continue loop
29         subne   r3,r3,#4             @ there may be bytes left
30 testb:  cmp     r3, r7               @ any bytes left to clear?
31         bge     idone                @ if not, then end loop
32         strb    r2, [r3],#1          @ clear byte, increment ptr
33         b       testb
34 idone:  ⋮
35
36 else:   ⋮
```

Listing 5.40
Very efficient initialization in assembly.

zero on each iteration of the loop, why not use a single store instruction to set four bytes to zero on each iteration? The only problem with this approach is that we must consider the possibility that the array may end in the middle of a word. However, this can be dealt with by using two consecutive loops. The first loop sets one *word* of the array to zero on each iteration, and the second loop finishes off any remaining bytes. Listing 5.40 shows the results of these additional improvements. This third implementation will run *much* faster than the previous implementations.

5.6 Chapter Summary

Spaghetti code is the bane of assembly programming, but it can easily be avoided. Although assembly language does not enforce structured programming, it does provide the low-level mechanisms required to write structured programs. The assembly programmer must be aware of, and assiduously practice, proper structured programming techniques. The burden of writing properly structured code blocks, with selection structures and iteration structures, lies with the programmer, and failure to apply structured programming techniques will result in code that is difficult to understand, debug, and maintain.

Subroutines provide a way to split programs into smaller parts, each of which can be written and debugged individually. This allows large projects to be divided among team members. In assembly language, defining and using subroutines is not as easy as in higher level languages. However, the benefits usually outweigh the costs. The C library provides a large number of functions. These can be accessed by an assembly program as long as it is linked with the C standard library.

Assembly provides the mechanisms to access aggregate data types. Arrays can be accessed using various addressing modes on the ARM processor. The pre-indexing and post-indexing modes allow array elements to be accessed using pointers, with the pointers being incremented after each element access. Fields in structured data records can be accessed using immediate offset addressing mode. The rich set of addressing modes available on the ARM processor allows the programmer to use aggregate data types more efficiently than on most processors.

Exercises

5.1   What does it mean for a register to be volatile? Which ARM registers are considered volatile according to the ARM function calling convention?

5.2   Fully explain the differences between static variables and automatic variables.

5.3   In ARM assembly language, write a function that is equivalent to the following C function.

```
1   int max(int a, int b)
2   {
3     if( a > b )
4       return a;
5     return b;
6   }
```

5.4 What are the two places where an automatic variable can be stored?

5.5 You are writing a function and you decided to use registers r4 and r5 within the function. Your function will not call any other functions; it is self-contained. Modify the following skeleton structure to ensure that r4 and r5 can be used within the function and are restored to comply with the ARM standards, but without unnecessary memory accesses.

```
1  myfunc: stmfd   sp!,{lr}
2              ⋮
3          @ function statements
4              ⋮
5          ldmfd   sp!,{lr}
```

5.6 Convert the following C program to ARM assembly, using a post-test loop:

```
1  int main()
2  {
3    for(i=0;i<10;i++)
4      printf("Hi!\n");
5    return 0;
6  }
```

5.7 Write a complete ARM function to shift a 64-bit value left by any given amount between 0 and 63 bits. The function should expect its arguments to be in registers r0, r1, and r2. The lower 32 bits of the value are passed in r0, the upper 32 bits of the value are passed in r1, and the shift amount is passed in r2.

5.8 Write a complete subroutine in ARM assembly that is equivalent to the following C subroutine.

```
1  /* This function copies 'count' bytes from 'src' to 'dest'.
2  void bytecopy(char dest[], char src[], int count)
3  {
4    count = count - 1;
5    while(count>=0)
6      {
7        dest[count] = src[count];
8        count = count - 1;
9      }
10 }
```

5.9 Write a complete function in ARM assembly that is equivalent to the following C function.

```
1  /* This function returns the minimum of six values.
2  int minsix(int a, int b, int c, int d, int e, int f)
3  {
4    if(b < a)
```

```
5       a = b;
6     if(d < c)
7       c = d;
8     if(c < a)
9       a = c;
10    if(f < e)
11      e = f;
12    if(e < a)
13      a = e;
14    return a;
15  }
```

5.10 Write an ARM assembly function to calculate the average of an array of integers, given a pointer to the array and the number of items in the array. Your assembly function must implement the following C function prototype:

```
int average(int *array, int number_of_items);
```

Assume that the processor does not support the div instruction, but there is a function available to divide two integers. You do not have to write this function, but you may need to call it. Its C prototype is:

```
int divide(int numerator, int denominator);
```

5.11 Write a complete function in ARM assembly that is equivalent to the following C function. Note that a and b must be allocated on the stack, and their addresses must be passed to scanf so that it can place their values into memory.

```
1  int read_and_add()
2  {
3    int a, b, sum;
4    scanf("%d",&a);
5    scanf("%d",&b);
6    sum = a + b;
7    return sum;
8  }
```

5.12 The factorial function can be defined as:

$$x! = \begin{cases} 1 & \text{if } x \leq 1, \\ x \times (x-1)! & \text{otherwise.} \end{cases}$$

The following C program repeatedly reads x from the user and calculates $x!$ It quits when it reads end-of-file or when the user enters a negative number or something that is not an integer.

Write this program in ARM assembly.

```
1   #include <stdio.h>
2
3   /* The factorial function calculates x! */
4   int factorial(int x)
5   {
6     if(x<2)
7       return 1;
8     return x * factorial(x-1);
9   }
10
11  /* main repeatedly asks for x, and prints x! */
12  int main()
13  {
14    int x,goodval;
15    do{
16        printf("Enter x: ");
17        goodval = fscanf("%d",&x);
18        if(goodval == 1)
19          printf("%d! = %d\n",x,factorial(x));
20      } while(goodval == 1);
21    return 0;
22  }
```

5.13   For large x, the factorial function is slow. However, a lookup table can be added to the function to improve average performance. This technique is commonly known as memoization or tabling, but is sometimes called dynamic programming. The following C implementation of the factorial function uses memoization. Modify your ARM assembly program from the previous problem to include memoization.

```
1   #define TABSIZE 50
2
3   /* The factorial function calculates x! */
4   int factorial(int x)
5   {
6     /* declare table and initialize to all zero */
7     static int table[TABSIZE] = {0};
8
9     /* handle base case */
10    if(x<2)
11      return 1;
12
13    /* if x! is not in the table and x is small enough,
14       then compute x! and put it in the table */
15    if((x < TABSIZE) && table[x] == 0))
```

```
16        table[x] =  x * factorial(x-1);
17
18    /* if x is small enough, then
19        return the value from the table */
20    if(x < TABSIZE)
21      return table[x];
22
23    /* if x is too large to be in the table, use
24      a recursive call */
25    return x * factorial(x-1);
26  }
```

Abstract Data Types

Chapter Outline

An abstract data type (ADT) is composed of data and the operations that work on that data. The ADT is one of the cornerstones of structured programming. Proper use of ADTs has many benefits. Most importantly, abstract data types help to support information hiding. A software module hides information by encapsulating the information into a module or other construct which presents an *interface*. The interface typically consists of the names of data types provided by the ADT and a set of subroutine definitions, or prototypes, for operating on the data types. The *implementation* of the ADT is hidden from the client code that uses the ADT.

A common use of information hiding is to hide the physical storage layout for data so that if it is changed, the change is restricted to a small subset of the total program. For example, if a three-dimensional point (x, y, z) is represented in a program with three floating point scalar variables, and the representation is later changed to a single array variable of size three, a module designed with information hiding in mind would protect the remainder of the program from such a change.

Information hiding reduces software development risk by shifting the code's dependency on an uncertain implementation onto a well-defined interface. Clients of the interface perform operations purely through the interface, which does not change. If the implementation changes, the client code does not have to change.

Encapsulating software and data structures behind an interface allows the construction of objects that mimic the behavior and interactions of objects in the real world. For example, a simple digital alarm clock is a real-world object that most people can use and understand. They can understand what the alarm clock does, and how to use it through the provided interface (buttons and display) without needing to understand every part inside of the clock. If the internal circuitry of the clock were to be replaced with a different implementation, people could continue to use it in the same way, provided that the interface did not change.

6.1 ADTs in Assembly Language

As with all other structured programming concepts, ADTs can be implemented in assembly language. In fact, most high-level compilers convert structured programming code into assembly during compilation. All that is required is that the programmer define the data structure(s), and the set of operations that can be used on the data. Listing 6.1 gives an example of an ADT *interface* in C. The type Image is not fully defined in the interface. This prevents client software from accessing the internal structure of the image data type. Therefore, programmers using the ADT can modify images only by using the provided

```c
1   #ifndef IMAGE_H
2   #define IMAGE_H
3   #include <stdio.h>
4
5   typedef unsigned char pval;
6
7   struct imageStruct;
8
9   typedef struct imageStruct Image;
10
11  Image *allocateImage();
12  void freeImage(Image *image);
13
14  int readImage(FILE *f, Image *image);
15  int writeImage(FILE *f, Image *image);
16
17  int setPixelRGB(Image *image, int row, int col, pval r, pval g, pval b);
18  int setPixelGray(Image *image, int row, int col, pval v);
19
20  pixel getPixelRGB(Image *image, int row, int col);
21  pval getPixelGray(Image *image, int row, int col);
22
23  #endif
```

Listing 6.1
Definition of an Abstract Data Type in a C header file.

```
 1  #ifndef IMAGE_PRIVATE_H
 2  #define IMAGE_PRIVATE_H
 3  #include <image.h>
 4
 5  typedef struct {
 6    pval r,g,b;
 7  } Pixel;
 8
 9  struct imageStruct;
10    int rows;      // number of rows in the image
11    int cols;      // number of columns in the image
12    Pixel *pixels; // array of pixel data
13    };
14  #endif
```

Listing 6.2
Definition of the image structure may be hidden in a separate header file.

functions. Other structured programming and object-oriented programming languages such as C++, Java, Pascal, and Modula 2 provide similar protection for data structures so that client code can access the data structure only through the provided interface. Note that only the `pval` definition is exposed, indicating to client programs that the red, green, and blue components of a pixel must be a number between 0 and 255. In C, as with other structured programming languages, the *implementation* of the subroutines can also be hidden by placing them in separate compilation modules. Those modules will have access to the internal structure of the `Image` data type.

Assembly language does not have the ability to define a data structure as such, but it does provide the mechanisms needed to specify the location of each field with respect to the beginning of a data structure, as well as the overall size of the data structure. With a little thought and effort, it is possible to implement ADTs in Assembly language. Listing 6.2 shows the private implementation of the `Image` data type, which is included by the C files which implement the `Image` data type. Listing 6.3 shows how the data structures from the previous listings can be defined in assembly language. With those definitions, any of the functions declared in Listing 6.1 can be written in assembly language.

6.2 Word Frequency Counts

Counting the frequency of words in written text has several uses. In digital forensics, it can be used to provide evidence as to the author of written communications. Different people have different vocabularies, and use words with differing frequency. Word counts can also be used

```
1    @@@ Definitions for pixel and image data structures
2
3        @@ pixel
4        .equ    p_red,    #0  @ offset to red value
5        .equ    p_green,  #1  @ offset to green value
6        .equ    p_blue,   #2  @ offset to blue value
7        .equ    p_size,   #3  @ size of the pixel data structure
8
9        @@ image
10       .equ    i_rows,   #0  @ offset to number of rows
11       .equ    i_cols,   #4  @ offset to number of columns
12       .equ    i_pixels, #8  @ offset to pointer to image data
13       .equ    i_size,   #12 @ size of the image data structure
```

Listing 6.3
Definition of an ADT in Assembly.

to classify documents by type. Scientific articles from different fields contain words specific to that field, and historical novels will differ from western novels in word frequency.

Listing 6.4 shows the main function for a simple C program which reads a text file and creates a list of all the words contained in a file, along with their frequency of occurrence. The program has been divided into two parts: the main program, and an ADT which is used to keep track the words and their frequencies, and to print a table of word frequencies.

```c
1    #include <stdlib.h>
2    #include <string.h>
3    #include <stdio.h>
4    #include <ctype.h>
5    #include <list.h>
6    /*********************************************************/
7    /* remove_punctuation copies the input string to a new   */
8    /* string, but omits any punctuation characters          */
9    char *remove_punctuation(char *word)
10   { char* newword = (char*)malloc(strlen(word)+1);
11     char* curdst = newword;
12     char* cursrc = word;
13     while( *cursrc != 0 )
14       {
15         if(strchr(",.\"!$();:{}\\[]", *cursrc) == NULL)
16           { /* Current character is not punctuation */
17             *curdst = tolower(*cursrc);
18             curdst++;
```

```
19        }
20      cursrc++;
21      }
22    *curdst=0;
23    return newword;
24 }
25
26 /******************************************************/
27 /* The main function reads whitespace separated words   */
28 /* from stdin, removes punctuation, and generates a word */
29 /* frequency list.                                      */
30 int main()
31 { int MaxWordLength = 1024;
32   char *nextword, *cleanword;
33   wordlist *list;
34   nextword = (char*)malloc(MaxWordLength*sizeof(char));
35   list = wl_alloc();
36   while(scanf("%s",nextword) == 1)
37     {
38       cleanword = remove_punctuation(nextword);
39       if(strlen(cleanword)>0)
40         wl_increment(list,cleanword);
41       free(cleanword);
42     }
43   printf("Alphabetical List\n");
44   wl_print(list);
45   printf("\nNumerical List\n");
46   wl_print_numerical(list);
47   wl_free(list);
48   return 0;
49 }
```

Listing 6.4
C program to compute word frequencies.

The interface for the ADT is shown in Listing 6.5. There are several ways that the ADT could be implemented. Note that the *interface* given in the header file does not show the internal fields of the word list data type. Thus, any file which includes this header is allowed to declare pointers to wordlist data types, but cannot access or modify any internal fields. The list of words could be stored in an array, a linked list, a binary tree, or some other data structure. The subroutines could be implemented in C or in some other language, including assembly. Listing 6.6 shows an implementation in C using a linked list. Note that the function for printing the word frequency list in numerical order has not been implemented. It will be written in assembly language. Since the program is split into multiple files, it is a good idea to use the make utility to build the executable program. A basic makefile is shown in Listing 6.7.

```
1   #ifndef LIST_H
2   #define LIST_H
3
4   /**********************************************************/
5   /* Define an opaque type, named wordlist                 */
6   typedef struct wlist wordlist;
7
8   /**********************************************************/
9   /* wl_alloc allocates and initializes a new word list.   */
10  wordlist* wl_alloc();
11
12  /**********************************************************/
13  /* wl_free frees all the storage used by a wordlist      */
14  void wl_free(wordlist* wl);
15
16  /**********************************************************/
17  /* wl_increment adds one to the count of the given word. */
18  /* If the word is not in the list, then it is added with */
19  /* a count of one.                                       */
20  void wl_increment(wordlist *list, char *word);
21
22  /**********************************************************/
23  /* wl_print_alphabetical prints a table showing the number*/
24  /* of occurrences for each word, followed by the word.   */
25  void wl_print(wordlist *list);
26
27  /**********************************************************/
28  /* wl_print_numerical prints a table showing the number  */
29  /* of occurrences for each word, followed by the word,   */
30  /* sorted in reverse order of occurrence.                */
31  void wl_print_numerical(wordlist *list);
32
33  #endif
```

Listing 6.5
C header for the wordlist ADT.

```
1    #include <stdlib.h>
2    #include <string.h>
3    #include <stdio.h>
4    #include <list.h>
5
6    /******************************************************/
7    /* The wordlistnode type is a linked list of words and  */
8    /* the number of times each word has occurred.          */
9    typedef struct wlist_node{
10     char *word;
11     int count;
12     struct wlist_node *next;
13   }wordlistnode;
14
15   /******************************************************/
16   /* The wordlist type holds a pointer to the linked list */
17   /* and keeps track of the number of nodes in the list   */
18   typedef struct wlist{
19     int nwords;
20     wordlistnode *head;
21   }wordlist;
22
23   /******************************************************/
24   /* wl_alloc allocates and initializes a new word list.  */
25   wordlist* wl_alloc()
26   { wordlist* tmp;
27     tmp = (wordlist*)malloc(sizeof(wordlist));
28     if(tmp == NULL)
29       {
30         printf("Unable to allocate wordlist\n");
31         exit(1);
32       }
33     tmp->nwords = 0;
34     tmp->head = NULL;
35     return tmp;
36   }
37
38   /******************************************************/
39   /* wl_free frees all the storage used by a wordlist     */
40   void wl_free(wordlist* wl)
41   {
42     wordlistnode *tmpa, *tmpb;
43     tmpa = wl->head;
44     while(tmpa != NULL)
45       {
```

```
46        tmpb = tmpa;
47        tmpa = tmpa->next;
48        free(tmpb->word);
49        free(tmpb);
50      }
51    free(wl);
52  }
53
54  /***********************************************************/
55  /* wln_lookup is used internally to search the list of    */
56  /* words. It returns a pointer to the wordlistnode. If    */
57  /* the word is not in the list, then it returns a pointer*/
58  /* to the place where the word should be inserted. If     */
59  /* the insertion point is the head of the list, then it   */
60  /* returns NULL.                                          */
61  wordlistnode* wln_lookup(wordlistnode* lst, char *word)
62  {
63    wordlistnode *prev = NULL;
64    while((lst != NULL)&&(strcmp(lst->word, word)<0))
65      {
66        prev = lst;
67        lst = lst->next;
68      }
69    if((lst != NULL)&&(strcmp(lst->word, word) == 0))
70      return lst;
71    else
72      return prev;
73  }
74
75  /***********************************************************/
76  /* wl_increment adds one to the count of the given word.  */
77  /* If the word is not in the list, then it is added with  */
78  /* a count of one.                                        */
79  void wl_increment(wordlist *list, char *word)
80  {
81    wordlistnode *newword;
82    wordlistnode *wlst = wln_lookup(list->head,word);
83    if((wlst == NULL)||(strcmp(wlst->word, word) != 0))
84      {
85        list->nwords++;
86        newword = (wordlistnode*)malloc(sizeof(wordlistnode));
87        if(newword == NULL)
88          {
89            printf("Unable to allocate wordlistnode\n");
90            exit(1);
91          }
```

```
92      newword->word = strdup(word);
93      newword->count = 1;
94      if(wlst == NULL)
95        {
96          newword->next = list->head;
97          list->head = newword;
98        }
99      else
100       {
101         newword->next = wlst->next;
102         wlst->next = newword;
103       }
104   }
105   else
106     wlst->count++;
107 }
108
109 /****************************************************/
110 /* wl_print_alphabetical prints a table showing the number*/
111 /* of occurrences for each word, followed by the word.    */
112 void wl_print(wordlist *list)
113 {
114   wordlistnode *wlist = list->head;
115   while(wlist != NULL) {
116     printf("%10d %s\n",wlist->count,wlist->word);
117     wlist=wlist->next;
118   }
119   printf("There are %d unique words in the document\n",
120           list->nwords);
121 }
122
123 /****************************************************/
124 /* wl_print_numerical prints a table showing the number   */
125 /* of occurrences for each word, followed by the word,    */
126 /* sorted in reverse order of occurrence.                 */
127 void wl_print_numerical(wordlist *list)
128 {
129   printf("wl_print_numerical has not been implemented");
130 }
```

Listing 6.6
C implementation of the wordlist ADT.

```
1   C_OBJS=wordfreq.o list.o
2   ASM_OBJS=
3   OBJS=$(C_OBJS) $(ASM_OBJS)
4
5   LFLAGS=-O2 -g
6   CFLAGS=-I. -O2 -g -Wall
7   SFLAGS=-I. -O2 -g -Wall
8   DEPENDFLAGS=-I. -M
9
10  wordfreq: $(OBJS)
11          gcc $(LFLAGS) -o wordfreq $(OBJS)
12
13  .S.o:
14          gcc $(SFLAGS) -c $<
15
16  .c.o:
17          gcc $(CFLAGS) -c $<
18
19  clean:
20          rm -f *.o *~ wordfreq
21
22  # make depend will create a file ".depend" with all the dependencies
23  depend:
24          rm -f .depend
25          $(CC) $(DEPENDFLAGS) $(C_OBJS:.o=.c) > .depend
26
27  # if we have a .depend file, include it
28  ifeq (.depend,$(wildcard .depend))
29  include .depend
30  endif
```

Listing 6.7
Makefile for the wordfreq program.

Suppose we wish to implement one of the functions from Listing 6.6 in ARM assembly language. We would delete the function from the C file, create a new file with the assembly version of the function, and modify the makefile so that the new file is included in the program. The header file and the main program file would not require any changes. The header file provides function prototypes that the C compiler uses to determine how parameters should be passed to the functions. As long as our new assembly function conforms to its C header definition, the program will work correctly.

6.2.1 Sorting by Word Frequency

The linked list is created in alphabetical order, but the `wl_print_numerical()` function is required to print it sorted in reverse order of number of occurrences. There are several ways in which this could be accomplished, with varying levels of efficiency. The possible approaches include, but are not limited to:

- Re-ordering the linked list using an insertion sort: This approach creates a complete new list by removing each item, one at a time, from the original list, and inserting it into a new list sorted by the number of occurrences rather than the words themselves. The time complexity for this approach would be $O(N^2)$, but would require no additional storage. However, if the list were later needed in alphabetical order, or any more words were to be added, then it would need to be re-sorted in the original order.
- Sorting the linked list using a merge sort algorithm: Merge sort is one of the most efficient sorting algorithms known and can be efficiently applied to data in files and linked lists. The merge sort works as follows:
  1. The sub-list size, i, is set to 1.
  2. The list is divided into sub-lists, each containing i elements. Each sub-list is assumed to be sorted. (A sub-list of length one is sorted by definition.)
  3. The sub-lists are merged together to create a list of sub-lists of size $2i$, where each sub-list is sorted.
  4. The sub-list size, i, is set to $2i$.
  5. The process is repeated from step 2 until $i \geq N$, where N is the number of items to be sorted.

  The time complexity for the merge sort algorithm is $N \log N$, which is far more efficient than the insertion sort. This approach would also require no additional storage. However, if the list were later needed in alphabetical order, or any more words were to be added, then it would need to be re-sorted in the original alphabetical order.
- Create an index, and sort the index rather than rebuilding the list. Since the number of elements in the list is known, we can allocate an array of pointers. Each pointer in the array is then initialized to point to one element in the linked list. The array forms an index, and the pointers in the array can be re-sorted in any desired order, using any common sorting method such as bubble sort ($O(N^2)$), in-place insertion sort ($O(N^2)$), quick sort ($O(N \log N)$), or others. This approach requires additional storage, but has the advantage that it does not need to modify the original linked list.

There are many other possibilities for re-ordering the list. Regardless of which method is chosen, the main program and the interface (header file) need not be changed. Different implementations of the sorting function can be substituted without affecting any other code.

The `wl_print_numerical()` function can be implemented in assembly as shown in
Listing 6.8. The function operates by re-ordering the linked list using an insertion sort as
described above. Listing 6.9 shows the change that must be made to the make file. Now, when
make is run, it compiles the two C files and the assembly file into object files, then links them
all together. The C implementation of `wl_print_numerical()` in `list.c` must be deleted or
commented out or the linker will emit an error indicating that it found two versions of
`wl_print_numerical()`.

```
1    @@@ Definitions for the wordlistnode type
2            .equ    wln_word,0      @ word field
3            .equ    wln_count,4     @ count filed
4            .equ    wln_next,8      @ next field
5            .equ    wln_size,12     @ sizeof(wordlistnode)
6    @@@ Definitions for the wordlist type
7            .equ    wl_nwords,0     @ number of words in list
8            .equ    wl_head,4       @ head of linked list
9            .equ    wl_size,8       @ sizeof(wordlist)
10   @@@ Define NULL
11           .equ    NULL,0
12
13   @@@ -------------------------------------------------------
14   @@@ The sort_numerical function sorts the list of words in
15   @@@ reverse by number of occurrences, and returns a
16   @@@ pointer to the head of the re-ordered list.
17   @@@ Records with identical counts will maintain their
18   @@@ original ordering with respect to each other.
19   @@@ r0 holds head of source list (head)
20   @@@ r1 holds destination list (dest)
21   @@@ r2 holds pointer to node currently being moved (node)
22   @@@ r3 holds pointer to current node in destination list (curr)
23   @@@ r4 holds pointer to previous node in destination list (prev)
24   @@@ r5 holds count of current node in destination list
25   @@@ r6 holds count of node currently being moved
26   sort_numerical:
27           stmfd   sp!,{r4-r6}
28           mov     r1,#NULL        @ initialize new list to NULL
29           @@ loop until source list is empty
30   loopa:  cmp     r0,#NULL
31           beq     endloopa
32           @@ detatch first node from source list
33           mov     r2,r0           @ node <- head
34           ldr     r5,[r2,#wln_count] @ load count for node
35           ldr     r0,[r0,#wln_next] @ head <- head->next
36           @@ find location to insert into destination list
37           mov     r3,r1           @ curr <- dest
```

```
38          mov     r4,#NULL          @ prev <- NULL
39  loopb:  cmp     r3,#NULL          @ Reached end of list?
40          beq     found
41          ldr     r6,[r3,#wln_count] @ load count for curr
42          cmp     r5,r6             @ compare with count for node
43          bgt     found
44          mov     r4,r3             @ previous <- current
45          ldr     r3,[r3,#wln_next] @ current <- current->next
46          b       loopb
47          @@ insert into destination list at current location
48  found:  str     r3,[r2,#wln_next] @ node-> next <- current
49          cmp     r4,#NULL          @ if prev == NULL
50          moveq   r1,r2             @ dest <- node
51          strne   r2,[r4,#wln_next] @ else prev->next <- node
52          @@ repeat with next list item
53          b       loopa
54  endloopa:
55          mov     r0,r1             @ return dest (sorted list)
56          ldmfd   sp!,{r4-r6}
57          mov     pc,lr
58
59  @@@ --------------------------------------------------------
60  @@@ wl_print_numerical prints a table showing the number
61  @@@ of occurrences for each word, followed by the word,
62  @@@ sorted in reverse order of occurrence.
63          .global wl_print_numerical
64  wl_print_numerical:
65          stmfd   sp!,{r0,lr}       @ save pointer to wordlist
66          ldr     r0,[r0,#wl_head]  @ load pointer to the linked list
67          bl      sort_numerical    @ re-sort the list
68          ldmfd   sp!,{r1}          @ load pointer to wordlist
69          str     r0,[r1,#wl_head]  @ update list pointer
70          mov     r0,r1             @ prepare to print
71          bl      wl_print          @ print the sorted list
72          ldmfd   sp!,{lr}          @ restore lr from stack
73          mov     pc,lr             @ return
```

Listing 6.8
ARM assembly implementation of `wl_print_numerical()`.

```
1  C_OBJS=wordfreq.o list.o
2  ASM_OBJS=wl_print_numerical.o
3  OBJS=$(C_OBJS) $(ASM_OBJS)
4
5  LFLAGS=-O2 -g -marm
```

```
 6  CFLAGS=-I. -O2 -g -Wall -marm
 7  SFLAGS=-I. -O2 -g -Wall -marm
 8  DEPENDFLAGS=-I. -M
 9
10  wordfreq: $(OBJS)
11          gcc $(LFLAGS) -o wordfreq $(OBJS)
12
13  .S.o:
14          gcc $(SFLAGS) -c $<
15
16  .C.o:
17          gcc $(CFLAGS) -c $<
18
19  clean:
20          rm -f *.o *~ wordfreq
21
22  # make depend will create a file ".depend" with all the dependencies
23  depend:
24          rm -f .depend
25          $(CC) $(DEPENDFLAGS) $(C_OBJS:.o=.c) > .depend
26
27  # if we have a .depend file, include it
28  ifeq (.depend,$(wildcard .depend))
29  include .depend
30  endif
```

Listing 6.9
Revised makefile for the wordfreq program.

6.2.2 Better Performance

The word frequency counter, as previously implemented, takes several minutes to count the frequency of words in the author's manuscript for this textbook on a Raspberry Pi. Most of the time is spent building the list of words and re-sorting the list in order of word frequency. Most of the time for both of these operations is spent in searching for the word in the list before incrementing its count or inserting it in the list. There are more efficient ways to build ordered lists of data.

Since the code is well modularized using an ADT, the internal mechanism of the list can be modified without affecting the main program. A major improvement can be made by changing the data structure from a linked list to a binary tree. Fig. 6.1 shows an example binary tree storing word frequency counts. The time required to insert into a linked list is $O(N)$, but the time required to insert into a binary tree is $O(\log_2 N)$. To give some perspective, the author's manuscript for this textbook contains about 125,000 words. Since $\log_2(125,000) < 17$, we would expect the linked list implementation to require about $\frac{125,000}{17} \approx 7353$ times as long as a binary tree implementation to process the author's manuscript for this textbook. In reality,

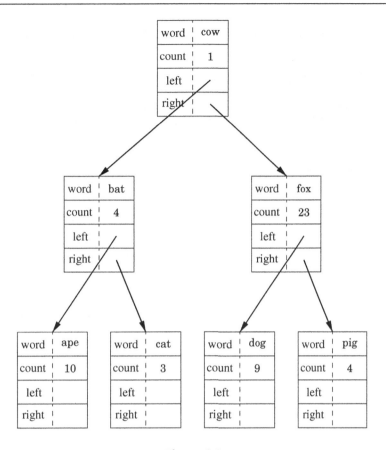

Figure 6.1
Binary tree of word frequencies.

there is some overhead to the binary tree implementation. Even with the extra overhead, we should see a significant speedup. Listing 6.10 shows the C implementation using a balanced binary tree instead of a linked list.

```
1   #include <stdlib.h>
2   #include <string.h>
3   #include <stdio.h>
4   #include <list.h>
5
6   #define MAX(x,y) (x<y?y:x)
7
8   /**********************************************************/
9   /* The wordlistnode type is a binary tree of words and   */
10  /* the number of times each word has occurred.           */
11  typedef struct wlist_node{
12    char *word;
```

```
13    int count;
14    struct wlist_node *left, *right;
15    int height;
16  }wordlistnode;
17
18  /*********************************************************/
19  /* wln_alloc allocates and initializes a wordlistnode    */
20  wordlistnode *wln_alloc(char *word)
21  {
22    wordlistnode *newword;
23    newword = (wordlistnode*)malloc(sizeof(wordlistnode));
24    if(newword == NULL)
25      {
26        printf("Unable to allocate wordlistnode\n");
27        exit(1);
28      }
29    newword->word = strdup(word);
30    newword->count = 1;
31    newword->left = NULL;
32    newword->right = NULL;
33    newword->height = 1;
34    return newword;
35  }
36
37  /*********************************************************/
38  /* wln_free frees the storage of a wordlistnode          */
39  void wln_free(wordlistnode *root)
40  {
41    if(root == NULL)
42      return;
43    free(root->word);
44    wln_free(root->left);
45    wln_free(root->right);
46    free(root);
47  }
48
49  /*********************************************************/
50  /* wln_lookup is used  to search the tree of words. It   */
51  /* returns a pointer to the wordlistnode. If the word is */
52  /* not in the list, then it returns NULL.                */
53  wordlistnode* wln_lookup(wordlistnode* root, char *word)
54  {
55    int cmp;
56    if(root != NULL)
57      {
58        cmp = strcmp(word,root->word);
```

```
59      if(cmp < 0)
60        root = wln_lookup(root->left,word);
61      else
62        if(cmp>0)
63          root = wln_lookup(root->right,word);
64    }
65    return root;
66  }
67
68  /***********************************************************/
69  /* wln_height finds the height of a node and returns       */
70  /* zero if the pointer is NULL.                            */
71  int wln_height(wordlistnode *node)
72  {
73    if(node == NULL)
74      return 0;
75    return node->height;
76  }
77
78  /***********************************************************/
79  /* wln_balance finds the balance factor of a node and      */
80  /* returns zero if the pointer is NULL.                    */
81  int wln_balance(wordlistnode *node)
82  {
83      if (node == NULL)
84          return 0;
85      return wln_height(node->left) - wln_height(node->right);
86  }
87
88  /***********************************************************/
89  /* wln_rotate_left rotates counterclockwise                */
90  wordlistnode* wln_rotate_left(wordlistnode* rt)
91  {
92    wordlistnode* nrt = rt->right;
93    rt->right = nrt->left;
94    nrt->left = rt;
95    rt->height =
96      MAX(wln_height(rt->left),wln_height(rt->right)) + 1;
97    nrt->height =
98      MAX(wln_height(nrt->left),wln_height(nrt->right)) + 1;
99    return nrt;
100 }
101
102 /***********************************************************/
103 /* wln_rotate_left rotates clockwise                       */
104 wordlistnode* wln_rotate_right(wordlistnode* rt)
```

```
105  {
106    wordlistnode* nrt = rt->left;
107    rt->left = nrt->right;
108    nrt->right = rt;
109    rt->height =
110      MAX(wln_height(rt->left),wln_height(rt->right)) + 1;
111    nrt->height =
112      MAX(wln_height(nrt->left),wln_height(nrt->right)) + 1;
113    return nrt;
114  }
115
116  /******************************************************/
117  /* wln_insert performs a tree insertion, and re-balances  */
118  wordlistnode * wln_insert(wordlistnode *root, wordlistnode *node)
119  {
120    int balance;
121    if (root == NULL)
122      /* handle case where tree is empty, or we reached a leaf */
123      root = node;
124    else
125      {
126        /* Recursively search for insertion point, and perform the
127           insertion. */
128        if (strcmp(node->word,root->word) < 0)
129          root->left  = wln_insert(root->left, node);
130        else
131          root->right = wln_insert(root->right, node);
132
133        /* As we return from the recursive calls, recalculate the heights
134           and perform rotations as necessary to re-balance the tree */
135        root->height = MAX(wln_height(root->left),
136                           wln_height(root->right)) + 1;
137
138        /* Calculate the balance factor */
139        balance = wln_balance(root);
140        if (balance > 1)
141          {
142            /* the tree is deeper on the left than on the right) */
143            if(strcmp(node->word,root->left->word) <= 0)
144              root = wln_rotate_right(root);
145            else
146              {
147                root->left = wln_rotate_left(root->left);
148                root = wln_rotate_right(root);
149              };
150          }
```

```
151        else
152          if(balance < -1)
153            {
154               /* the tree is deeper on the right than on the left) */
155               if(strcmp(node->word,root->right->word) >= 0)
156                 root = wln_rotate_left(root);
157               else
158                 {
159                   root->right = wln_rotate_right(root->right);
160                   root = wln_rotate_left(root);
161                 }
162            }
163        }
164    return root;
165  }
166
167  /*********************************************************/
168  /* The wordlist type holds a pointer to the binary tree  */
169  /* and keeps track of the number of nodes in the list    */
170  typedef struct wlist{
171    int nwords;
172    wordlistnode *root;
173  }wordlist;
174
175  /*********************************************************/
176  /* wl_alloc allocates and initializes a new word list.   */
177  wordlist* wl_alloc()
178  { wordlist* tmp;
179    tmp = (wordlist*)malloc(sizeof(wordlist));
180    if(tmp == NULL)
181      {
182        printf("Unable to allocate wordlist\n");
183        exit(1);
184      }
185    tmp->nwords = 0;
186    tmp->root = NULL;
187    return tmp;
188  }
189
190  /*********************************************************/
191  /* wl_free frees all the storage used by a wordlist      */
192  void wl_free(wordlist* wl)
193  {
194    wln_free(wl->root);
195    free(wl);
196  }
```

```
197
198   /*****************************************************/
199   /* wl_increment adds one to the count of the given word.  */
200   /* If the word is not in the list, then it is added with  */
201   /* a count of one.                                       */
202   void wl_increment(wordlist *list, char *word)
203   {
204     wordlistnode *newword;
205     wordlistnode *wlst = wln_lookup(list->root,word);
206     if((wlst == NULL)||(strcmp(wlst->word, word) != 0))
207       { /* create a new node */
208         list->nwords++;
209         newword = wln_alloc(word);
210         list->root = wln_insert(list->root,newword);
211       }
212     else
213       wlst->count++;
214   }
215
216   /*****************************************************/
217   /* wln_print is an interal function to print a table    */
218   /* showing the number of occurrences for each word,     */
219   /* followed by the word.                                */
220   void wln_print(wordlistnode *list)
221   {
222     if(list != NULL)
223       {
224         wln_print(list->left);
225         printf("%10d '%s'\n",list->count,list->word);
226         wln_print(list->right);
227       }
228   }
229
230   /*****************************************************/
231   /* wl_print_alphabetical prints a table showing the number*/
232   /* of occurrences for each word, followed by the word.   */
233   void wl_print(wordlist *list)
234   {
235     wln_print(list->root);
236     printf("There are %d unique words in the document\n",
237           list->nwords);
238   }
239
240   #ifndef USE_ASM
241   /*****************************************************/
242   /* wl_print_numerical prints a table showing the number  */
```

```
243   /* of occurrences for each word, followed by the word,    */
244   /* sorted in reverse order of occurrence.                  */
245   void wl_print_numerical(wordlist *list)
246   {
247     printf("wl_print_numerical has not been implemented\n");
248   }
249   #endif
```

Listing 6.10
C implementation of the wordlist ADT using a tree.

With the tree implementation, `wl_print_numerical()` could build a new tree, sorted on the word frequency counts. However, it may be more efficient to build a separate index, and sort the index by word frequency counts. The assembly code will allocate an array of pointers, and set each pointer to one of the nodes in the tree, as shown in Fig. 6.2. Then, it will use a quick sort to sort the pointers into descending order by word frequency count, as shown in Fig. 6.3. This implementation is shown in Listing 6.11.

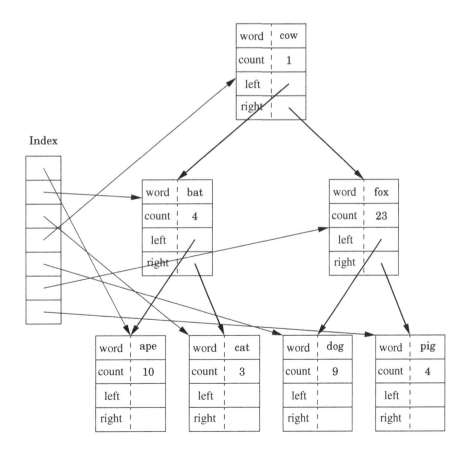

Figure 6.2
Binary tree of word frequencies with index added.

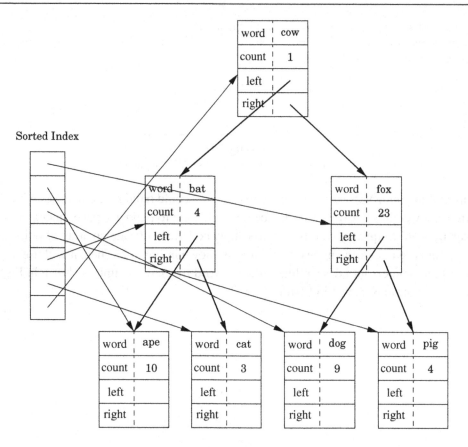

Figure 6.3
Binary tree of word frequencies with sorted index.

```
1   @@@ Definitions for the wordlistnode type
2           .equ    wln_word,0     @ word field
3           .equ    wln_count,4    @ count filed
4           .equ    wln_left,8     @ left field
5           .equ    wln_right,12   @ right field
6           .equ    wln_height,16  @ height of this node
7           .equ    wln_size,20    @ sizeof(wordlistnode)
8   @@@ Definitions for the wordlist type
9           .equ    wl_nwords,0    @ number of words in list
10          .equ    wl_head,4      @ head of linked list
11          .equ    wl_size,8      @ sizeof(wordlist)
12  @@@ Define NULL
13          .equ    NULL,0
14          .data
15  failstr:.asciz  "Unable to allocate index\n"
```

```
16    fmtstr: .asciz  "%10d '%s'\n"
17            .text
18    @@@ ------------------------------------------------------------
19    @@@ wordlistnode **getptrs(wordlistnode *ptrs[],wordlistnode *node)
20    @@@ this function recursively traverses the tree, filling in the
21    @@@ array of pointers.
22    @@@ r0 is incremented as each pointer is stored, so it returns
23    @@@ a pointer to the next pointer in the array that needs to
24    @@@ be set.
25    getptrs:
26            cmp     r1,#0               @ if node == NULL
27            moveq   pc,lr               @ exit immediately
28            stmfd   sp!,{r4,lr}
29            mov     r4,r1               @ save address of node
30            ldr     r1,[r4,#wln_left]   @ get ptr to left child
31            bl      getptrs             @ process left child
32            str     r4,[r0],#4          @ Store address of node
33            ldr     r1,[r4,#wln_right]  @ get ptr to right child
34            bl      getptrs             @ process right child
35            ldmfd   sp!,{r4,pc}
36
37    @@@ ------------------------------------------------------------
38    @@@ wl_print_numerical prints a table showing the number
39    @@@ of occurrences for each word, followed by the word,
40    @@@ sorted in reverse order of occurrence.
41            .global wl_print_numerical
42    wl_print_numerical:
43            stmfd   sp!,{r4-r6,lr}      @ save registers
44            mov     r4,r0               @ copy original pointer
45            ldr     r5,[r0,#wl_nwords]  @ load nwords
46            lsl     r0,r5,#2            @ multiply by four
47            bl      malloc              @ allocate storage
48            cmp     r0,#0               @ check return value
49            bne     malloc_ok
50            ldr     r0,=failstr         @ load pointer to string
51            bl      printf
52            mov     r0,#1
53            bl      exit                @ exit(1)
54    malloc_ok:
55            mov     r6,r0               @ save pointer to array
56            ldr     r1,[r4,#wl_head]    @ get pointer to tree
57            bl      getptrs             @ fill in the pointers
58            mov     r0,r6               @ get pointer to array
59            add     r1,r0,r5,lsl #2     @ get pointer to end of array
60            sub     r1,r1,#4
61            bl      wl_quicksort        @ re-sort the array of pointers
```

```
62          @@ Print the word frequency list.
63          mov     r4,#0           @ do a for loop
64   loop:  cmp     r4,r5
65          bge     done
66          ldr     r0,=fmtstr
67          ldr     r3,[r6,r4,lsl #2] @ get next pointer
68          add     r4,r4,#1
69          ldr     r1,[r3,#wln_count]@load count
70          ldr     r2,[r3,#wln_word] @load ptr to word
71          bl      printf
72          b       loop
73   done:  ldmfd   sp!,{r4-r6,pc}  @ restore & return
74
75   @@@ --------------------------------------------------------
76   @@@ function wl_quicksort(wln **left,wln **right) quicksorts
77   @@@ the array of pointers in order of the word counts
78   wl_quicksort:
79          cmp     r0,r1
80          movge   pc,lr           @ return if length<=1
81          stmfd   sp!,{r4-r7,lr}
82          ldr     r12,[r0]        @ use count of first item as
83          ldr     r12,[r12,#wln_count] @ pivot value in r12
84          mov     r4,r0           @ current left
85          mov     r5,r1           @ current right
86          mov     r6,r0           @ original left(first)
87          mov     r7,r1           @ original right(last)
88   loopa: cmp     r4,r5           @ while left <= right &&
89          bgt     finale
90          ldr     r0,[r4]         @ (*left)->count > pivot
91          ldr     r1,[r0,#wln_count]
92          cmp     r1,r12
93          ble     loopb
94          add     r4,r4,#4        @ increment left
95          b       loopa
96   loopb: cmp     r4,r5           @ while left < right &&
97          bgt     finale
98          ldr     r2,[r5]         @ (*right)->count < pivot
99          ldr     r3,[r2,#wln_count]
100         cmp     r3,r12
101         bge     cmp
102         sub     r5,r5,#4        @ decrement right
103         b       loopb
104  cmp:   cmp     r4,r5           @ if( left <= right )
105         bgt     finale
106         str     r0,[r5],#-4     @ swap pointers and
107         str     r2,[r4],#4      @ change indices
```

```
108        b       loopa
109 finale: mov     r0,r6       @ quicksort array from
110        mov     r1,r5       @ first to current right
111        bl      wl_quicksort
112        mov     r0,r4       @ quicksort array from
113        mov     r1,r7       @ current left to last)
114        bl      wl_quicksort
115        ldmfd   sp!,{r4-r7,pc}
```

Listing 6.11
ARM assembly implementation of wl_print_numerical() with a tree.

The tree-based implementation gets most of its speed improvement through using two $O(N \log N)$ algorithms to replace $O(N^2)$ algorithms. These examples show how a small part of a program can be implemented in assembly language, and how to access C data structures from assembly language. The functions could just as easily have been written in C rather than assembly, without greatly affecting performance. Later chapters will show examples where the assembly implementation *does* have significantly better performance than the C implementation.

6.3 Ethics Case Study: Therac-25

The Therac-25 was a device designed for radiation treatment of cancer. It was produced by Atomic Energy of Canada Limited (AECL), which had previously produced the Therac-6 and Therac-20 units in partnership with CGR of France. It was capable of treating tumors close to the skin surface using electron beam therapy, but could also be configured for Megavolt X-ray therapy to treat deeper tumors. The X-ray therapy required the use of a tungsten radiation shield to limit the area of the body that was exposed to the potentially lethal radiation produced by the device.

The Therac-25 used a double pass accelerator, which provided more power, in a smaller space, at less cost, compared to its predecessors. The second major innovation was that computer control was a central part of the design, rather than an add-on component as in its predecessors. Most of the hardware safety interlocks that were integral to the designs of the Therac-6 and Therac-20 were seen as unnecessary, because the software would perform those functions. Computer control was intended to allow operators to set up the machine more quickly, allowing them to spend more time communicating with patients and to treat more patients per day. It was also seen as a way to reduce production costs by relying on software, rather than hardware, safety interlocks.

There were design issues with both the software and the hardware. Although this machine was built with the goal of saving lives, between 1985 and 1986, three deaths and other injuries were attributed to the hardware and software design of this machine. Death due to radiation exposure is usually slow and painful, and the problem was not identified until the damage had been done.

6.3.1 History of the Therac-25

AECL was required to obtain US Food and Drug Administration (FDA) approval before releasing the Therac-25 to the US market. They obtained approval quickly by declaring "pre-market equivalence," effectively claiming that the new machine was not significantly different from its predecessors. This practice was common in 1984, but was overly optimistic, considering that most of the safety features had been changed from hardware to software implementations. With FDA approval, AECL made the Therac-25 commercially available and performed a Fault Tree Analysis to evaluate the safety of the device.

Fault Tree Analysis, as its name implies, requires building a tree to describe every possible fault and assigning probabilities to those faults. After building the tree, the probabilities of hazards, such as overdose, can be calculated. Unfortunately, the engineers assumed that the software (much of which was re-used from the previous Therac models) would operate correctly. This turned out not to be the case, because the hardware interlocks present in the previous models had hidden some of the software faults. The analysts did consider some possible computer faults, such as an error being caused by cosmic rays, but assigned extremely low probabilities to those faults. As a result, the assessment was very inaccurate.

When the first report of an overdose was reported to AECL in 1985, they sent an engineer to the site to investigate. They also filed a report with the FDA and the Canadian Radiation Protection Board (CRPB). AECL also notified all users of the fact that there had been a report and recommended that operators should visually confirm hardware settings before each treatment. The AECL engineers were unable to reproduce the fault, but suspected that it was due to the design and placement of a microswitch. They redesigned the microswitch and modified all of the machines that had been deployed. They also retracted their recommendation that operators should visually confirm hardware settings before each treatment.

Later that year, a second incident occurred. In this case, there is no evidence that AECL took any action. In January of 1986, AECL received another incident report. An employee at AECL responded by denying that the Therac-25 was at fault, and stated that no other similar incidents had been reported. Another incident occurred in March of that year. AECL sent an engineer to investigate. The engineer was unable to determine the cause, and suggested that it was due to an electrical problem, which may have caused an electrical shock. An independent engineering firm was called to examine the machine and reported that it was very unlikely that the machine could have delivered an electrical shock to the patient. In April of 1986, another incident was reported. In this case, the AECL engineers, working with the medical physicist at the hospital, were able to reproduce the sequence of events that lead to the overdose.

As required by law, AECL filed a report with the FDA. The FDA responded by declaring the Therac-25 defective. AECL was ordered to notify all of the sites where the Therac-25 was in use, investigate the problem, and file a corrective action plan. AECL notified all sites, and recommended removing certain keys from the keyboard on the machines. The FDA responded by requiring them to send another notification with more information about the defect and the consequent hazards. Later in 1986, AECL filed a revised corrective action plan.

Another overdose occurred in January 1987, and was attributed to a different software fault. In February, the FDA and CRPB both ordered that all Therac-25 units be shut down, pending effective and permanent modifications. AECL spent six months developing a new corrective action plan, which included a major overhaul of the software, the addition of mechanical safety interlocks, and other safety-related modifications.

6.3.2 Overview of Design Flaws

The Therac-25 was controlled by a DEC PDP-11 computer, which was the most popular minicomputer ever produced. Around 600,000 were produced between 1970 and 1990 and used for a variety of purposes, including embedded systems, education, and general data processing. It was a 16-bit computer and was far less powerful than a Raspberry Pi. The Therac-25 computer was programmed in assembly language by one programmer and the source code was not documented. Documentation for the hardware components was written in French. After the faults were discovered, a commission concluded that the primary problems with the Therac-25 were attributable to poor software design practices, and not due to any one of several specific coding errors. This is probably the best known case where a poor overall software design, and insufficient testing, led to loss of life.

The worst problems in the design and engineering of the machine were:

- The code was not subjected to independent review.
- The software design was not considered during the assessment of how the machine could fail or malfunction.
- The operator could ignore malfunctions and cause the machine to proceed with treatment.
- The hardware and software were designed separately and not tested as a complete system until the unit was assembled at the hospitals where it was to be used.
- The design of the earlier Therac-6 and Therac-20 machines included hardware interlocks which would ensure that the X-ray mode could not be activated unless the tungsten radiation shield was in place. The hardware interlock was replaced with a software interlock in the Therac-25.
- Errors were displayed as numeric codes, and there was no indication of the severity of the error condition.

The operator interface consisted of a keyboard and text-mode monitor, which was common in the early 1980s. The interface had a data entry area in the middle of the screen and a command line at the bottom. The operator was required to enter parameters in the data entry area, then move the cursor to the command line to initiate treatment. When the operator moved the cursor to the command line, internal variables were updated and a flag variable was set to indicate that data entry was complete. That flag was cleared when a command was entered on the command line. If the operator moved the cursor back to the data entry area without entering a command, then the flag was not cleared, and any subsequent changes to the data entry area did not affect the internal variables.

A global variable was used to indicate that the magnets were currently being adjusted. This variable was modified by two functions, and did not always contain the correct value. Adjusting the magnets required about eight seconds, and the flag was correct for only a small period at the beginning of this time period.

Due to the errors in the design and implementation of the software, the following sequence of events could result in the machine causing injury to, or even the death of, the patient:

1. The operator mistakenly specified high-power mode during data entry.
2. The operator moved the cursor to the command line area.
3. The operator noticed the mistake, and moved the cursor back to the data entry area without entering a command.
4. The operator corrected the mistake and moved the cursor back to the command line.
5. The operator entered the command line area, left it, made the correction, and returned within the eight-second window required for adjusting the magnets.

If the above sequence occurred, then the operator screen could indicate that the machine was in low power mode, although it was actually set in high-power mode. During a final check before initiating the beam, the software would find that the magnets were set for high-power mode but the operator setting was for low power mode. It displayed a numeric error code and prevented the machine from starting. The operator could clear the error code by resetting the computer (which only required one key to be pressed on the keyboard). This caused the tungsten shield to withdraw but left the machine in X-ray mode. When the operator entered the command to start the beam, the machine could be in high-power mode without having the tungsten shield in place. X-rays were applied to the unprotected patient.

It took some time for this critical flaw to appear. The failure only occurred when the operator initially made a one-keystroke mistake in entering the prescription data, moved to the command area, and then corrected the mistake within eight seconds. Initially, operators were slow to enter data, and spent a lot of time making sure that the prescription was correct before initiating treatment. As they became more familiar with the machine, they were able to enter data and correct mistakes more quickly. Eventually, operators became familiar enough with the machine that they could enter data, make a correction, and return to the command area within the critical eight-second window. Also, the operators became familiar with the machine reporting numeric error codes without any indication of the severity of the code. The operators were given a table of codes and their meanings. The code reported was "no dose" and indicated "treatment pause." There is no reason why the operator should consider that to be a serious problem; they had become accustomed to frequent malfunctions that did not have any consequences to the patient.

Although the code was written in assembly language, that fact was not cited as an important factor. The fundamental problems were poor software design and overconfidence. The reuse of code in an application for which it was not initially designed also may have contributed to the system flaws. A proper design using established software design principles, including structured programming and abstract data types, would almost certainly have avoided these fatalities.

6.4 Chapter Summary

The abstract data type is a structured programming concept which contributes to software reliability, eases maintenance, and allows for major revisions to be performed in a safe way. Many high-level languages enforce, or at least facilitate, the use of ADTs. Assembly language does not. However, the ethical assembly language programmer will make the extra effort to write code that conforms to the standards of structured programming and use abstract data types to help ensure safety, reliability, and maintainability.

ADTs also facilitate the implementation of software modules in more than one language. The interface specifies the components of the ADT, but not the implementation. The implementation can be in any language. As long as assembly programmers and compiler authors generate code that conforms to a well-known standard, their code can be linked with code written in other languages.

Poor coding practices and poor design can lead to dire consequences, including loss of life. It is the responsibility of the programmer, regardless of the language used, to make ethical decisions in the design and implementation of software. Above all, the programmer must be aware of the possible consequences of the decisions they make.

Exercises

6.1   What are the advantages of designing software using abstract data types?

6.2   Why is the internal structure of the `Pixel` data type hidden from client code in Listing 6.2?

6.3   High-level languages provide mechanisms for information hiding, but assembly does not. Why should the assembly programmer not simply bypass all information hiding and access the internal data structures of any ADT directly?

6.4   The assembly code in `wl_print_numerical()` accesses the internal structure of the `wordlistnode` data type. Why is it allowed to do so? *Should* it be allowed to do so?

6.5   Given the following definitions for a stack ADT:

```
1    /* File: stack.h */
2
3    typedef struct IntStackStruct *IntStack;
4
5    /* create an empty stack and return a pointer to it */
6    IntStack InitStack ();
7
8    /* Push value onto the stack.
9       Return 1 for success. Return 0 if stack is full. */
10   int Push (IntStack stack, int k);
11
12   /* Remove value from top of stack.
13      Return 1 for success. Return 0 if stack was empty. */
14   int Pop (IntStack stack);
15
16   /* Return the value that is at the top of the stack. */
17   int Top (IntStack stack);
18
19   /* Print the elements of the stack. */
20   extern PrintStack (IntStack stack);
```

```
1    /* File: stack.c */
2
3    #define STACKSIZE 100
4
5    /* The stack is implemented as  an array of items and
6       the index of the item at the top */
7    struct IntStackStruct {
8      int stackItems [STACKSIZE];
9      int top;
10   };
11
12   typedef struct IntStackStruct *IntStack;
```

Write the `InitStack()` function in ARM assembly language.

6.6   Referring to the previous question, write the `Push()` function in ARM assembly language.

6.7   Referring to the previous two questions, write the `Pop()` function in ARM assembly language.

6.8   Referring to the previous three questions, write the `Top()` function in ARM assembly language.

6.9   Referring to the previous three questions, write the `PrintStack()` function in ARM assembly language.

6.10 Re-implement all of the previous stack functions using a linked list rather than a static array.

6.11 The "Software Engineering Code of Ethics And Professional Practice" states that a responsible software engineer should "Approve software only if they have well-founded belief that it is safe, meets specifications, passes appropriate tests. . ." (sub-principle 1.03) and "Ensure adequate testing, debugging, and review of software. . .on which they work." (sub-principle 3.10). Unfortunately, defects did make their way into the system. The software engineering code of ethics also states that a responsible software engineer should "Treat all forms of software maintenance with the same professionalism as new development."

    (a)   Explain how the Software Engineering Code of Ethics And Professional Practice were violated by the Therac 25 developers.

    (b)   How should the engineers and managers at AECL have responded when problems were reported?

    (c)   What other ethical and non-ethical considerations may have contributed to the deaths and injuries?

PART II

Performance Mathematics

Integer Mathematics

There are some differences between the way calculations are performed in a computer versus the way most of us were taught as children. The first difference is that calculations are performed in binary instead of base ten. Another difference is that the computer is limited to a fixed number of binary digits, which raises the possibility of having a result that is too large to fit in the number of bits available. This occurrence is referred to as *overflow*. The third difference is that subtraction is performed using complement addition.

Addition in base b is very similar to base ten addition, except that the result of each column is limited to $b - 1$. For example, binary addition works exactly the same as decimal addition, except that the result of each column is limited to 0 or 1. The following figure shows an addition in base ten and the equivalent addition in base two.

$$
\begin{array}{cc}
\begin{array}{r}
1\ \ \ \ \\
7\ 5 \\
+\ \ 1\ 9 \\
\hline
9\ 4
\end{array}
&
=
\begin{array}{r}
1\ \ 1\ \ \ \ \ \ \\
0\ 1\ 0\ 0\ 1\ 0\ 1\ 1 \\
+\ \ 0\ 0\ 0\ 1\ 0\ 0\ 1\ 1 \\
\hline
0\ 1\ 0\ 1\ 1\ 1\ 1\ 0
\end{array}
\end{array}
$$

Modern Assembly Language Programming with the ARM Processor. http://dx.doi.org/10.1016/B978-0-12-803698-3.00007-3

The carry from one column to the next is shown as a small number above the column that it is being carried into. Note that carries from one column to the next are done the same way in both bases. The only difference is that there are more columns in the base two addition because it takes more digits to represent a number in binary than it does in decimal.

7.1 Subtraction by Addition

Finding the complement was explained in Section 1.3.3. Subtraction can be computed by adding the radix complement of the subtrahend to the menuend. Example 7.1 shows a complement subtraction with positive results. When $x < y$, the result will be negative. In the complement method, this means that there will be a '1' in the most significant bit, and in order to convert the result to base ten, we must take the radix complement. Example 7.2 shows complement subtraction with negative results. Example 7.3 shows several more signed addition and subtraction operations in base ten and binary.

Example 7.1 Ten's Complement Subtraction

Suppose we wish to calculate $384_{10} - 56_{10}$ using the complements method. After extending both numbers to the same number of digits, we have $384_{10} - 056_{10}$. From Eq. (1.1), the ten's complement of 056_{10} is $10^4 - 056_{10} = 944_{10}$. Adding gives us $384_{10} + 944_{10} = 1328_{10}$. After discarding the leading "1", we have 328, which is the correct result. Both methods of subtraction are shown below:

$$
\begin{array}{r}
{}^{7}\ {}^{14}\\
3\ \cancel{8}\ \cancel{4}\\
-\ \ 5\ 6\\
\hline
3\ 2\ 8
\end{array}
\quad = \quad
\begin{array}{r}
{}^{1}\\
3\ 8\ 4\\
+9\ 4\ 4\\
\hline
\cancel{1}3\ 2\ 8
\end{array}
$$

Example 7.2 Ten's Complement Subtraction With a Negative Result

Suppose we want to calculate $284 - 481$. Both numbers have three digits, so it is not necessary to pad with leading zeros. Adding the ten's complement of y to x gives $284 + 519 = 803$. This is obviously the wrong answer, since the expected answer is -197. But all is not lost, because 803 happens to be the ten's complement of 197. The fact that the first digit of the result is greater than four indicates that we must take the ten's complement of the result and add a negative sign.

7.2 Binary Multiplication

Many processors have hardware multiply instructions. However hardware multipliers require a large number of transistors, and consume significant power. Processors designed for extremely

Example 7.3   Signed Addition and Subtraction in Decimal and Binary

2 3		0 0 0 1 0 1 1 1
+ 1 5	=	+ 0 0 0 0 1 1 1 1
3 8		0 0 1 0 0 1 1 0

2 3		0 0 0 1 0 1 1 1
− 1 5	=	+ 1 1 1 1 0 0 0 1
8		1|0 0 0 0 1 0 0 0

− 2 3		1 1 1 0 1 0 0 1
+ 1 5	=	+ 0 0 0 0 1 1 1 1
− 8		1 1 1 1 1 0 0 0

− 2 3		1 1 1 0 1 0 0 1
− 1 5	=	+ 1 1 1 1 0 0 0 1
− 3 8		1|1 1 0 1 1 0 1 0

low power consumption or very small size usually do not implement a multiply instruction, or only provide multiply instructions that are limited to a small number of bits. On these systems, the programmer must implement multiplication using basic data processing instructions.

7.2.1 Multiplication by a Power of Two

If the multiplier is a power of two, then multiplication can be accomplished with a shift to the left. Consider the 4-bit binary number $x = x_3 \times 2^3 + x_2 \times 2^2 + x_1 \times 2^1 + x_0 \times 2^0$, where x_n denotes bit n of x. If x is shifted left by one bit, introducing a zero into the least significant bit, then it becomes $x_3 \times 2^4 + x_2 \times 2^3 + x_1 \times 2^2 + x_0 \times 2^1 + 0 \times 2^0 = 2\left(x_3 \times 2^3 + x_2 \times 2^2 + x_1 \times 2^1 + x_0 \times 2^0 + 0 \times 2^{-1}\right)$. Therefore, a shift of one bit to the left is equivalent to multiplication by two. This argument can be extended to prove that a shift left by n bits is equivalent to multiplication by 2^n.

7.2.2 Multiplication of Two Variables

Most techniques for binary multiplication involve computing a set of partial products and then summing the partial products together. This process is similar to the method taught to primary schoolchildren for conducting long multiplication on base ten integers, but has been modified here for application to binary. The method typically taught in school for multiplying decimal numbers is based on calculating partial products, shifting them to the left and then adding them together. The most difficult part is to obtain the partial products, as that involves multiplying a long number by one base ten digit. The following example shows how the partial products are formed when multiplying 123 by 456.

The first partial product can be written as $123 \times 6 \times 10^0 = 738$. The second is $123 \times 5 \times 10^1 = 6150$, and the third is $123 \times 4 \times 10^2 = 49200$. In practice, we usually leave out the trailing zeros. The procedure is the same in binary, but is simpler because the partial

```
              1 2 3
        ×     4 5 6
              7 3 8     (this is 123 × 6)
            6 1 5 0     (this is 123 × 5, shifted one position to the left)
    +     4 9 2 0 0     (this is 123 × 4, shifted two positions to the left)
          5 6 0 8 8
```

product involves multiplying a long number by a single base 2 digit. Since the multiplier is always either zero or one, the partial product is very easy to compute. The product of multiplying any binary number x by a single binary digit is always either 0 or x. Therefore, the multiplication of two binary numbers comes down to shifting the multiplicand left appropriately for each non-zero bit in the multiplier, and then adding the shifted numbers together.

Suppose we wish to multiply two four-bit numbers, 1011 and 1010:

```
            1 0 1 1     this is 11_10
        ×   1 0 1 0     this is 10_10
            0 0 0 0     1011 × 0
          1 0 1 1       1011 × 1, shifted one position to the left
        0 0 0 0         1011 × 0, shifted two positions to the left
    +   1 0 1 1         1011 × 1, shifted three positions to the left
        1 1 0 1 1 1 0   this is 110_10
```

Notice in the previous example that each partial sum is either zero or x shifted by some amount. A slightly quicker way to perform the multiplication is to leave out any partial sum which is zero. Example 7.4 shows the results of multiplying 101_{10} by 89_{10} in decimal and binary using this shorter method. For implementation in hardware and software, it is easier to *accumulate* the partial products, by adding each to a running sum, rather than building a circuit to add multiple binary numbers at once.

Example 7.4   Equivalent Multiplication in Decimal and Binary

```
                                        0 1 1 0 0 1 0 1
            1 0 1                    ×   0 1 0 1 1 0 0 1
        ×     8 9
                                        0 1 1 0 0 1 0 1
            9 0 9     =                0 1 1 0 0 1 0 1
          8 0 8                      0 1 1 0 0 1 0 1
          8 9 8 9                  0 1 1 0 0 1 0 1
                            0 0 1 0 0 0 1 1 0 0 0 1 1 1 0 1
```

Binary multiplication can be implemented as a sequence of shift and add instructions. Given two registers, x and y, and an *accumulator* register a, the product of x and y can be computed

$a \leftarrow 0$
while $y \neq 0$ **do**
  if $LSB(y) = 1$ **then**
    $a \leftarrow a + x$
  end if
  $x \leftarrow x$ shifted left 1 bit
  $y \leftarrow y$ shifted right 1 bit
end while

Algorithm 1
Algorithm for binary multiplication.

using Algorithm 1. When applying the algorithm, it is important to remember that, in the general case, the result of multiplying an n bit number by an m bit number is (at most) an $n + m$ bit number. For instance $11_2 \times 11_2 = 1001_2$. Therefore, when applying Algorithm 1, it is necessary to know the number of bits in x and y. Since x is shifted left on each iteration of the loop, the registers used to store x and a must both be at least as large as the number of bits in x plus the number of bits in y.

Assume we wish to multiply two numbers, $x = 01101001$ and $y = 01011010$. Applying Algorithm 1 results in the following sequence:

a	x	y	Next operation
0000000000000000	0000000001101001	01011010	shift only
0000000000000000	0000000011010010	00101101	add, then shift
0000000011010010	0000000110100100	00010110	shift only
0000000011010010	0000001101001000	00001011	add, then shift
0000010000011010	0000011010010000	00000101	add, then shift
0000101010101010	0000110100100000	00000010	shift only
0000101010101010	0001101001000000	00000001	add, then shift
0010010011101010	0011010010000000	00000000	shift only
	$105 \times 90 = 9450$		

To multiply two n bit numbers, you must be able to add two $2n$-bit numbers. On the ARM processor, n is usually assumed to be 32-bits, because that is the natural word size for the ARM processor. Adding 64-bit numbers requires two add instructions and the carry from the least-significant 32 bits must be added to the sum of the most-significant 32 bits. The ARM processor provides a convenient way to perform the add with carry. Assume we have two 64 bit numbers, x and y. We have x in r0, r1 and y in r2, r3, where the high order words of each number are in the higher-numbered registers, and we want to calculate $x = x + y$. Listing 7.1 shows a two instruction sequence for the ARM processor. The first instruction adds the two least-significant words together and sets (or clears) the carry bit and other flags in the CPSR. The second instruction adds the two most significant words along with the carry bit.

On the ARM processor, the algorithm to multiply two 32-bit unsigned integers is very efficient. Listing 7.2 shows one possible algorithm for multiplying two 32-bit numbers to obtain a 64-bit result. The code is a straightforward implementation of the algorithm, and some modifications can be made to improve efficiency. For example, if we only want a 32-bit result, we do not need to perform 64-bit addition. This significantly simplifies the code, as shown in Listing 7.3.

```
1      adds    r0,r0,r2    @ add the low-order words, and
2                          @ set flags in CPSR
3      adc     r1,r1,r3    @ add the high-order words plus
4                          @ the carry flag
```

Listing 7.1
ARM assembly code for adding two 64 bit numbers.

```
1           mov    r0, #0   @ r0 = low-order word (LOW) of result
2           mov    r1, #0   @ r1 = high-order word (HOW) of result
3           ldr    r2, =x   @ load pointer to multiplicand
4           ldr    r2, [r2] @ r2<-low-order word (LOW) of multiplicand
5           mov    r3, #0   @ r3<-high-order word (HOW) of multiplicand
6           ldr    ip, =y   @ load pointer to multiplier
7           ldr    ip, [ip] @ ip<-multiplier
8   loop: tst    ip, #1   @ is y odd?
9           addnes r0,r0,r2 @ add and set carry flag if y is odd
10          tst    ip, #1   @ previous add may have changed flags
11          adcne  r1,r1,r3 @ add and use carry flag  if y is odd
12          lsls   r2,r2,#1 @ shift LOW of x left into carry bit
13          lsl    r3,r3,#1 @ make room for the carry bit in HOW of x
14          adc    r3,r3,#0 @ add carry bit to HOW of x
15          lsrs   ip,ip,#1 @ shift y right
16          bne    loop     @ if y==0, we are done
```

Listing 7.2
ARM assembly code for multiplication with a 64 bit result.

```
1        mov     r0, #0   @ r0 is result
2        ldr     ip, =y   @ ip is multiplier
3        ldr     ip,[ip]
4        ldr     r2, =x   @ r2 is multiplicand
5        ldr     r2,[r2]
6        lsrs    ip,ip,#1 @ shift y right into carry flag
7  loop:
8        addcs   r0,r0,r2 @ add if carry is set
9        lsl     r2,r2,#1 @ shift multiplicand left
10       lsrs    ip,ip,#1 @ shift y right into carry flag
11       bne     loop     @ if y==0, we are done
```

Listing 7.3
ARM assembly code for multiplication with a 32 bit result.

7.2.3 Multiplication of a Variable by a Constant

If x or y is a constant, then a loop is not necessary. The multiplication can be directly translated into a sequence of shift and add operations. This will result in much more efficient code than the general algorithm. If we inspect the constant multiplier, we can usually find a pattern to exploit that will save a few instructions. For example, suppose we want to multiply a variable x by 10_{10}. The multiplier $10_{10} = 1010_2$, so we only need to add x shifted left 1 bit to x shifted left 3 bits as shown below:

```
1        ldr     r0, =x
2        ldr     r0,[r0]          @ load x  (r0 <- x)
3        lsl     r0,r0,#1         @ shift x left one bit (r0 <- 2x)
4        add     r0,r0,r0,lsl #2 @ (r0 <- 2x + 8x)
```

Now suppose we want to multiply a number x by 11_{10}. The multiplier $11_{10} = 1011_2$, so we will add x to x shifted left one bit plus x shifted left 3 bits as in the following:

```
1        ldr     r1, =x
2        ldr     r1,[r1]          @ load x (r1 <- x)
3        add     r0,r1,r1,lsl #1 @ shift one bit and add (r0 <- x + 2x)
4        add     r0,r0,r1,lsl #3 @ (r0 <- 3x + 8x)
```

If we wish to multiply a number x by 1000_{10}, we note that $1000_{10} = 1111101000_2$ It looks like we need one shift plus five add/shift operations, or six add/shift operations. With a little thought, the number of operations can be reduced from six to five as shown below:

```
1    ldr    r1, =x
2    ldr    r1,[r1]         @ load x
3    add    r0,r1,r1,lsl #1 @ shift and add: r2 <- x*3
4    add    r0,r0,r0,lsl #2 @ r2 <- x*3 + x*3*4 (x*15)
5    add    r0,r1,r0,lsl #1 @ r2 <- x + x*15*2 (x*31)
6    lsl    r0,#5           @ r2 <- x*31*32  (x*992)
7    add    r0,r0,r1,lsl #3 @ r1 <- x*992 + x*8 = x * 1000
```

Applying the basic multiplication algorithm to multiply a number x by 255_{10} would result in seven add/shift operations, but we can do it with only three operations and use only one register, as shown below:

```
1    ldr    r0, =x
2    ldr    r0,[r0]         @ load x
3    add    r0,r0,r0,lsl #1 @ shift and add: r0 <- x*3
4    add    r0,r0,r0,lsl #2 @ r0 <- x*3 + x*3*4 (x*15)
5    add    r0,r0,r0,lsl #4 @ r0 <- x*15 + x*15*16 (x*255)
```

Most modern systems have assembly language instructions for multiplication, but hardware multiply units require a relatively large number of transistors. For that reason, processors intended for small embedded applications often do not have a multiply instruction. Even when a hardware multiplier is available, on some processors it is often more efficient to use shift, add, and subtract operations when multiplying by a constant. The hardware multiplier units that are available on most ARM processors are very powerful. They can typically perform multiplication with a 32-bit result in as little as one clock cycle. The long multiply instructions take between three and five clock cycles, depending on the size of the operands. Using the multiply instruction on an ARM processor to multiply by a constant usually requires loading the constant into a register before performing the multiply. Therefore, if the multiplication can be performed using three or fewer shift, add, and subtract instructions, then it will be equal to or better than using the multiply instruction.

7.2.4 Signed Multiplication

Consider the two multiplication problems shown in Figs. 7.1 and 7.2. Note that the result of a multiply depends on whether the numbers are interpreted as unsigned numbers or signed numbers. For this reason, most computer CPUs have two different multiply operations for signed and unsigned numbers.

If the CPU provides only an unsigned multiply, then a signed multiply can be accomplished by using the unsigned multiply operation along with a conditional complement. The following procedure can be used to implement signed multiplication.

$$
\begin{array}{r}
-39 \\
\times \quad 73 \\
\hline
657 \\
219 \\
\hline
-2847
\end{array}
\quad = \quad
\begin{array}{r}
1\ 1\ 0\ 1\ 1\ 0\ 0\ 1 \\
\times \quad 0\ 1\ 0\ 0\ 1\ 0\ 0\ 1 \\
\hline
1\ 1\ 1\ 1\ 1\ 1\ 1\ 1\ 1\ 0\ 1\ 1\ 0\ 0\ 1 \\
1\ 1\ 1\ 1\ 1\ 1\ 0\ 1\ 1\ 0\ 0\ 1 \\
1\ 1\ 1\ 1\ 0\ 1\ 1\ 0\ 0\ 1 \\
\hline
1\ 1\ 1\ 1\ 0\ 1\ 0\ 0\ 1\ 1\ 1\ 0\ 0\ 0\ 0\ 1
\end{array}
$$

Figure 7.1

In signed 8-bit math, 11011001_2 is -39_{10}.

$$
\begin{array}{r}
217 \\
\times \quad 73 \\
\hline
511 \\
73 \\
146 \\
\hline
15841
\end{array}
\quad = \quad
\begin{array}{r}
1\ 1\ 0\ 1\ 1\ 0\ 0\ 1 \\
\times \quad 0\ 1\ 0\ 0\ 1\ 0\ 0\ 1 \\
\hline
0\ 0\ 0\ 0\ 0\ 0\ 0\ 0\ 1\ 1\ 0\ 1\ 1\ 0\ 0\ 1 \\
0\ 0\ 0\ 0\ 0\ 1\ 1\ 0\ 1\ 1\ 0\ 0\ 1 \\
0\ 0\ 1\ 1\ 0\ 1\ 1\ 0\ 0\ 1 \\
\hline
0\ 0\ 1\ 1\ 1\ 1\ 0\ 1\ 1\ 1\ 1\ 0\ 0\ 0\ 0\ 1
\end{array}
$$

Figure 7.2

In unsigned 8-bit math, 11011001_2 is 217_{10}.

1. if the multiplier is negative, take the two's complement,
2. if the multiplicand is negative, take the two's complement,
3. perform unsigned multiply, and
4. if the multiplier or multiplicand was negative (but not both), then take two's complement of result.

Example 7.5 demonstrates this method using one negative number.

7.2.5 Multiplying Large Numbers

Consider the method used for multiplying two digit numbers is base ten, using only the one-digit multiplication tables. Fig. 7.3 shows how a two digit number $a = a_1 \times 10^1 + a_0 \times 10^0$ is multiplied by another two digit number $b = b_1 \times 10^1 + b_0 \times 10^0$ to produce a four digit result using basic multiplication operations which only take one digit from a and one digit from b at each step.

Example 7.5   Signed Multiplication Using Unsigned Math

$$73 \times -39 = 73 \times 39 \times -1$$

```
                                    0 0 1 0 0 1 1 1
          7 3                   ×    0 1 0 0 1 0 0 1
     ×      3 9               ─────────────────────────
     ─────────                       0 0 1 0 0 1 1 1
          6 5 7                   0 0 1 0 0 1 1 1
        2 1 9        =        0 0 1 0 0 1 1 1
     ─────────               ─────────────────────────
        2 8 4 7              0 0 0 0 1 0 1 1 0 0 0 1 1 1 1 1
     ×     − 1               one's complement:
     ─────────               1 1 1 1 0 1 0 0 1 1 1 0 0 0 0 0
      − 2 8 4 7              two's complement:
                            1 1 1 1 0 1 0 0 1 1 1 0 0 0 0 1
```

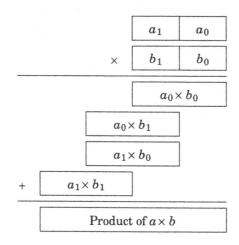

Figure 7.3
Multiplication of large numbers.

This technique can be used for numbers in any base and for any number of digits. Recall that one hexadecimal digit is equivalent to exactly four binary digits. If a and b are both 8-bit numbers, then they are also 2-digit hexadecimal numbers. In other words 8-bit numbers can be divided into groups of four bits, each representing one digit in base sixteen. Given a multiply operation that is capable of producing an 8-bit result from two 4-bit inputs, the technique shown above can then be used to multiply two 8-bit numbers using only 4-bit multiplication operations.

Carrying this one step further, suppose we are given two 16-bit numbers, but our computer only supports multiplying eight bits at a time and producing a 16-bit result. We can consider each 16-bit number to be a two digit number in base 256, and use the above technique to perform four eight bit multiplies with 16-bit results, then shift and add the 16-bit results to obtain the final 32-bit result. This approach can be extended to implement efficient multiplication of arbitrarily large numbers, using a fixed-sized multiplication operation.

7.3 Binary Division

Binary division can be implemented as a sequence of shift and subtract operations. When performing binary division by hand, it is convenient to perform the operation in a manner very similar to the way that decimal division is performed. As shown in Fig. 7.4, the operation is identical, but takes more steps in binary.

7.3.1 Division by a Power of Two

If the divisor is a power of two, then division can be accomplished with a shift to the right. Using the same approach as was used in Section 7.2.1, it can be shown that a shift right by n bits is equivalent to division by 2^n. However, care must be taken to ensure that an arithmetic shift is used if the numerator is a signed two's complement number, and a logical shift is used if the numerator is unsigned.

```
              1110110101
       1101) 11000000111001
             1101000000000
             ─────────────
             1011000111001
              110100000000
             ─────────────
              100100111001
               11010000000
             ─────────────
               1010111001
                110100000
             ─────────────
                100011001
                 11010000
             ─────────────
                  1001001
                   110100
             ─────────────
                    10101
                     1101
                    ─────
                     1000
```

```
          949
    13) 12345
        11700
        ─────
          645
          520
         ────
          125
          117
          ───
            8
```

Figure 7.4
Longhand division in decimal and binary.

7.3.2 Division by a Variable

The algorithm for dividing binary numbers is somewhat more complicated than the algorithm for multiplication. The algorithm consists of two main phases:

1. shift the divisor left until it is greater than dividend and count the number of shifts, then
2. repeatedly shift the divisor back to the right and subtract whenever possible.

Fig. 7.5 shows the algorithm in more detail. Because of the complexity of the algorithm, division in hardware requires a significant number of transistors. The ARM architecture did not introduce a divide instruction until ARMv7, and even then it was not implemented on all processors. Many ARM systems (including the Raspberry Pi) do not have hardware division. However, the ARM processor instruction set makes it possible to write very efficient code for division.

Before we introduce the ARM code, we will take some time to step through the algorithm using an example. Let us begin by dividing 94 by 7. The result is shown below:

$$94 \div 7 =$$

$$
\begin{array}{r}
1101 \\
\hline
111 \,\overline{)\, 1011110} \\
111000 \\
\hline
100110 \\
11100 \\
\hline
1010 \\
111 \\
\hline
11
\end{array}
$$

To implement the algorithm, we need three registers, one for the dividend, one for the divisor, and one for a counter. The dividend and divisor are loaded into their registers and the counter is initialized to zero as shown below:

Dividend	0	1	0	1	1	1	1	0
Divisor	0	0	0	0	0	1	1	1
Counter	0	0	0	0	0	0	0	0

Next, the divisor is shifted left and the counter incremented repeatedly until the divisor is greater than the dividend. This is shown in the following sequence:

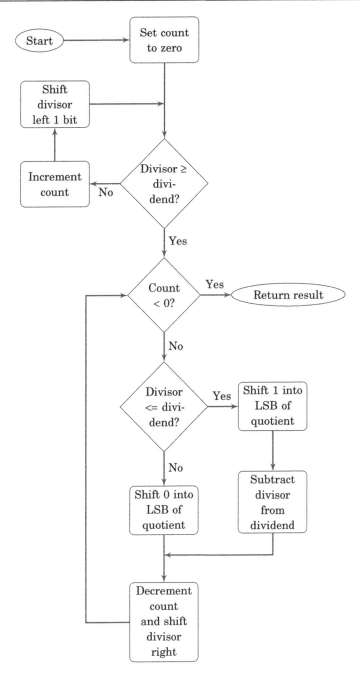

Figure 7.5
Flowchart for binary division.

Dividend	0	1	0	1	1	1	1	0
Divisor	0	0	0	0	1	1	1	0
Counter	0	0	0	0	0	0	0	1

Dividend	0	1	0	1	1	1	1	0
Divisor	0	0	0	1	1	1	0	0
Counter	0	0	0	0	0	0	1	0

Dividend	0	1	0	1	1	1	1	0
Divisor	0	0	1	1	1	0	0	0
Counter	0	0	0	0	0	0	1	1

Dividend	0	1	0	1	1	1	1	0
Divisor	0	1	1	1	0	0	0	0
Counter	0	0	0	0	0	1	0	0

Next, we allocate a register for the quotient and initialize it to zero. Then, according to the algorithm, we repeatedly subtract if possible, shift to the right, and decrement the counter. This sequence continues until the counter becomes negative. For our example this results in the following sequence:

Quotient	0	0	0	0	0	0	0	0
Dividend	0	1	0	1	1	1	1	0
Divisor	0	1	1	1	0	0	0	0
Counter	0	0	0	0	0	1	0	0

Divisor > Dividend: No subtract, shift 0 into Quotient, decrement Counter, shift Divisor right

Quotient	0	0	0	0	0	0	0	0
Dividend	0	1	0	1	1	1	1	0
Divisor	0	0	1	1	1	0	0	0
Counter	0	0	0	0	0	0	1	1

Divisor <= Dividend: Subtract, shift 1 into Quotient, decrement Counter, shift Divisor right

Quotient	0	0	0	0	0	0	0	1
Dividend	0	0	1	0	0	1	1	0
Divisor	0	0	0	1	1	1	0	0
Counter	0	0	0	0	0	0	1	0

Divisor <= Dividend: Subtract, shift 1 into Quotient, decrement Counter, shift Divisor right

Quotient	0	0	0	0	0	0	1	1
Dividend	0	0	0	0	1	0	1	0
Divisor	0	0	0	0	1	1	1	0
Counter	0	0	0	0	0	0	0	1

Divisor > Dividend: No subtract, shift 0 into Quotient, decrement Counter, shift Divisor right

Quotient	0	0	0	0	0	1	1	0
Dividend	0	0	0	0	1	0	1	0
Divisor	0	0	0	0	0	1	1	1
Counter	0	0	0	0	0	0	0	0

Divisor <= Dividend: Subtract, shift 1 into Quotient, decrement Counter, shift Divisor right

Quotient	0	0	0	0	1	1	0	1
Dividend	0	0	0	0	0	0	1	1
Divisor	0	0	0	0	0	0	1	1
Counter	1	1	1	1	1	1	1	1

Counter < 0: We are finished

When the algorithm terminates, the quotient register contains the result of the division, and the modulus (remainder) is in the dividend register. Thus, one algorithm is used to compute both the quotient and the modulus at the same time. There are variations on this algorithm. For example, one variation is to shift a single bit left in a register, rather than incrementing a count. This variation has the same two phases as the previous algorithm, but counts in powers of two rather than by ones. The following sequence shows what occurs after each iteration of the first loop in the algorithm.

Dividend	0	1	0	1	1	1	1	0
Divisor	0	0	0	0	0	1	1	1
Power:	0	0	0	0	0	0	0	1

Dividend	0	1	0	1	1	1	1	0
Divisor	0	0	0	0	1	1	1	0
Power:	0	0	0	0	0	0	1	0

Dividend	0	1	0	1	1	1	1	0
Divisor	0	0	0	1	1	1	0	0
Power:	0	0	0	0	0	1	0	0

Dividend	0	1	0	1	1	1	1	0
Divisor	0	0	1	1	1	0	0	0
Power:	0	0	0	0	1	0	0	0

Dividend	0	1	0	1	1	1	1	0
Divisor	0	1	1	1	0	0	0	0
Power:	0	0	0	1	0	0	0	0

The divisor is greater than the dividend, so the algorithm proceeds to the second phase. In this phase, if the divisor is less than or equal to the dividend, then the power register is added to the quotient and the divisor is subtracted from the dividend. Then, the power and Divisor registers are shifted to the right. The process is repeated until the power register is zero. The following sequence shows what the registers will contain at the end of each iteration of the second loop.

Quotient	0	0	0	0	0	0	0	0	Divisor > Dividend: shift Power right, shift Divisor right
Dividend	0	1	0	1	1	1	1	0	
Divisor	0	1	1	1	0	0	0	0	
Power:	0	0	0	1	0	0	0	0	

Quotient	0	0	0	0	0	0	0	0	Divisor ≤ Dividend: Dividend -= Divisor, Quotient += Power, shift Power right, shift Divisor right
Dividend	0	1	0	1	1	1	1	0	
Divisor	0	0	1	1	1	0	0	0	
Power:	0	0	0	0	1	0	0	0	

Quotient	0	0	0	0	1	0	0	0	Divisor ≤ Dividend: Dividend -= Divisor, Quotient += Power, shift Power right, shift Divisor right
Dividend	0	0	1	0	0	1	1	0	
Divisor	0	0	0	1	1	1	0	0	
Power:	0	0	0	0	0	1	0	0	

Quotient	0	0	0	0	1	1	0	0	Divisor > Dividend: shift Power right, shift Divisor right
Dividend	0	0	0	0	1	0	1	0	
Divisor	0	0	0	0	1	1	1	0	
Power:	0	0	0	0	0	0	1	0	

Quotient	0	0	0	0	1	1	0	0	Divisor ≤ Dividend: Dividend -= Divisor, Quotient += Power, shift Power right, shift Divisor right
Dividend	0	0	0	0	1	0	1	0	
Divisor	0	0	0	0	0	1	1	1	
Power:	0	0	0	0	0	0	0	1	

Quotient	0	0	0	0	1	1	0	1	Power = 0: We are finished
Dividend	0	0	0	0	0	0	1	1	
Divisor	0	0	0	0	0	0	1	1	
Power:	0	0	0	0	0	0	0	0	

As with the previous version, when the algorithm terminates, the quotient register contains the result of the division, and the modulus (remainder) is in the dividend register. Listing 7.4 shows the ARM assembly code to implement this version of the division algorithm for 32-bit numbers, and the counting method for 64-bit numbers.

```
1    @@@ -------------------------------------------------------------
2    @@@ divide.S
3    @@@ Author: Larry Pyeatt
4    @@@ Date: 10/16/2014
5    @@@
6    @@@ Division functions in ARM assembly language
7    @@@ -------------------------------------------------------------
8
9            .text
10           .align  2
11   @@@ -------------------------------------------------------------
12           @@ udiv32 takes a 32-bit unsigned dividend in r0 and
13           @@ divides it by a 32-bit unsigned divisor in r1.
14           @@ Returns the quotient in r0 and remainder in r1
15           @@ It calls no other functions and only
16           @@ uses r0-r3. We don't need to use the stack
17           .global udiv32
18   udiv32: cmp     r1,#0           @ if divisor == zero
19           beq     quitudiv32      @   exit immediately
20           mov     r2,r1           @ move divisor to r2
21           mov     r1,r0           @ move dividend to r1
22           mov     r0,#0           @ clear r0 to accumulate result
23           mov     r3,#1           @ set "current" bit in r3
24   divstrt:cmp     r2,#0           @ WHILE ((msb of r2 != 1)
25           blt     divloop
26           cmp     r2,r1           @ && (divisor < dividend))
27           lslls   r2,r2,#1        @  shift divisor left
28           lslls   r3,r3,#1        @  shift "current" bit left
29           bls     divstrt         @ end WHILE
30   divloop:cmp     r1,r2           @ if dividend >= divisor
31           subhs   r1,r1,r2        @    subtract divisor from dividend
32           addhs   r0,r0,r3        @    set "current" bit in the result
33           lsr     r2,r2,#1        @ shift divisor right
34           lsrs    r3,r3,#1        @ Shift current bit right into carry
35           bcc     divloop         @ If carry not clear, R3 has
36                                   @ shifted one bit past where it
37                                   @ started, and we are done.
38   quitudiv32:
39           mov     pc,lr
40
41   @@@ -------------------------------------------------------------
42           @@ sdiv32 takes a 32-bit signed dividend in r0 and
43           @@ divides it by a 32-bit signed divisor in r1.
44           @@ Returns the quotient in r0 and remainder in r1
45           @@ It calls udiv32 to do the real work
```

```
46          .global sdiv32
47   sdiv32: stmfd   sp!,{r4,lr}
48          @@ If dividend is negative
49          cmp     r0,#0
50          rsblt   r0,r0,#0        @ complement it
51          movlt   r4,#1           @ and set sign bit for result
52          movge   r4,#0           @ else clear sign bit for result
53          @@ If divisor is negative
54          cmp     r1,#0
55          rsblt   r1,r1,#0        @ complement it
56          eorlt   r4,#1           @ complement sign bit for result
57          bl      udiv32          @ perform division
58          @@ complement result if needed
59          cmp     r4,#0
60          rsbne   r0,r0,#0
61          ldmfd   sp!,{r4,pc}
62
63   @@@ ------------------------------------------------------------
64          @@ udiv64 takes a 64 bit unsigned dividend in r1:r0
65          @@ and divides it by a 64 bit unsigned divisor in r3:r2
66          @@ Returns a 64-bit result in r1:r0 and
67          @@ 64-bit modulus in r3:r2
68          .global udiv64
69   udiv64:
70          @@ check for divisor of zero
71          cmp     r2,#0
72          cmpeq   r3,#0
73          beq     quitudiv64
74          stmfd   sp!,{r4-r6}
75          mov     r4,r2           @ move divisor to r5:r4
76          mov     r5,r3
77          mov     r2,r0           @ move dividend to r3:r2
78          mov     r3,r1
79          mov     r0,#0           @ clear r1:r0 to accumulate result
80          mov     r1,#0
81          mov     r6,#0           @ set counter to zero
82   divstrt64:
83          @@ shift divisor left until its msb is set, or
84          @@   until divisor>=dividend
85          cmp     r5,#0           @ is msb of divisor set?
86          blt     divloop64       @ end loop if msb of divisor is set
87          cmp     r5,r3           @ compare high words
88          cmpeq   r4,r2           @ conditionally compare low words
89          bhs     divloop64       @ end loop if divisor >= dividend
90          lsl     r5,#1           @   shift r5:r4 (divisor) left
91          lsls    r4,#1
```

```
 92          orrcs   r5,r5,#1
 93          add     r6,r6,#1          @   increment count
 94          b       divstrt64        @ end WHILE
 95  divloop64:
 96          lsl     r1,#1            @ shift quotient left
 97          lsls    r0,#1
 98          orrcs   r1,#1
 99          cmp     r5,r3            @ compare divisor to dividend
100          cmpeq   r4,r2            @ conditionally compare low words
101          bhi     NoSub            @ IF (divisor<=dividend) Unsigned!
102          subs    r2,r2,r4         @   subtract divisor from dividend
103          sbc     r3,r3,r5
104          orr     r0,r0,#1         @   set lsb of quotient
105  NoSub:  lsr     r4,#1            @ shift divisor right
106          lsrs    r5,#1
107          orrcs   r4,#0x80000000
108          subs    r6,#1            @ decrement count
109          bge     divloop64        @ continue until count is negative
110          ldmfd   sp!,{r4-r6}
111  quitudiv64:
112          mov     pc,lr
113
114  @@@ --------------------------------------------------------------
115          @@ sdiv64 takes a 64 bit signed dividend in r1:r0
116          @@ and divides it by a 64 bit signed divisor in r3:r2
117          @@ Returns a 64-bit result in r1:r0 and
118          @@ 64-bit modulus in r3:r2
119          .global sdiv64
120  sdiv64:
121          stmfd   sp!,{r4,lr}
122          mov     r4,#0            @ r4 holds the sign of the result
123          @@ Complement dividend if it is negative
124          cmp     r1,#0
125          bge     NotNeg1
126          mvn     r0,r0            @ complement if negative
127          mvn     r1,r1
128          adds    r0,r0,#1         @ add one to get two's complement
129          adc     r1,r1,#0
130          eor     r4,r4,#1         @ keep track of sign
131  NotNeg1:
132          @@ Complement divisor if it is negative
133          cmp     r3,#0
134          bge     NotNeg2
135          mvn     r2,r2            @ complement if negative
136          mvn     r3,r3
137          adds    r2,r2,#1         @ add one to get two's complement
```

```
138          adc      r3,r3,#0
139          eor      r4,r4,#1        @ keep track of sign
140   NotNeg2:
141          bl       udiv64          @ do unsigned division
142          @@ Complement result if sign bit is set
143          cmp      r4,#0
144          beq      NoComplement
145          mvn      r0,r0           @ complement if negative
146          mvn      r1,r1
147          adds     r0,r0,#1        @ add one to get 2's complement
148          adc      r1,r1,#0
149   NoComplement:
150          ldmfd    sp!,{r4,pc}
```

Listing 7.4
ARM assembly implementation of signed and unsigned 32-bit and 64-bit division functions

7.3.3 Division by a Constant

In general, division is slow. Newer ARM processors provide a hardware divide instruction which requires between two and twelve clock cycles to produce a result, depending on the size of the operands. Older processors must perform division using software, as previously described. In either case, division is by far the slowest of the basic mathematical operations. However, division by a constant c can be converted to a multiply by the reciprocal of c. It is obviously much more efficient to use a multiply instead of a divide wherever possible. Efficient division of a variable by a constant is achieved by applying the following equality:

$$x \div c = x \times \frac{1}{c}. \tag{7.1}$$

The only difficulty is that we have to do it in binary, using only integers. If we modify the right-hand side by multiplying and dividing by some power of two (2^n), we can rewrite Eq. (7.1) as follows:

$$x \div c = x \times \frac{2^n}{c} \times 2^{-n}. \tag{7.2}$$

Recall that, in binary, multiplying by 2^n is the same as shifting left by n bits, while multiplying by 2^{-n} is done by shifting right by n bits. Therefore, Eq. (7.2) is just Eq. (7.1) with two shift operations added. The two shift operations cancel each other out. Now, let

$$m = \frac{2^n}{c}. \tag{7.3}$$

We can rewrite Eq. (7.2) as:

$$x \div c = x \times m \times 2^{-n}. \tag{7.4}$$

We now have a method for dividing by a constant c which involves multiplying by a different constant, m, and shifting the result. In order to achieve the best precision, we want to choose n such that m is as large as possible with the number of bits we have available.

Suppose we want efficient code to calculate $x \div 23$ using 8-bit *signed* integer multiplication. Our first task is to find $m = \frac{2^n}{c}$ such that $01111111_2 \geq m \geq 01000000_2$. In other words, we want to find the value of n where the most significant bit of m is zero, and the next most significant bit of m is one. If we choose $n = 11$, then

$$m = \frac{2^{11}}{23} \approx 89.0434782609.$$

Rounding to the nearest integer gives $m = 89$. In 8 bits, m is 01011001_2 or 59_{16}. We now have values for m and n, and therefore we can apply Eq. (7.4) to divide any number x by 23. The procedure is simple: calculate $y = x \times m$, then shift y right by 11 bits.

However, there are two more considerations. First, when the divisor is positive, the result for some values of x may be incorrect due to rounding error. It is usually sufficient to increment the reciprocal value by one in order to avoid these errors. In the previous example, the number would be changed from 59_{16} to $5A_{16}$. When implementing this technique for finding the reciprocal, the programmer should always verify that the results are correct for all input values. The second consideration is when the dividend is negative. In that case it is necessary to subtract one from the final result.

For example, to calculate $101_{10} \div 23_{10}$ in binary, with eight bits of precision, we first perform the multiplication as follows:

```
              0 1 1 0 0 1 0 1
          ×   0 1 0 1 1 0 1 0
          ─────────────────────
              0 1 1 0 0 1 0 1
            0 1 1 0 0 1 0 1
          0 1 1 0 0 1 0 1
        0 1 1 0 0 1 0 1
      ─────────────────────────
      1 0 0 0 1 1 0 0 0 1 1 1 1 0
```

Then shift the result right by 11 bits. 10001100011101_2 shifted right 11_{10} bits is: $100_2 = 4_{10}$. If the modulus is required, it can be calculated as $101 \mod 23 = 101 - (4 \times 23) = 9$, which once again requires multiplication by a constant.

In the previous example the shift amount of 11 bits provided the best precision possible. But how was that number chosen? The shift amount, n, can be directly computed as

$$n = p + \lfloor \log_2 c \rfloor - 1, \tag{7.5}$$

where p is the desired number of bits of precision. The value of m can then be computed as

$$m = \begin{cases} \frac{2^n}{c} + 1 & c > 0, \\ \frac{2^n}{c} & \text{otherwise.} \end{cases} \tag{7.6}$$

For example, to divide by the constant 33, with 16 bits of precision, we compute n as

$$n = 16 + \lfloor \log_2 33 \rfloor - 1 = 16 + \lfloor 5.044394 \rfloor - 1 = 16 + 5 - 1 = 20,$$

and then we compute m as

$$m = \frac{2^{20}}{33} + 1 = 31776.030303 \approx 31776 = 7C20_{16}.$$

Therefore, multiplying a 16 bit number by $7C20_{16}$ and then shifting right 20 bits is equivalent to dividing by 33.

Example 7.6   Division by Constant 193

To divide by a constant 193, with 32 bits of precision, the multiplier is computed using Eqs. (7.5) and 7.6 with $p = 32$ as follows:

$$m = \frac{2^{32+7-1}}{193} + 1 = \frac{2^{38}}{193} + 1 = 1424237860.81 \approx 1424237860 = 54E42524_{16}.$$

The shift amount, n, is 38 bits.

Example 7.6 shows how to calculate m and n for division by 193. On the ARM processor, division by a constant can be performed very efficiently. Listing 7.5 shows how division by 193 can be implemented using only a few lines of code. In the listing, the numbers are 32 bits

```
1    @ The following code will calculate r2/193
2    @ It will leave the quotient in r0 and the remainder in r1
3    @ It will also use registers r2 and r3 for temporary variables
4    ldr    r3,=0x54E42524 @ load 1/193 shifted left by 38 bits
5    smull r0,r1,r3,r2    @ multiply (3 to 7 clock cycles)
6    mov    r3,r2,asr #31 @ get sign of numerator (0 or -1)
7    rsb    r0,r3,r1,asr#6 @ shift right and adjust for sign
8                         @ now get the modulus, if needed
9    mov    r1,#193       @ move denominator to r1
10   mul    r1,r1,r0      @ multiply denominator by quotient
11   sub    r1,r2,r1      @ subtract that from numerator
```

Listing 7.5
ARM assembly code for division by constant 193.

in length, so the constant m is much larger than in the example that was multiplied by hand, but otherwise the method is the same.

On processors without the multiply instruction, we can use the technique of shifting and adding shown previously. If we wish to divide by 23 using 32 bits of precision, we compute the multiplier as

$$m = \frac{2^{32+4-1}}{23} + 1 = \frac{2^{35}}{23} + 1 = 1493901669.17 \approx 1493901669 = 590B2165_{16}.$$

That is $01011001000010110010000101100101_2$. Note that there are only 12 non-zero bits, and the pattern 1011001 appears three times in the 32-bit multiplier. The multiply can be implemented as $2^{24}(2^6x + 2^4x + 2^3x + 2^0x) + 2^{13}(2^6x + 2^4x + 2^3x + 2^0x) + 2^2(2^6x + 2^4x + 2^3x + 2^0x) + 2^0x$. So the following code sequence can be used on processors that do not have the multiply instruction:

```
1    @ The following code will calculate r2/23
2    @ It will leave the quotient in r0 and the remainder in r1
3    @ It will also use registers r2, r3, r4, and r5
4    movs   r4, r2            @ r4:r5 <- r2 extended to 64 bits
5    movlt  r5, #0xFFFFFFFF   @ extend the sign if r2 < 0
6    movge  r5, #0            @
7    mov    r0, r4            @ copy to r0 and r1
8    mov    r1, r5
9    @ calculate 2^6x+2^4x+2^3x+2^0x
10   adds   r4,r4,r0 lsl #3
11   adc    r5,r5,r0 asr #(32-3)
12   adds   r4,r4,r0 lsl #4
13   adc    r5,r5,r0 asr #(32-4)
14   adds   r4,r4,r0 lsl #6
15   adc    r5,r5,r0 asr #(32-6)
16   @ now perform three 64-bit shift-add operations
17   lsl    r5,r5,#2
18   orr    r5,r5,r4 lsr #(32-2)
19   lsl    r4,r4,#2
20   adds   r0,r0,r4
21   adc    r1,r1,r5
22   lsl    r5,r5,#11
23   orr    r5,r5,r4 lsr #(32-11)
24   lsl    r4,r4,#11
25   adds   r0,r0,r4
26   adc    r1,r1,r5
27   lsl    r5,r5,#11
28   orr    r5,r5,r4 lsr #(32-11)
29   lsl    r4,r4,#11
30   adds   r0,r0,r4
```

```
31    adc     r1,r1,r5

32

33    mov     r3,r2,asr #31  @ get sign of numerator (0 or -1)
34    rsb     r0,r3,r1,asr#3 @ shift right and adjust for sign
35                           @ now get the modulus, if needed
36    mov     r1,#23         @ move denominator to r1
37    mul     r1,r1,r0       @ multiply denominator by quotient
38    sub     r1,r2,r1       @ subtract that from numerator
```

Listing 7.6
ARM assembly code for division of a variable by a constant without using a multiply instruction.

7.3.4 Dividing Large Numbers

Section 7.2.5 showed how large numbers can be multiplied by breaking them into smaller numbers and using a series of multiplication operations. There is no similar method for synthesizing a large division operation with an arbitrary number of digits in the dividend and divisor. However, there is a method for dividing a large dividend by a divisor given that the division operation can operate on numbers with at least the same number of digits as in the divisor.

Suppose we wish to perform division of an arbitrarily large dividend by a one digit divisor using a basic division operation that can divide a two digit dividend by a one digit divisor. The operation can be performed in multiple steps as follows:

1.   Divide the most significant digit of the dividend by the divisor. The result is the most significant digit of the quotient.
2.   Prepend the remainder from the previous division step to the next digit of the dividend, forming a two-digit number, and divide that by the divisor. This produces the next digit of the result.
3.   Repeat from step 2 until all digits of the dividend have been processed.
4.   Take the final remainder as the modulus.

The following example shows how to divide 6189 by 7 using only 2-digits at a time:

$$
\begin{array}{cccc}
\dfrac{0}{7\overline{)6}} &
\begin{array}{r} 8 \\ \hline 7\overline{)61} \\ 56 \\ \hline 5 \end{array} &
\begin{array}{r} 8 \\ \hline 7\overline{)58} \\ 56 \\ \hline 2 \end{array} &
\begin{array}{r} 4 \\ \hline 7\overline{)29} \\ 28 \\ \hline 1 \end{array}
\end{array}
$$

$$x = 6189 \div 7 = 0884 \text{ remainder } 1$$

This method can be applied in any base and with any number of digits. The only restriction is that the basic division operation must be capable of dividing a $2n$ digit number by an n digit number and producing a $2n$ digit quotient and an n digit remainder. for example, the `div` instruction available on Cortex M3 and newer processors is capable of dividing a 32-bit dividend by a 32-bit divisor, producing a 32-bit quotient. The remainder can be calculated by multiplying the quotient by the divisor and subtracting the product from the dividend. Using this division operation it is possible to divide an arbitrarily large number by a 16-bit divisor.

We have seen that, given a divide operation capable of dividing an n digit number by an n digit number, it is possible to divide a dividend with any number of digits by a divisor with $\frac{n}{2}$ digits. Unfortunately, there is no similar method to deal with an arbitrarily large divisor, or to divide an arbitrarily large dividend by a divisor with more than $\frac{n}{2}$ digits. In those cases the division must be performed using a general division algorithm as shown previously.

7.4 Big Integer ADT

For some programming tasks, it may be helpful to deal with arbitrarily large integers. For example, the factorial function and Ackerman's function grow very quickly and will overflow a 32-bit integer for small input values. In this section, we will outline an abstract data type which provides basic operations for arbitrarily large integer values. Listing 7.7 shows the C header for this ADT, and Listing 7.8 shows the C implementation. Listing 7.9 shows a small program that uses the bigint ADT to create a table of $x!$ for all x between 0 and 100.

```
1  #ifndef BIGINT_H
2  #define BIGINT_H
3
4  struct bigint_struct;
5
6  /* define bigint to be a pointer to a bigint_struct */
7  typedef struct bigint_struct* bigint;
8
9  /* there are three ways to create a bigint */
10 bigint bigint_from_str(char *s);
11 bigint bigint_from_int(int i);
12 bigint bigint_copy(bigint source);
13
14 /* bigints can be converted to integers */
15 /* if it won't fit in an integer, the program exits */
16 int bigint_to_int(bigint b);
17
18 /* to print a bigint, you must convert it to a string */
```

```
19  char *bigint_to_str(bigint b);
20
21  /* this function frees the memory used by a bigint */
22  void bigint_free(bigint b);
23
24  /* there are five arithmetic operations */
25  bigint bigint_add(bigint l, bigint r);
26  bigint bigint_sub(bigint l, bigint r);
27  bigint bigint_mul(bigint l, bigint r);
28  bigint bigint_div(bigint l, bigint r);
29  bigint bigint_negate(bigint b);
30
31  /* There are seven comparison operations.
32     They return 1 for true, or 0 for false. */
33  inline int bigint_is_zero(bigint b);
34  inline int bigint_le(bigint l, bigint r);
35  inline int bigint_lt(bigint l, bigint r);
36  inline int bigint_ge(bigint l, bigint r);
37  inline int bigint_gt(bigint l, bigint r);
38  inline int bigint_eq(bigint l, bigint r);
39  inline int bigint_ne(bigint l, bigint r);
40
41  #endif
```

Listing 7.7
Header file for a big integer abstract data type.

```
1   #include <bigint.h>
2   #include <string.h>
3   #include <math.h>
4   #include <stdlib.h>
5   #include <stdio.h>
6   #include <ctype.h>
7   #include <stdint.h>
8
9   #ifdef EIGHT_BIT
10  typedef uint8_t chunk;
11  typedef int8_t schunk;
12  typedef uint16_t bigchunk;
13  #define CHUNKMASK 0xFF
14  #else
15  #ifdef SIXTEEN_BIT
16  typedef uint16_t chunk;
17  typedef int16_t schunk;
18  typedef uint32_t bigchunk;
19  #define CHUNKMASK 0xFFFF
```

```
20   #else
21   typedef uint32_t chunk;
22   typedef int32_t schunk;
23   typedef uint64_t bigchunk;
24   #define CHUNKMASK 0xFFFFFFFF
25   #endif
26   #endif
27
28   #define BITSPERCHUNK ((sizeof(chunk)<<3))
29
30   /* A bigint is an array of chunks of bits */
31   struct bigint_struct{
32     chunk *blks;          /* array of bit chunks */
33     int size;             /* number of chunks in the array */
34   };
35
36   #define MAX(a,b) ((a<b)?b:a)
37
38   bigint bigint_adc(bigint l, bigint r, chunk carry);
39
40   /**********************************************************/
41   /*   Utility functions                                  */
42   /**********************************************************/
43   void alloc_err()
44   {
45     printf("error allocating\n");
46     exit(1);
47   }
48
49   bigint bigint_alloc(int chunks)
50   {
51     bigint r;
52     if((r = (bigint)malloc(sizeof(struct bigint_struct))) == NULL)
53       {
54         perror("bigint_alloc");
55         exit(1);
56       }
57     r->size = chunks;
58     r->blks = (chunk*)malloc(chunks * sizeof(chunk));
59     if(r->blks == NULL)
60       {
61         perror("bigint_alloc");
62         exit(1);
63       }
64     return r;
65   }
```

```
66
67    /*****************************************************/
68    void bigint_free(bigint b)
69    {
70      if(b != NULL)
71        {
72          if(b->blks != NULL)
73            free(b->blks);
74          free(b);
75        }
76    }
77
78    /*****************************************************/
79    void bigint_dump(bigint b)
80    {
81      int i;
82      printf("%d chunks:",b->size);
83      for(i=b->size-1;i>=0;i--)
84        printf(" %02X",b->blks[i]);
85      printf("\n");
86    }
87
88    /*****************************************************/
89    bigint bigint_trim(bigint b)
90    {
91      bigint d;
92      int i;
93      for(i=b->size-1; (i>0) && (!b->blks[i] || (b->blks[i]==CHUNKMASK)) ; i--);
94      if( i < (b->size-1) &&
95          ((b->blks[i]>>(BITSPERCHUNK-1) && b->blks[i+1]==0)||
96           (!b->blks[i]>>(BITSPERCHUNK-1) && b->blks[i+1]==CHUNKMASK)))
97        ++i;
98      ++i;
99      if(i < b->size)
100       {
101         d = bigint_alloc(i);
102         memcpy(d->blks,b->blks,d->size*sizeof(chunk));
103       }
104     else
105       d = bigint_copy(b);
106     return d;
107   }
108
109   /*****************************************************/
110   /* smallmod divides a bigint by a small number
111      and returns the modulus. b changes as a SIDE-EFFECT.
```

```
112      This is used by the to_str function. */
113   unsigned bigint_smallmod(bigint b,chunk num)
114   {
115     bigchunk tmp;
116     int i;
117     if(num >= (1<<(BITSPERCHUNK-1)))
118       {
119         fprintf(stderr,"bigint_smallmod: divisor out of range\n");
120         exit(1);
121       }
122     /* start with most significant chunk and work down, taking
123        two overlapping chunks at a time */
124     tmp = b->blks[b->size-1];
125     for(i=b->size-1; i>0;i--)
126       {
127         b->blks[i] = tmp/num;
128         tmp = ((tmp % num) << BITSPERCHUNK) | b->blks[i-1];
129       }
130     b->blks[0] = tmp/num;
131     tmp = (tmp % num);
132     return tmp;
133   }
134
135   /*********************************************************/
136   /* bigint_cmp compares two bigints
137      returns -1 if l<r
138      returns 0 if l==r
139      returns 1 if l>r
140   */
141   int bigint_cmp(bigint l, bigint r)
142   {
143     int i=l->size-1;
144     int j=r->size-1;
145     while(i>j)
146       if(l->blks[i--])
147         return 1;
148     while(j>i)
149       if(r->blks[j--])
150         return -1;
151     while(i>=0)
152       {
153         if(l->blks[i]<r->blks[i])
154           return -1;
155         if(l->blks[i]>r->blks[i])
156           return 1;
157         i--;
```

```
158        }
159      return 0;
160    }
161
162    /*************************************************************/
163    inline int bigint_is_zero(bigint b)
164    {
165      int i;
166      for(i=0;i<b->size;i++)
167        if(b->blks[i])
168          return 0;
169      return 1;
170    }
171
172    /*************************************************************/
173    bigint bigint_shift_left_chunk(bigint l, int chunks)
174    {
175        bigint tmp;
176        int i;
177        tmp=bigint_alloc(l->size+chunks);
178        for(i=-chunks;i<l->size;i++)
179          {
180            if(i<0)
181              tmp->blks[i+chunks]=0;
182            else
183              tmp->blks[i+chunks]=l->blks[i];
184          }
185        return tmp;
186    }
187
188    /*************************************************************/
189    bigint bigint_shift_right_chunk(bigint l, int chunks)
190    {
191        bigint tmp;
192        int i;
193        tmp=bigint_alloc(l->size-chunks);
194        for(i=0;i<tmp->size;i++)
195          {
196            if(i<chunks)
197              tmp->blks[i]=0;
198            else
199              tmp->blks[i]=l->blks[i-chunks];
200          }
201        return tmp;
202    }
203
```

```
204    /**********************************************************/
205    /* Conversion and copy functions                          */
206    /**********************************************************/
207    bigint bigint_copy(bigint source)
208    {
209      bigint r;
210      r = bigint_alloc(source->size);
211      memcpy(r->blks,source->blks,r->size*sizeof(chunk));
212      return r;
213    }
214
215    /**********************************************************/
216    bigint bigint_complement(bigint b)
217    {
218      int i;
219      bigint r = bigint_copy(b);
220      for(i=0;i<r->size;i++)
221        r->blks[i] ^= CHUNKMASK;
222      return r;
223    }
224
225    /**********************************************************/
226    bigint bigint_negate(bigint b)
227    {
228      bigint tmp1,tmp2;
229      bigint r = bigint_complement(b);
230      tmp1=bigint_from_int(1);
231      tmp2=bigint_adc(r,tmp1,0);
232      bigint_free(tmp1);
233      bigint_free(r);
234      return tmp2;
235    }
236
237    /**********************************************************/
238    char *bigint_to_str(bigint b)
239    {
240      int chars,i,negative=0;
241      unsigned remainder;
242      char *s,*r;
243      bigint tmp,tmp2;
244      /* rough estimate of the number of characters needed */
245      chars = log10(pow(2.0,(b->size * BITSPERCHUNK)))+3;
246      i = chars-1;
247      if((s = (char*)malloc(1 + chars * sizeof(char))) == NULL)
248        {
249          perror("bigint_str");
```

```
250        exit(1);
251      }
252    s[i]=0;
253    tmp = bigint_copy(b);
254    if(tmp->blks[tmp->size-1] & (1<< (BITSPERCHUNK-1)))
255      {
256        negative=1;
257        tmp2 = bigint_negate(tmp);
258        bigint_free(tmp);
259        tmp=tmp2;
260      }
261    if(bigint_is_zero(tmp))
262      s[--i] = '0';
263    else
264      do
265        {
266          remainder = bigint_smallmod(tmp,10);
267          s[--i] = remainder + '0';
268        } while(!bigint_is_zero(tmp));
269    if(negative)
270      s[--i] = '-';
271    r = strdup(s+i);
272    bigint_free(tmp);
273    free(s);
274    return r;
275  }
276
277  /********************************************************/
278  bigint bigint_from_str(char *s)
279  {
280    bigint d;
281    bigint power;
282    bigint ten;
283    bigint tmp;
284    bigint currprod;
285    int i,negative=0;
286    d = bigint_from_int(0);
287    ten =   bigint_from_int(10);
288    power = bigint_from_int(1);
289    if(*s == '-')
290      {
291        negative = 1;
292        s++;
293      }
294    for(i=strlen(s)-1; i>=0;i--)
295      {
```

```
296        if(!isdigit(s[i]))
297          {
298            fprintf(stderr,"Cannot convert string to bigint\n");
299            exit(1);
300          }
301        tmp = bigint_from_int(s[i]-'0');
302        currprod = bigint_mul(tmp,power);
303        bigint_free(tmp);
304        tmp = bigint_adc(currprod,d,0);
305        bigint_free(d);
306        d=tmp;
307        bigint_free(currprod);
308        if(i>0)
309          {
310            tmp = bigint_mul(power,ten);
311            bigint_free(power);
312            power = tmp;
313          }
314      }
315    if(negative)
316      {
317        tmp=bigint_negate(d);
318        bigint_free(d);
319        d=tmp;
320      }
321    return d;
322  }
323
324  /************************************************/
325  int bigint_to_int(bigint b)
326  {
327    int i,negative=0,result=0;
328    bigint tmp1, tmp2;
329    tmp1 = bigint_trim(b); /* make a trimmed copy */
330    if(tmp1->size*sizeof(chunk) > sizeof(int))
331      {
332        fprintf(stderr,
333                "Cannot convert bigint to int\n%ld bytes\n",
334                (long)tmp1->size*sizeof(chunk));
335        exit(1);
336      }
337    /* check sign and negate if necessary */
338    if(tmp1->blks[tmp1->size-1] & (1<<(BITSPERCHUNK-1)))
339      {
340        negative=1;
341        tmp2=bigint_negate(tmp1);
```

```
342        bigint_free(tmp1);
343        tmp1=tmp2;
344      }
345    for(i=tmp1->size-1;i>=0;i--)
346      result |= (tmp1->blks[i]<<(i*BITSPERCHUNK));
347    bigint_free(tmp1);
348    if(negative)
349      result = -result;
350    return result;
351  }
352
353  /**********************************************************/
354  bigint bigint_from_int(int val)
355  {
356    bigint d,tmp;
357    int i;
358    int nchunks = sizeof(int)/sizeof(chunk);
359    d = bigint_alloc(nchunks);
360    for(i=0;i<d->size;i++)
361      d->blks[i] = (val >> (i*BITSPERCHUNK)) & CHUNKMASK;
362    tmp = bigint_trim(d);
363    bigint_free(d);
364    return tmp;
365  }
366
367  /**********************************************************/
368  bigint  bigint_extend(bigint b,int nchunks)
369  {
370    bigint tmp;
371    int i,negative;
372    negative=0;
373    if(b->blks[b->size-1] & (1<<(BITSPERCHUNK-1)))
374      negative=1;
375    tmp = bigint_alloc(nchunks);
376    for(i=0;i<nchunks;i++)
377      if(i < b->size)
378        tmp->blks[i] = b->blks[i];
379      else
380        if(negative)
381          tmp->blks[i] = CHUNKMASK;
382        else
383          tmp->blks[i] = 0;
384    return tmp;
385  }
386
387  #ifndef USE_ASM
```

```
388   /**********************************************************/
389   /* this is the internal add function.  It includes a     */
390   /* carry. Several other functions use it.                */
391   bigint bigint_adc(bigint l, bigint r, chunk carry)
392   {
393     bigint sum,tmpl,tmpr;
394     int i,nchunks;
395     bigchunk tmpsum;
396     /* allocate one extra chunk to make sure overflow
397        cannot occur */
398     nchunks = MAX(l->size,r->size)+1;
399     /* make sure both operands are the same size */
400     tmpl = bigint_extend(l,nchunks);
401     tmpr = bigint_extend(r,nchunks);
402     /* allocate space for the result */
403     sum = bigint_alloc(nchunks);
404     /* perform the addition */
405     for(i=0 ;i < nchunks ; i++)
406       {
407         /* add the current block of bits */
408         tmpsum = tmpl->blks[i] + tmpr->blks[i] + carry;
409         sum->blks[i] = tmpsum & CHUNKMASK;
410         /* calculate the carry bit for the next block */
411         carry = (tmpsum >> BITSPERCHUNK)&CHUNKMASK;
412       }
413     bigint_free(tmpl);
414     bigint_free(tmpr);
415     tmpl = bigint_trim(sum);
416     bigint_free(sum);
417     return tmpl;
418   }
419   #endif
420
421   /**********************************************************/
422   /* Mathematical operations                               */
423   /**********************************************************/
424
425   /**********************************************************/
426   /* The add function calls adc to perform an add with     */
427   /* initial carry of zero                                 */
428   bigint bigint_add(bigint l, bigint r)
429   {
430     return bigint_adc(l,r,0);
431   }
432
433   /**********************************************************/
```

```
434  bigint bigint_sub(bigint l, bigint r)
435  {
436    bigint tmp1,tmp2;
437    tmp1 = bigint_complement(r);
438    tmp2 = bigint_adc(l,tmp1,1);
439    bigint_free(tmp1);
440    return tmp2;
441  }
442
443  /***************************************************/
444  bigint bigint_shift_left(bigint l, int shamt)
445  {
446    int extra,i;
447    bigint tmp;
448    l = bigint_extend(l,l->size+1);
449    extra = shamt % BITSPERCHUNK;
450    shamt = shamt / BITSPERCHUNK;
451    if(shamt)
452      {
453        tmp = l;
454        l = bigint_shift_left_chunk(l,shamt);
455        bigint_free(tmp);
456      }
457    if(extra)
458      {
459        for(i=l->size-1;i>0;i--)
460          {
461            l->blks[i] = (l->blks[i]<<extra) |
462              (l->blks[i-1]>>(BITSPERCHUNK-extra));
463          }
464        l->blks[0] = (l->blks[0]<<extra);
465      }
466    tmp = bigint_trim(l);
467    bigint_free(l);
468    return tmp;
469  }
470
471  /***************************************************/
472  bigint bigint_shift_right(bigint l, int shamt)
473  {
474    int extra,i;
475    bigint tmp;
476    extra = shamt % BITSPERCHUNK;
477    shamt = shamt / BITSPERCHUNK;
478    l = bigint_shift_right_chunk(l,shamt);
479    if(extra)
```

```
480        {
481         for(i=0;i<l->size;i++)
482           {
483              l->blks[i] = (l->blks[i]>>extra) |
484                (l->blks[i+1]<<(BITSPERCHUNK-extra));
485           }
486        }
487    tmp = bigint_trim(l);
488    bigint_free(l);
489    return tmp;
490  }
491
492  /*****************************************************/
493  bigint bigint_mul_uint(bigint l, unsigned r)
494  {
495    bigint sum;
496    bigint tmp1,tmp2;
497    int i,negative=0;
498    bigchunk tmpchunk;
499    sum = bigint_from_int(0);
500    /* make sure the right operand is not too large */
501    if(r > CHUNKMASK)
502      {
503        fprintf(stderr,"bigint_mul_uint: Integer too large\n");
504        exit(1);
505      }
506    /* make sure the left operand is not negative */
507    if(l->blks[l->size-1]&(1<<(BITSPERCHUNK-1)))
508      {
509        negative ^= 1;
510        l = bigint_negate(l);
511      }
512    /* perform the multiply */
513    for(i=0;i<l->size;i++)
514      {
515        tmpchunk = (bigchunk)l->blks[i] * r;
516        tmp1 = bigint_alloc(3);
517        tmp1->blks[0] = tmpchunk & CHUNKMASK;
518        tmp1->blks[1] = (tmpchunk>>BITSPERCHUNK) & CHUNKMASK;
519        tmp1->blks[2] = 0;
520        tmp2 = bigint_shift_left_chunk(tmp1,i);
521        bigint_free(tmp1);
522        tmp1=bigint_adc(sum,tmp2,0);
523        bigint_free(sum);
524        bigint_free(tmp2);
525        sum = tmp1;
```

```
526        }
527      /* result may need to be negated */
528      if(negative)
529        {
530          tmp1 = sum;
531          sum = bigint_negate(sum);
532          bigint_free(tmp1);
533        }
534      return sum;
535    }
536
537    /**********************************************************/
538    /* bigint_mul uses the algorithm from Section 7.2.5      */
539    bigint bigint_mul(bigint l, bigint r)
540    {
541      bigint sum;
542      bigint tmp1,tmp2;
543      int i,negative=0;
544      /* the result may require the sum
545         of the number of chunks in l and r */
546      sum = bigint_from_int(0);
547      /* make sure the right operand is not negative */
548      if(r->blks[r->size-1] & (1<<(BITSPERCHUNK-1)))
549        {
550          negative=1;
551          r=bigint_negate(r); /* make negated copy of r */
552        }
553      for(i=0;i<r->size;i++)
554        {
555          tmp1 = bigint_mul_uint(l,r->blks[i]);
556          tmp2 = bigint_shift_left_chunk(tmp1,i);
557          bigint_free(tmp1);
558          tmp1 = sum;
559          sum = bigint_adc(sum,tmp2,0);
560          bigint_free(tmp1);
561          bigint_free(tmp2);
562        }
563      if(negative)
564        {
565          tmp1 = sum;       /* copy original */
566          sum = bigint_negate(sum); /* create complement */
567          bigint_free(tmp1); /* free original */
568          bigint_free(r);
569        }
570      return sum;
571    }
```

```
572
573   /*****************************************************/
574   bigint bigint_div(bigint l, bigint r)
575   {
576     bigint lt,rt,tmp,q;
577     int shift,chunkshift,negative=0;
578     q = bigint_from_int(0);
579     lt = bigint_trim(l);
580     rt = bigint_trim(r);
581     if(lt->size >= rt->size)
582       {
583         /* make sure the right operand is not negative */
584         if(r->blks[r->size-1]&(1<<(BITSPERCHUNK-1)))
585           {
586             negative = 1;  /* track sign of result */
587             tmp = rt;
588             rt = bigint_negate(rt);
589             bigint_free(tmp);
590           }
591         /* make sure the left operand is not negative */
592         if(l->blks[l->size-1]&(1<<(BITSPERCHUNK-1)))
593           {
594             negative ^= 1;  /* track sign of result */
595             tmp = lt;
596             lt = bigint_negate(lt);
597             bigint_free(tmp);
598           }
599         /* do shift by chunks */
600         chunkshift = lt->size - rt->size - 1;
601         if(chunkshift>0)
602           {
603             tmp = rt;
604             rt = bigint_shift_left_chunk(rt,chunkshift);
605             bigint_free(tmp);
606           }
607         /* do remaining shift bit-by-bit */
608         shift = 0;
609         while((shift < 31) && bigint_lt(rt,lt))
610           {
611             shift++;
612             tmp = rt;
613             rt = bigint_shift_left(rt,1);
614             bigint_free(tmp);
615           }
616         shift += (chunkshift * BITSPERCHUNK); /* total shift */
617         /* loop to shift right and subtract */
```

```
618        while(shift >= 0)
619          {
620            tmp = q;
621            q = bigint_shift_left(q,1);
622            bigint_free(tmp);
623            if(bigint_le(rt,lt))
624              {
625                /* perform subtraction */
626                tmp = lt;
627                lt = bigint_sub(lt,rt);
628                bigint_free(tmp);
629                /* change lsb from zero to one */
630                q->blks[0] |= 1;
631              }
632            tmp = rt;
633            rt = bigint_shift_right(rt,1);
634            bigint_free(tmp);
635            shift --;
636          }
637        /* correct the sign of the result */
638        if(negative)
639          {
640            tmp = bigint_negate(q);
641            bigint_free(q);
642            q = tmp;
643          }
644      }
645    bigint_free(rt);
646    bigint_free(lt);
647    return q;
648  }
649
650  /****************************************************/
651  /* Test and compare functions                       */
652  /****************************************************/
653  inline int bigint_le(bigint l, bigint r)
654  {
655    return (bigint_cmp(l, r) < 1);
656  }
657
658  /****************************************************/
659  inline int bigint_lt(bigint l, bigint r)
660  {
661    return (bigint_cmp(l, r) == -1);
662  }
663
```

```
664   /***************************************************/
665   inline int bigint_ge(bigint l, bigint r)
666   {
667     return (bigint_cmp(l, r) > -1);
668   }
669
670   /***************************************************/
671   inline int bigint_gt(bigint l, bigint r)
672   {
673     return (bigint_cmp(l, r) == 1);
674   }
675
676   /***************************************************/
677   inline int bigint_eq(bigint l, bigint r)
678   {
679     return (!bigint_cmp(l, r));
680   }
681
682   /***************************************************/
683   inline int bigint_ne(bigint l, bigint r)
684   {
685     return abs(bigint_cmp(l, r));
686   }
```

Listing 7.8
C source code file for a big integer abstract data type.

```
1    #include <bigint.h>
2    #include <stdio.h>
3    #include <stdlib.h>
4
5    bigint bigfact(bigint x)
6    {
7      bigint tmp1 = bigint_from_int(1);
8      bigint tmp2;
9      if(bigint_le(x,tmp1))
10        return tmp1;
11     tmp2 = bigint_sub(x,tmp1);
12     bigint_free(tmp1);
13     tmp1 = bigfact(tmp2);
14     bigint_free(tmp2);
15     tmp2 = bigint_mul(x,tmp1);
16     bigint_free(tmp1);
17     return tmp2;
18   }
19
```

```
20   int main()
21   {
22     bigint a;
23     char *s;
24     int i,j;
25     bigint factable[151];
26
27     for(i=0;i<151;i++)
28       {
29         a = bigint_from_int(i);
30         factable[i] = bigfact(a);
31         bigint_free(a);
32       }
33
34     for(i=0;i<151;i++)
35       {
36         s = bigint_to_str(factable[i]);
37         printf("%4d %s\n",i,s);
38         free(s);
39       }
40
41     return 0;
42   }
```

Listing 7.9
Program using the bigint ADT to calculate the factorial function.

The implementation could be made more efficient by writing some of the functions in assembly language. One opportunity for improvement is in the add function, which must calculate the carry from one chunk of bits to the next. In assembly, the programmer has direct access to the carry bit, so carry propagation should be much faster.

When attempting to speed up a C program by converting selected parts of it to assembly language, it is important to first determine where the most significant gains can be made. A profiler, such as gprof, can be used to help identify the sections of code that will matter most. It is also important to make sure that the result is not just highly optimized C code. If the code cannot benefit from some features offered by assembly, then it may not be worth the effort of re-writing in assembly. The code should be re-written from a pure assembly language viewpoint.

It is also important to avoid premature assembly programming. Make sure that the C algorithms and data structures are efficient before moving to assembly. if a better algorithm can give better performance, then assembly may not be required at all. Once the assembly is written, it is more difficult to make major changes to the data structures and algorithms. Assembly language optimization is the *final* step in optimization, not the first one.

Well-written C code is modularized, with many small functions. This helps readability, promotes code reuse, and may allow the compiler to achieve better optimization. However, each function call has some associated overhead. If optimal performance is the goal, then calling many small functions should be avoided. For instance, if the piece of code to be optimized is in a loop body, then it may be best to write the entire loop in assembly, rather than writing a function and calling it each time through the loop. Writing in assembly is not a guarantee of performance. Spaghetti code is slow. Load/store instructions are slow. Multiplication and division are slow. The secret to good performance is avoiding things that are slow. Good optimization requires rethinking the code to take advantage of assembly language.

The `bigint_adc` function was re-written in assembly, as shown in Listing 7.10. This function is used internally by several other functions in the bigint ADT to perform addition and subtraction. The profiler indicated that it is used more than any other function. If assembly language can make this function run faster, then it should have a profound effect on the program.

```
 1
 2          .equ    bi_blks,0       @ offset to block pointer
 3          .equ    bi_size,4       @ offset to size int
 4          .equ    bi_struct_sz,8  @ size of the bigint struct
 5
 6          .equ    NULL,0
 7
 8  @@@ bigint bigint_adc(bigint l, bigint r, int carry)
 9  @@@ This function adds two big integers along with a carry bit.
10  @@@ NOTE: labels beginning with "." are ignored by the profiler.
11          .text
12          .global bigint_adc
13          .type   bigint_adc, %function
14  bigint_adc:
15          stmfd   sp!,{r4-r11,lr} @ store everything
16          @@ They may have different lengths. Put longest on left.
17          ldr     r8,[r0,#bi_size]
18          ldr     r9,[r1,#bi_size]
19          cmp     r8,r9
20          bge     .noswap
21          @@ skip next part if they are already ordered.
22          mov     r3,r0           @ swap pointers
23          mov     r0,r1
24          mov     r1,r3
25          mov     r3,r8           @ swap sizes
26          mov     r8,r9
27          mov     r9,r3
28  .noswap:@@ r4  : pointer to blocks for longest bigint
29          @@ r5  : pointer to blocks for shortest bigint
```

```
30          @@ r6  : pointer to blocks for result
31          @@ r7  : loop counter
32          @@ r8  : size of longest bigint
33          @@ r9  : size of shortest bigint
34          @@ r10 : CPSR flags
35          @@ r11 : tmp1
36          @@ r12 : tmp2
37          ldr     r4,[r0,#bi_blks]  @ load pointer
38          lsl     r10,r2,#29        @ initialize carry bit
39          ldr     r5,[r1,#bi_blks]  @ load pointer
40          lsl     r0,r8,#2          @ calculate result size
41          bl      malloc            @ allocate storage
42          cmp     r0,#NULL          @ check for NULL
43          bleq    alloc_err
44          mov     r6,r0
45          mov     r7,#0             @ initialize loop counter
46  .loopa: ldr     r11,[r4,r7, lsl #2]@ load current chunk
47          ldr     r12,[r5,r7, lsl #2]@ load current chunk
48          msr     CPSR_f,r10        @ restore flags (carry bit)
49          adcs    r11,r11,r12       @ add chunks
50          str     r11,[r6,r7, lsl #2]@ store result
51          add     r7,r7,#1          @ increment count
52          mrs     r10,CPSR          @ save flags in r10
53          cmp     r7,r9             @ cmp will change flags
54          blt     .loopa
55          @@ We may have chunks remaining in the longest bigint
56          cmp     r7,r8             @ are there any chunks remaining?
57          bge     .finish
58          tst     r12,#0x80000000   @ fill r12 with the sign of
59          moveq   r12,#0            @ the shorter bigint
60          movne   r12,#0xFFFFFFFF
61  .loopb: ldr     r11,[r4,r7, lsl #2]@ load current chunk
62          msr     CPSR_f,r10        @ restore flags (carry bit)
63          adcs    r11,r11,r12       @ add chunks
64          str     r11,[r6,r7, lsl #2]@ store result
65          add     r7,r7,#1          @ increment count
66          mrs     r10,CPSR          @ save flags in r10
67          cmp     r7,r8             @ cmp will change flags
68          blt     .loopb
69  .finish:@@ if there was overflow on the final add, then
70          @@ extend the result and copy the sign (carry) bit
71          msr     CPSR_f,r10        @ restore flags
72          bvc     .noext
73          movcc   r11,#0
74          movcs   r11,#0xFFFFFFFF
75          lsl     r0,r8,#2          @ calculate number of bytes
```

```
 76          add     r0,r0,#4            @ increase storage space
 77          bl      malloc
 78          cmp     r0,#NULL           @ check for NULL
 79          bleq    alloc_err
 80          mov     r4,r0              @ protect new pointer
 81          mov     r1,r6              @ get pointer to source
 82          lsl     r2,r8,#2           @ calculate number of bytes
 83          bl      memcpy             @ copy the data
 84          mov     r0,r6              @ free the old storage
 85          bl      free
 86          str     r11,[r4,r8,lsl#2]  @ store extended bits
 87          add     r8,r8,#1           @ calculate new size
 88          mov     r6,r4
 89          b       .return
 90 .noext: @@ if we did not extend, then we may need to trim
 91          @@ r4  : current
 92          @@ r5  : next
 93          @@ r6  : pointer to blocks for result
 94          @@ r7  : i
 95          @@ r8  : size of result
 96          @@ r9  : new size of result
 97          subs    r7,r8,#1           @ i = size - 1
 98          ble     .return            @ can't be trimmed
 99          mov     r9,r8              @ newsize = size;
100          ldr     r5,[r6,r7,lsl #2]  @ load next
101          tst     r5,#0x80000000
102          bne     .nloop
103 .ploop: mov     r4,r5              @ current = next
104          subs    r7,r7,#1           @ decrement index
105          blt     .trimit            @ done if it is <= 0
106          ldr     r5,[r6,r7,lsl #2]  @ load next chunk
107          cmp     r4,#0x00000000     @ done if not leading zeros
108          bne     .trimit
109          tst     r5,#0x80000000     @ done if next sign bit is set
110          bne     .trimit
111          sub     r9,r9,#1           @ current can be trimmed
112          b       .ploop
113 .nloop: mov     r4,r5              @ current = next
114          subs    r7,r7,#1           @ decrement index
115          blt     .trimit            @ done if it is <= 0
116          ldr     r5,[r6,r7,lsl #2]  @ load next chunk
117          cmp     r4,#0xFFFFFFFF     @ done if not leading ones
118          bne     .trimit
119          tst     r5,#0x80000000     @ done if next sign bit not set
120          beq     .trimit
121          sub     r9,r9,#1           @ current can be trimmed
```

```
122         b         .nloop
123 .trimit:cmp       r8,r9
124         beq       .return
125         lsl       r0,r9,#2
126         bl        malloc
127         cmp       r0,#NULL         @ check for NULL
128         bleq      alloc_err
129         mov       r10,r0
130         mov       r1,r6
131         lsl       r2,r9,#2
132         bl        memcpy
133         mov       r0,r6
134         bl        free
135         mov       r6,r10
136         mov       r8,r9
137 .return:mov       r0,#bi_struct_sz
138         bl        malloc
139         cmp       r0,#NULL         @ check for NULL
140         bleq      alloc_err
141         str       r8,[r0,#bi_size] @ store size
142         str       r6,[r0,#bi_blks] @ store pointer to blocks
143         ldmfd     sp!,{r4-r11,pc}  @ return
144         .size     bigint_adc, .-bigint_adc
```

Listing 7.10
ARM assembly implementation if the `bigint_adc` function.

The `bigfact` main function was executed 50 times on a Raspberry Pi, using the C version of `bigint_adc` and then with the assembly version. The total time required using the C version was 27.65 seconds, and the program spent 54.0% of its time (14.931 seconds) in the `bigint_adc` function. The assembly version ran in 15.07 seconds, and the program spent 15.3% of its time (2.306 seconds) in the `bigint_adc` function. Therefore the assembly version of the function achieved a speedup of 6.47 over the C implementation. Overall, the program achieved a speedup of 1.83 by writing one function in assembly.

Running `gprof` on the improved program reveals that most of the time is now spent in the `bigint_mul` function (63.2%) and two functions that it calls: `bigint_mul_uint` (39.1%) and `bigint_shift_left_chunk` (21.6%). It seems clear that optimizing those two functions would further improve performance.

7.5 Chapter Summary

Complement mathematics provides a method for performing all basic operations using only the complement, add, and shift operations. Addition and subtraction are fast, but multiplication and division are relatively slow. In particular, division should be avoided whenever possible. The exception to this rule is division by a power of the radix, which can be implemented as a shift. Good assembly programmers replace division by a constant *c* with multiplication by the reciprocal of *c*. They also replace the multiply instruction with a series of shifts and add or subtract operations when it makes sense to do so. These optimizations can make a big difference in performance.

Writing sections of a program in assembly can result in better performance, but it is not guaranteed. The chance of achieving significant performance improvement is increased if the following rules are used:

1. Only optimize the parts that really matter.
2. Design data structures with assembly in mind.
3. Use efficient algorithms and data structures.
4. Write the assembly code last.
5. Ignore the C version and write good, clean, assembly.
6. Reduce function calls wherever it makes sense.
7. Avoid unnecessary memory accesses.
8. Write good code. The compiler will beat poor assembly every time, but good assembly will beat the compiler every time.

Understanding the basic mathematical operations can enable the assembly programmer to work with integers of any arbitrary size with efficiency that cannot be matched by a C compiler. However, it is best to focus the assembly programming on areas where the greatest gains can be made.

Exercises

7.1 Multiply -90 by 105 using *signed* 8-bit binary multiplication to form a signed 16-bit result. Show all of your work.

7.2 Multiply 166 by 105 using *unsigned* 8-bit binary multiplication to form an unsigned 16-bit result. Show all of your work.

7.3 Write a section of ARM assembly code to multiply the value in r1 by 13_{10} using only shift and add operations.

7.4 The following code will multiply the value in r0 by a constant *C*. What is *C*?

```
1       add     r1,r0,r0,lsl #1
2       add     r0,r1,r0,lsl #2
```

7.5 Show the optimally efficient instruction(s) necessary to multiply a number in register r0 by the constant 67_{10}.

7.6 Show how to divide 78_{10} by 6_{10} using binary long division.

7.7   Demonstrate the division algorithm using a sequence of tables as shown in Section 7.3.2 to divide 155_{10} by 11_{10}.

7.8   When dividing by a constant value, why is it desirable to have m as large as possible?

7.9   Modify your program from Exercise 5.13 in Chapter 5 to produce a 64-bit result, rather than a 32-bit result.

7.10   Modify your program from Exercise 5.13 in Chapter 5 to produce a 128-bit result, rather than a 32-bit result. How would you do this in C?

7.11   Write the `bigint_shift_left_chunk` function from Listing 7.8 in ARM assembly, and measure the performance improvement.

7.12   Write the `bigint_mul_uint` function in ARM assembly, and measure the performance improvement.

7.13   Write the `bigint_mul` function in ARM assembly, and measure the performance improvement.

Non-Integral Mathematics

Chapter Outline

Chapter 7 introduced methods for performing computation using integers. Although many problems can be solved using only integers, it is often necessary (or at least more convenient) to perform computation using real numbers or even complex numbers. For our purposes, a non-integral number is any number that is not an integer. Many systems are only capable of

Modern Assembly Language Programming with the ARM Processor. http://dx.doi.org/10.1016/B978-0-12-803698-3.00008-5

performing computation using binary integers, and have no hardware support for non-integral calculations. In this chapter, we will examine methods for performing non-integral calculations using only integer operations.

8.1 Base Conversion of Fractional Numbers

Section 1.3.2 explained how to convert integers in a given base into any other base. We will now extend the methods to convert fractional values. A fractional number can be viewed as consisting of an integer part, a *radix point*, and a fractional part. In base 10, the radix point is also known as the decimal point. In base 2, it is called the binimal point. For base 16, it is the heximal point, and in base 8 it is an octimal point. The term *radix point* is used as a general term for a location that divides a number into integer and fractional parts, without specifying the base.

8.1.1 Arbitrary Base to Decimal

The procedure for converting fractions from a given base b into base ten is very similar to the procedure used for integers. The only difference is that the digit to the left of the radix point is weighted by b^0 and the exponents become increasingly negative for each digit right of the radix point. The basic procedure is the same for any base b. For example, the value 101.0101_2 can be converted to base ten by expanding it as follows:

$$1 \times 2^2 + 0 \times 2^1 + 1 \times 2^0 + 0 \times 2^{-1} + 1 \times 2^{-2} + 0 \times 2^{-3} + 1 \times 2^{-4}$$
$$= 4 + 0 + 1 + 0 + \frac{1}{4} + 0 + \frac{1}{16}$$
$$= 5.3125_{10}$$

Likewise, the hexadecimal fraction 4F2.9A0 can be converted to base ten by expanding it as follows:

$$4 \times 16^2 + 15 \times 16^1 + 2 \times 16^0 + 9 \times 16^{-1} + 10 \times 16^{-2} + 0 \times 16^{-3}$$
$$= 1024 + 240 + 2 + \frac{9}{16} + \frac{10}{256} + \frac{0}{4096}$$
$$= 1266.60156250_{10}$$

8.1.2 Decimal to Arbitrary Base

When converting from base ten into another base, the integer and fractional parts are treated separately. The base conversion for the integer part is performed in exactly the same way as in

Section 1.3.2, using repeated division by the base b. The fractional part is converted using repeated multiplication. For example, to convert the decimal value 5.6875_{10} to a binary representation:

1. Convert the integer portion, 5_{10} into its binary equivalent, 101_2.
2. Multiply the decimal fraction by two. The integer part of the result is the first binary digit to the right of the radix point.
    Because $x = 0.6875 \times 2 = 1.375$, the first binary digit to the right of the point is a 1. So far, we have $5.625_{10} = 101.1_2$
3. Multiply the fractional part of x by 2 once again.
    Because $x = 0.375 \times 2 = 0.75$, the second binary digit to the right of the point is a 0. So far, we have $5.625_{10} = 101.10_2$
4. Multiply the fractional part of x by 2 once again.
    Because $x = 0.75 \times 2 = 1.50$, the third binary digit to the right of the point is a 1. So now we have $5.625 = 101.101$
5. Multiply the fractional part of x by 2 once again.
    Because $x = 0.5 \times 2 = 1.00$, the fourth binary digit to the right of the point is a 1. So now we have $5.625 = 101.1011$
6. Since the fractional part is now zero, we know that all remaining digits will be zero.

The procedure for obtaining the fractional part can be accomplished easily using a tabular method, as shown below:

	Result	
Operation	**Integer**	**Fraction**
$0.6875 \times 2 = 1.375$	1	0.375
$0.375 \times 2 = 0.75$	0	0.75
$0.75 \times 2 = 1.5$	1	0.5
$0.5 \times 2 = 1.0$	1	0.0

Putting it all together, $5.6875_{10} = 101.1011_2$. After converting a fraction from base 10 into another base, the result should be verified by converting back into base 10. The results from the previous example can be expanded as follows:

$$1 \times 2^2 + 0 \times 2^1 + 1 \times 2^0 + 1 \times 2^{-1} + 0 \times 2^{-2} + 1 \times 2^{-3} + 1 \times 2^{-4}$$

$$= 4 + 0 + 1 + \frac{1}{2} + 0 + \frac{1}{8} + \frac{1}{16}$$

$$= 5.6875_{10}$$

Converting decimal fractions to base sixteen is accomplished in a very similar manner. To convert 842.2343750_{10} into base 16, we first convert the integer portion by repeatedly dividing by 16 to yield 34A. We then repeatedly multiply the fractional part, extracting the integer portion of the result each time as shown in the table below:

Operation	Result	
	Integer	Fraction
$0.234375 \times 16 = 3.75$	3	0.75
$0.75 \times 16 = 12.0$	12	0.0

In the second line, the integer part is 12, which must be replaced with a hexadecimal digit. The hexadecimal digit for 12_{10} is C, so the fractional part is 3C. Therefore, $842.234375_{10} = 34A.3C_{16}$ The result is verified by converting it back into base 10 as follows:

$$3 \times 16^2 + 4 \times 16^1 + 10 \times 16^0 + 3 \times 16^{-1} + 12 \times 16^{-2}$$

$$= 768 + 64 + 10 + \frac{3}{16} + \frac{12}{256}$$

$$= 842.234375_{10}$$

Bases that are powers-of-two

Converting fractional values between binary, hexadecimal, and octal can be accomplished in the same manner as with integer values. However, care must be taken to align the radix point properly. As with integers, converting from hexadecimal or octal to binary is accomplished by replacing each hex or octal digit with the corresponding binary digits from the appropriate table shown in Fig. 1.3.

For example, to convert $5AC.43B_{16}$ to binary, we just replace "5" with "0101," replace "A" with "1010," replace "C" with "1100," replace "4" with "0100," replace "3" with "0011," replace "B" with "1011," So, using the table, we can immediately see that $5AC.43B_{16} = 010110101100.010000111011_2$. This method works exactly the same way for converting from octal to binary, except that it uses the table on the right side of Fig. 1.3.

Converting fractional numbers from binary to hexadecimal or octal is also very easy when using the tables. The procedure is to split the binary string into groups of bits, working outwards from the radix point, then replace each group with its hexadecimal or octal equivalent. For example, to convert 01110010.10101111_2 to hexadecimal, just divide the number into groups of four bits, starting at the radix point and working outwards in both directions. It may be necessary to pad with zeroes to make a complete group on the left or right, or both. Our example is grouped as follows: $|0000|0111|0010.1010|1110|_2$. Now each group of four bits is converted to hexadecimal by looking up the corresponding hex digit in the table on the left side of Fig. 1.3. This yields $072.AE_{16}$. For octal, the binary number would be grouped as follows: $|001|110|010.101|011|100|_2$. Now each group of three bits is converted to octal by looking up the corresponding digit in the table on the right side of Fig. 1.3. This yields 162.534_8.

8.2 Fractions and Bases

One interesting phenomenon that is often encountered is that fractions which terminate in one base may become non-terminating, repeating fractions in another base. For example, the binary representation of the decimal fraction $\frac{1}{10}$ is a repeating fraction, as shown in Example 8.1. The resulting fractional part from the last step performed is exactly the same as in the second step. Therefore, the sequence will repeat. If we continue, we will repeat the sequence of steps 2–5 forever. Hence, the final binary representation will be:

$$0.1_{10} = 0.00011001100110011\ldots_2$$
$$= 0.0\overline{0011}_2$$

Because of this phenomenon, it is impossible to exactly represent 1.10_{10} (and many other fractional quantities) as a binary fraction in a finite number of bits.

The fact that some base 10 fractions cannot be exactly represented in binary has lead to many subtle software bugs and round-off errors, when programmers attempt to work with currency (and other quantities) as real-valued numbers. In this section, we explore the idea that the representation problem can be avoided by working in some base other than base 2. If that is the case, then we can simply build hardware (or software) to work in that base, and will be able to represent any fractional value precisely using a finite number of digits. For brevity, we will refer to a binary fractional quantity as a *binimal* and a decimal fractional quantity as a *decimal*. We would like to know whether there are more non-terminating decimals than binimals, more non-terminating binimals than decimals, or neither. Since there are an infinite number of non-terminating decimals and an infinite number of non-terminating binimals, we could be tempted to conclude that they are equal. However, that is an oversimplification. If we ask the question differently, we can discover some important information. A better way to ask the question is as follows:

Question: Is the set of terminating decimals a subset of the set of terminating binimals, or vice versa, or neither?

Example 8.1   A Non-Terminating, Repeating Binimal

$$.1 \times 2 = 0.2$$
$$.2 \times 2 = 0.4$$
$$.4 \times 2 = 0.8$$
$$.8 \times 2 = 1.6$$
$$.6 \times 2 = 1.2$$
$$.2 \times 2 = 0.4$$

We start by introducing a lemma which can be used to predict whether or not a terminating fraction in one base will terminate in another base. We introduce the notation $x \mid y$ (read as "x divides y") to indicate that y can be evenly divided by x.

Lemma 8.2.1. *If x, $0 < x < 1$, terminates in some base B (a product of primes), then $x = \frac{N_x}{D_x}$, and $D_x = p_1^{k_1} p_2^{k_2} \ldots p_n^{k_n}$, where the p_i are the prime factors of B.*

Proof. Let $x = \frac{N_x}{D_x}$, and $D_x = p_1^{k_1} p_2^{k_2} \ldots p_n^{k_n}$, where the p_i are the prime factors of B. Then $D_x \mid N_x \times B^{k_{max}}$, where $k_{max} = max(k_1, k_2, \ldots k_n)$, so $x = \frac{N_x}{D_x x}$ terminates after k_{max} or fewer divisions.

Let $x = \frac{N_x}{D_x}$ terminate after k divisions. Then $D_x \mid N_x \times B^k$. Since D_x does not evenly divide N_x, D_x must be composed of some combination of the prime factors of B. Thus, D_x can be expressed as $p_1^{k_1} p_2^{k_2} \ldots p_n^{k_n}$. $\square$

Theorem 8.2.1. *The set of terminating binimals is a subset of the set of terminating Decimals.*

Proof. Let b be a terminating binimal. Then, by Lemma 8.2.1, $b = \frac{N_b}{D_b}$, such that $D_b = 2^k$, for some $k \geq 0$. Therefore, $D_b = 2^k 5^m$, for some $k, m > 0$, and again by the Lemma, b is also a terminating decimal. $\square$

Theorem 8.2.2. *The set of terminating decimals is not a subset of the set of terminating binimals.*

Proof. Let d be a terminating decimal such that $d = \frac{N_d}{D_d}$, where $D_d = 2^k 5^m$. If $m > 0$, then by the Lemma, d is a non-terminating binimal. $\square$

Answer: The set of terminating binimals is a subset of the set of terminating decimals, but the set of terminating decimals is not a subset of the set of terminating binimals.

Implications

Theorem 8.2.1 implies that any binary fraction can be expressed exactly as a decimal fraction, but Theorem 8.2.2 implies that there are decimal fractions which cannot be expressed exactly in binary. Every fraction (when expressed in lowest terms) which has a non-zero power of five in its denominator cannot be represented in binary with a finite number of bits. Another implication is that some fractions cannot be expressed exactly in either binary or decimal. For example, let $B = 30 = 2 * 3 * 5$. Then any number with denominator $2^{k_1} 3^{k_2} 5^{k_3}$ terminates in base 30. However if $k_2 \neq 0$, then the fraction will terminate in neither base two nor base ten, because three is not a prime factor of ten or two.

Another implication of the theorem is that the more prime factors we have in our base, the more fractions we can express exactly. For instance, the smallest base that has two, three, and five as prime factors is base 30. Using that base, we can exactly express fractions in radix notation that cannot be expressed in base ten or in base two with a finite number of digits. For example, in base 30, the fraction $\frac{11}{15}$ will terminate after one division since $15 = 3^1 5^1$. To see what the number will look like, let us extend the hexadecimal system of using letters to represent digits beyond 9. So we get this chart for base 30:

$$0_{10} \rightarrow 0_{30} \quad 1_{10} \rightarrow 1_{30} \quad 2_{10} \rightarrow 2_{30} \quad 3_{10} \rightarrow 3_{30} \quad 4_{10} \rightarrow 4_{30}$$
$$5_{10} \rightarrow 5_{30} \quad 6_{10} \rightarrow 6_{30} \quad 7_{10} \rightarrow 7_{30} \quad 8_{10} \rightarrow 8_{30} \quad 9_{10} \rightarrow 9_{30}$$
$$10_{10} \rightarrow A_{30} \quad 11_{10} \rightarrow B_{30} \quad 12_{10} \rightarrow C_{30} \quad 13_{10} \rightarrow D_{30} \quad 14_{10} \rightarrow E_{30}$$
$$15_{10} \rightarrow F_{30} \quad 16_{10} \rightarrow G_{30} \quad 17_{10} \rightarrow H_{30} \quad 18_{10} \rightarrow I_{30} \quad 19_{10} \rightarrow J_{30}$$
$$20_{10} \rightarrow K_{30} \quad 21_{10} \rightarrow L_{30} \quad 22_{10} \rightarrow M_{30} \quad 23_{10} \rightarrow N_{30} \quad 24_{10} \rightarrow O_{30}$$
$$25_{10} \rightarrow P_{30} \quad 26_{10} \rightarrow Q_{30} \quad 27_{10} \rightarrow R_{30} \quad 28_{10} \rightarrow S_{30} \quad 29_{10} \rightarrow T_{30}$$

Since $\frac{11}{15} = \frac{22}{30}$, the fraction can be expressed precisely as $0.M_{30}$. Likewise, the fraction $\frac{13}{45}$ is $0.2\overline{8}_{10}$ but terminates in base 30. Since $45 = 3^3 5^1$, this number will have three or fewer digits following the radix point. To compute the value, we will have to raise it to higher terms. Using 30^2 as the denominator gives us:

$$\frac{13}{45} = \frac{260}{900}$$

Now we can convert it to base 30 by repeated division. $\frac{260}{30} = 8$ with remainder 20. Since $20 < 30$, we cannot divide again. Therefore, $\frac{13}{45}$ in base 30 is $0.8K$.

Although base 30 can represent all fractions that can be expressed in bases two and ten, there are still fractions that cannot be represented in base 30. For example, $\frac{1}{7}$ has the prime factor seven in its denominator, and therefore will only terminate in bases were seven is a prime factor of the base. The fraction $\frac{1}{7}$ will terminate in base 7, base 14, base 21, base 42 and many others, but not in base 30. Since there are an infinite number of primes, no number system is immune from this problem. No matter what base the computer works in, there are fractions that cannot be expressed exactly with a finite number of digits. Therefore, it is incumbent upon programmers and hardware designers to be aware of round-off errors and take appropriate steps to minimize their effects.

For example, there is no reason why the hardware clocks in a computer should work in base ten. They can be manufactured to measure time in base two. Instead of counting seconds in tenths, hundredths or thousandths, they could be calibrated to measure in fourths, eighths, sixteenths, 1024ths, etc. This would eliminate the round-off error problem in keeping track of time.

8.3 Fixed-Point Numbers

As shown in the previous section, given a finite number of bits, a computer can only *approximately* represent non-integral numbers. It is often necessary to accept that limitation and perform computations involving approximate values. With due care and diligence, the results will be accurate within some acceptable error tolerance. One way to deal with real-valued numbers is to simply treat the data as *fixed-point* numbers. Fixed-point numbers are treated as integers, but the programmer must keep track of the radix point during each operation. We will present a systematic approach to designing fixed-point calculations.

When using fixed-point arithmetic, the programmer needs a convenient way to describe the numbers that are being used. Most languages have standard data types for integers and floating point numbers, but very few have support for fixed-point numbers. Notable exceptions include PL/1 and Ada, which provide support for fixed-point binary and fixed-point decimal numbers. We will focus on fixed-point binary, but the techniques presented can also be applied to fixed-point numbers in any base.

8.3.1 Interpreting Fixed-Point Numbers

Each fixed-point binary number has three important parameters that describe it:

1. whether the number is signed or unsigned,
2. the position of the radix point in relation to the right side of the sign bit (for signed numbers) or the position of the radix point in relation to the most significant bit (for unsigned numbers), and
3. the number of fractional bits stored.

Unsigned fixed-point numbers will be specified as $U(i, f)$, where i is the position of the radix point in relation to the *left* side of the most significant bit, and f is the number of bits stored in the fractional part.

For example, $U(10, 6)$ indicates that there are six bits of precision in the fractional part of the number, and the radix point is ten bits to the right of the most significant bit stored. The layout for this number is shown graphically as:

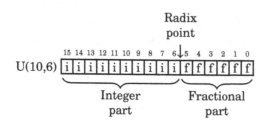

where i is an integer bit and f is a fractional bit. Very small numbers with no integer part may have a negative *i*. For example, U(−8, 16) specifies an unsigned number with no integer part, eight leading zero bits which are not actually stored, and 16 bits of fractional precision. The layout for this number is shown graphically as:

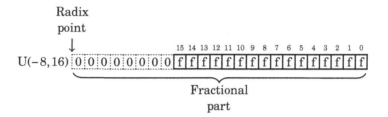

Likewise, signed fixed-point numbers will be specified using the following notation: S(*i,f*), where *i* is the position of the radix point in relation to the *right* side of the sign bit, and *f* is the number of fractional bits stored. As with integer two's-complement notation, the sign bit is always the leftmost bit stored. For example, S(9, 6) indicates that there are six bits in the fractional part of the number, and the radix point is nine bits to the right of the sign bit. The layout for this number is shown graphically as:

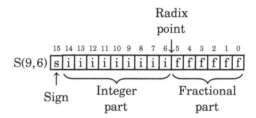

where i is an integer bit and f is a fractional bit. Very small numbers with no integer part may have a negative *i*. For example, S(−7, 16) specifies a signed number with no integer part, six leading sign bits which are not actually stored, a sign bit that *is* stored and 15 bits of fraction. The layout for this number is shown graphically as:

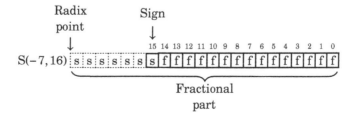

Note that the "hidden" bits in a signed number are assumed to be copies of the sign bit, while the "hidden" bits in an unsigned number are assumed to be zero.

The following figure shows an unsigned fixed-point number with seven bits in the integer part and nine bits in the fractional part. It is a $U(7,9)$ number. Note that the total number of bits is $7 + 9 = 16$

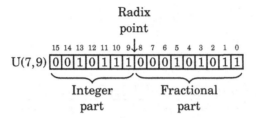

The value of this number in base 10 can be computed by summing the values of each non-zero bit as follows:

$$2^{13-9} + 2^{11-9} + 2^{10-9} + 2^{9-9} + 2^{5-9} + 2^{3-9} + 2^{1-9} + 2^{0-9}$$
$$= 2^4 + 2^2 + 2^1 + 2^0 + 2^{-4} + 2^{-6} + 2^{-8} + 2^{-9}$$
$$= 16 + 4 + 2 + 1 + \frac{1}{16} + \frac{1}{64} + \frac{1}{256} + \frac{1}{512}$$
$$= 23.083984375_{10}$$

Likewise, the following figure shows a signed fixed-point number with nine bits in the integer part and six bits in the fractional part. It is as $S(9,6)$ number. Note that the total number of bits is $9 + 6 + 1 = 16$.

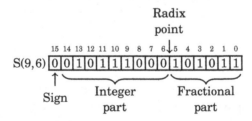

The value of this number in base 10 can be computed by summing the values of each non-zero bit as follows:

$$2^{13-6} + 2^{11-6} + 2^{10-6} + 2^{9-6} + 2^{5-6} + 2^{3-6} + 2^{1-6} + 2^{0-6}$$
$$= 2^7 + 2^5 + 2^4 + 2^3 + 2^{-1} + 2^{-3} + 2^{-5} + 2^{-6}$$
$$= 128 + 32 + 16 + 8 + \frac{1}{2} + \frac{1}{8} + \frac{1}{32} + \frac{1}{64}$$
$$= 184.671875_{10}$$

Note that in the above two examples, the pattern of bits are identical. The value of a number depends upon how it is interpreted. The notation that we have introduced allows us to easily specify exactly how a number is to be interpreted. For signed values, if the first bit is non-zero, then the two's complement should be taken before the number is evaluated. For example, the following figure shows an S(8, 7) number that has a negative value.

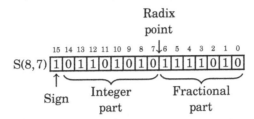

The value of this number in base 10 can be computed by taking the two's complement, summing the values of the non-zero bits, and adding a negative sign to the result. The two's complement of 1011010101111010 is 0100101010000101 + 1 = 0100101010000110. The value of this number is:

$$- \left(2^{14-7} + 2^{11-7} + 2^{9-7} + 2^{7-7} + 2^{2-7} + 2^{1-7} \right)$$
$$= - \left(2^7 + 2^4 + 2^2 + 2^0 + 2^{-5} + 2^{-6} \right)$$
$$= - \left(128 + 16 + 4 + 1 + \frac{1}{32} + \frac{1}{64} \right)$$
$$= -149.046875_{10}$$

For a final example we will interpret this bit pattern as an S(−5, 16). In that format, the layout is:

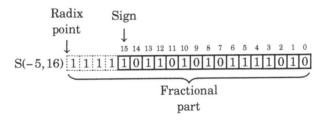

The value of this number in base ten can be computed by taking the two's complement, summing the values of the non-zero bits, and adding a negative sign to the result. The two's complement is:

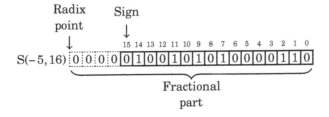

The value of this number interpreted as an $S(-5, 16)$ is:

$$- \left(2^{-6} + 2^{-9} + 2^{-11} + 2^{-13} + 2^{-18} + 2^{-19} \right) = -0.0181941986083984375$$

8.3.2 Q Notation

Fixed-point number formats can also be represented using Q notation, which was developed by Texas Instruments. Q notation is equivalent to the S/U format used in this book, except that the integer portion is not always fully specified. In general, Q formats are specified as Qm, n where m is the number of integer bits, and n is the number of fractional bits. If a fixed word size w is being used then m may be omitted, and is assumed to be $w - n$. For example, a Q10 number has 10 fractional bits, and the number of integer bits is not specified, but is assumed to be the number of bits required to complete a word of data. A Q2,4 number has two integer bits and four fractional bits in a 6-bit word. There are two conflicting conventions for dealing with the sign bit. In one convention, the sign bit is included as part of m, and in the other convention, it is not. When using Q notation, it is important to state which convention is being used. Additionally, a U may be prefixed to indicate an unsigned value. For example UQ8.8 is equivalent to $U(8, 8)$, and Q7,9 is equivalent to $S(7, 9)$.

8.3.3 Properties of Fixed-Point Numbers

Once the decision has been made to use fixed-point calculations, the programmer must make some decisions about the specific representation of each fixed-point variable. The combination of size and radix will affect several properties of the numbers, including:

Precision: the maximum number of non-zero bits representable,

Resolution: the smallest non-zero magnitude representable,

Accuracy: the magnitude of the maximum difference between a true real value and its approximate representation,

Range: the difference between the largest and smallest number that can be represented, and

Dynamic range: the ratio of the maximum absolute value to the minimum positive absolute value representable.

Given a number specified using the notation introduced previously, we can determine its properties. For example, an $S(9, 6)$ number has the following properties:

Precision: $P = 16$ bits

Resolution: $R = 2^{-6} = 0.015625$

Accuracy: $A = \frac{R}{2} = 0.0078125$

Range: Minimum value is $1000000000.000000 = -512$
Maximum value is $0111111111.111111 = 1023.9921875$
Range is $G = 1023.9921875 + 512 = 1535.9921875$

Dynamic range: For a signed fixed-point rational representation, $S(i, f)$, the dynamic range is

$$D = 2 \times \frac{2^i}{2^{-f}} = 2^{i+f+1} = 2^P.$$

Therefore, the dynamic range of an $S(9, 6)$ is $2^{16} = 65536$.

Being aware of these properties, the programmer can select fixed-point representations that fit the task that they are trying to solve. This allows the programmer to strive for very efficient code by using the smallest fixed-point representation possible, while still guaranteeing that the results of computations will be within some limits for error tolerance.

8.4 Fixed-Point Operations

Fixed-point numbers are actually stored as integers, and all of the integer mathematical operations can be used. However, some care must be taken to track the radix point at each stage of the computation. The advantages of fixed-point calculations are that the operations are very fast and can be performed on any computer, even if it does not have special hardware support for non-integral numbers.

8.4.1 Fixed-Point Addition and Subtraction

Fixed-point addition and subtraction work exactly like their integer counterparts. Fig. 8.1 gives some examples of fixed-point addition with signed numbers. Note that in each case, the numbers are aligned so that they have the same number of bits in their fractional part. This requirement is the only difference between integer and fixed-point addition. In fact, integer arithmetic is just fixed-point arithmetic with no bits in the fractional part. The arithmetic that was covered in Chapter 7 was fixed-point arithmetic using only $S(i, 0)$ and $U(i, 0)$ numbers. Now we are simply extending our knowledge to deal with numbers where $f \neq 0$. There are some rules which must be followed to ensure that the results are correct. The rules for

2.25	00010.010
+ 1.50 =	+ 00001.100
3.75	00011.110

11.125	01011.001
− 5.625 =	+ 11010.011
5.500	00101.100

−12.375	10011.101
+ 5.250 =	+ 00101.010
− 7.125	11000.111

Figure 8.1
Examples of fixed-point signed arithmetic.

subtraction are the same as the rules for addition. Since we are using two's complement math, subtraction is performed using addition.

Suppose we want to add an $S(7, 8)$ number to an $S(7, 4)$ number. The radix points are at different locations, so we cannot simply add them. Instead, we must shift one of the numbers, changing its format, until the radix points are aligned. The choice of which one to shift depends on what format we desire for the result. If we desire eight bits of fraction in our result, then we would shift the $S(7, 4)$ left by four bits, converting it into an $S(7, 8)$. With the radix points aligned, we simply use an integer addition operation to add the two numbers. The result will have it's radix point in the same location as the two numbers being added.

8.4.2 Fixed Point Multiplication

Recall that the result of multiplying an n bit number by an m bit number is an $n + m$ bit number. In the case of fixed-point numbers, the size of the fractional part of the result is the sum of the number of fractional bits of each number, and the total size of the result is the sum of the total number of bits in each number. Consider the following example where two $U(5, 3)$ numbers are multiplied together:

```
              0 0 0 1 1 . 1 1 0
        ×     0 0 0 1 0 . 1 0 0
          0 0 0 1 . 1 1 1 0
        0 0 0 1 1 1 . 1 0
0 0 0 0 0 0 1 0 0 1 . 0 1 1 0 0 0
```

The result is a $U(10, 6)$ number. The number of bits in the result is the sum of all of the bits of the multiplicand and the multiplier. The number of fractional bits in the result is the sum of the number of fractional bits in the multiplicand and the multiplier. There are three simple rules to predict the resulting format when multiplying any two fixed-point numbers.

Unsigned Multiplication

The result of multiplying two unsigned numbers $U(i_1, f_1)$ and $U(i_2, f_2)$ is a $U(i_1 + i_2, f_1 + f_2)$ number.

Mixed Multiplication

The result of multiplying a signed number $S(i_1, f_1)$ and an unsigned number $U(i_2, f_2)$ is an $S(i_1 + i_2, f_1 + f_2)$ number.

Signed Multiplication

The result of multiplying two signed numbers $S(i_1, f_1)$ and $S(i_2, f_2)$ is an $S(i_1 + i_2 + 1, f_1 + f_2)$ number.

Note that this rule works for integers as well as fixed-point numbers, since integers are really fixed-point numbers with $f = 0$. If the programmer desires a particular format for the result, then the multiply is followed by an appropriate shift.

Listing 8.1 gives some examples of fixed-point multiplication using the ARM multiply instructions. In each case, the result is shifted to produce the desired format. It is the responsibility of the programmer to know what type of fixed-point number is produced after each multiplication and to adjust the result by shifting if necessary.

```
1    @@ Multiply two S(10,5) numbers and produce an S(10,5) result.
2    mul    r0,r1,r2      @ x = a * b -> S(21,10)
3    asr    r0,r0,#5      @ shift back to S(10,5)
4
5    @@ Multiply two U(12,4) numbers and produce a U(22,6) result.
6    mul    r3,r4,r5      @ x = a * b -> U(24,8)
7    lsr    r3,r3,#2      @ shift back to U(22,6)
8
9    @@ Multiply two S(16,15) numbers and produce an S(16,15) result.
10   smull r0,r1,r2,r3    @ x = a * b -> S(33,30)
11   lsr    r0,r0,#17     @ get 15 bits from r0
12   orr    r0,r1,lsl #15 @ combine with 17 bits from r1
13
14   @@ Multiply two U(10,22) numbers and produce a U(10,22) result.
15   umull r0,r1,r2,r3    @ x = a * b -> U(20,44)
16   lsr    r0,r0,#10     @ get 22 bits from r0
17   orr    r0,r1,lsl #22 @ combine with 10 bits from r1
```

Listing 8.1

Examples of fixed-point multiplication in ARM assembly.

8.4.3 Fixed Point Division

Derivation of the rule for determining the format of the result of division is more complicated than the one for multiplication. We will first consider only unsigned division of a dividend with format $U(i_1, f_1)$ by a divisor with format $U(i_2, f_2)$.

Results of fixed point division

Consider the results of dividing two fixed-point numbers, using integer operations with limited precision. The value of the least significant bit of the dividend N is 2^{-f_i} and the value of the least significant bit of the divisor D is 2^{-f_2}. In order to perform the division using integer operations, it is necessary to multiply N by 2^{f_i} and multiply D by 2^{f_2} so that both numbers are integers. Therefore, the division operation can be written as:

$$Q = \frac{N \times 2^{f_1}}{D \times 2^{f_2}} = \frac{N}{D} \times 2^{f_1 - f_2}.$$

Note that no multiplication is actually performed. Instead, the programmer mentally shifts the radix point of the divisor and dividend, then computes the radix point of the result. For example, given two $U(5, 3)$ numbers, the division operation is accomplished by converting them both to integers, performing the division, then computing the location radix point:

$$Q = \frac{N \times 2^3}{D \times 2^3} = \frac{N}{D} \times 2^0.$$

Note that the result is an integer. If the programmer wants to have some fractional bits in the result, then the dividend must be shifted to the left before the division is performed.

If the programmer wants to have f_q fractional bits in the quotient, then the amount that the dividend must be shifted can easily be computed as

$$s = f_q + f_1 - f_2.$$

For example, suppose the programmer wants to divide 01001.011 stored as a $U(28, 3)$ by 00011.110 which is also stored as a $U(28, 3)$, and wishes to have six fractional bits in the result. The programmer would first shift 01001.011 to the left by six bits, then perform the division and compute the position of the radix in the result as shown:

$$01001.011 \div 00011.110 = (0000001001011000000 \div 00011110) \times 2^{-6-3+3}$$

$$\times 2^{-6} = 10.100000$$

$$\begin{array}{r} 10100000 \\ \hline 11110 \overline{)\,1001011000000} \\ 111100000000 \\ \hline 1111000000 \\ 1111000000 \\ \hline 0 \end{array}$$

Since the divisor may be between zero and one, the quotient may actually require *more* integer bits than there are in the dividend. Consider that the largest possible value of the dividend is $N_{max} = 2^{i_1} - 2^{-f_1}$, and the smallest positive value for the divisor is $D_{min} = 2^{-f_2}$. Therefore, the maximum quotient is given by:

$$Q_{max} = \frac{2^{i_1} - 2^{-f_1}}{2^{-f_2}} = 2^{i_1 + f_2} - 2^{f_1 - f_2}.$$

Taking the limit of the previous equation,

$$\lim_{f_1 - f_2 \to -\infty} Q_{max} = 2^{i_1 + f_2},$$

provides the following bound on how many bits are required in the integer part of the quotient:

$$Q_{max} < 2^{i_1 + f_2}.$$

Therefore, in the worst case, the quotient will require $i_1 + f_2$ integer bits. For example, if we divide a U(3, 5), $a = 111.11111 = 7.96875_{10}$, by a U(5, 3), $b = 00000.001 = 0.125_{10}$, we end up with a U(6, 2) $q = 111111.11 = 63.75_{10}$.

The same thought process can be used to determine the results for signed division as well as mixed division between signed and unsigned numbers. The results can be reduced to the following three rules:

Unsigned Division
  The result of dividing an unsigned fixed-point number $U(i_1, f_1)$ by an unsigned number $U(i_2, f_2)$ is a $U(i_1 + f_2, f_1 - f_2)$ number.
Mixed Division
  The result of dividing two fixed-point numbers where one of them is signed and the other is unsigned is an $S(i_1 + f_2, f_1 - f_2)$ number.
Signed Division
  The result of dividing two signed fixed-point numbers is an $S(i_1 + f_2 + 1, f_1 - f_2)$ number.

Consider the results when a U(2, 3), $a = 00000.001 = 0.125_{10}$ is divided by a U(4, 1), $b = 1000.0 = 8.0_{10}$. The quotient is $q = 0.000001$, which requires six bits in the fractional part. However, if we simply perform the division, then according to the rules shown above, the result will be a U(8, −2). There is no such thing as a U(8, −2), so the result is meaningless.

When $f_2 > f_1$, blindly applying the rules will result in a negative fractional part. To avoid this, the dividend can be shifted left so that it has at least as many fractional bits as the divisor. This leads to the following rule: If $f_2 > f_1$ then convert the divisor to an $S(i_1, x)$, where $x \geq f_2$, then apply the appropriate rule. For example, dividing an S(5, 2) by a U(3, 12) would result in an S(17, −10). But shifting the S(5, 2) 16 bits to the left will result in an S(5, 18), and dividing that by a U(3, 12) will result in an S(17, 6).

Maintaining precision

Recall that integer division produces a result and a remainder. In order to maintain precision, it is necessary to perform the integer division operation in such a way that all of the significant bits are in the result and only insignificant bits are left in the remainder. The easiest way to accomplish this is by shifting the dividend to the left before the division is performed.

To find a rule for determining the shift necessary to maintain full precision in the quotient, consider the worst case. The minimum positive value of the dividend is $N_{min} = 2^{-f_1}$ and the largest positive value for the divisor is $D_{min} = 2^{i_2} - 2^{-f_2}$. Therefore, the minimum positive quotient is given by:

$$Q_{min} = \frac{2^{-f_1}}{2^{i_2} - 2^{-f_2}}$$

$$= \frac{\frac{1}{2^{f_1}}}{\frac{2^{i_2+f_2}}{2^{f_2}}}$$

$$= \frac{2^{f_2}}{2^{f_1+i_2+f_2}}$$

$$= \frac{1}{2^{f_1+i_2}}$$

$$= 2^{-(i_2+f_1)}$$

Therefore, in the worst case, the quotient will require $i_2 + f_1$ fractional bits to maintain precision. However, fewer bits can be reserved if full precision is not required.

Recall that the least significant bit of the quotient will be $2^{-(i_2+f_1)}$. Shifting the dividend left by $i_2 + f_2$ bits will convert it into a $U(i_1, i_2 + f_1 + f_2)$. Using the rule above, when it is divided by a $U(i_2, f_2)$, the result is a $U(i_1 + f_2, i_2 + f_1)$. This is the minimum size which is guaranteed to preserve all bits of precision. The general method for performing fixed-point division while maintaining maximum precision is as follows:

1. shift the dividend left by $i_2 + f_2$, then
2. perform integer division.

The result will be a $U(i_1 + f_2, i_2 + f_1)$ for unsigned division, or an $S(i_1 + f_2 + 1, i_2 + f_1)$ for signed division. The result for mixed division is left as an exercise for the student.

8.4.4 Division by a Constant

Section 7.3.3 introduced the idea of converting division by a constant into multiplication by the reciprocal of that constant. In that section it was shown that by pre-multiplying the reciprocal by a power of two (a shift operation), then dividing the final result by the same power of two (a shift operation), division by a constant could be performed using only integer operations with a more efficient multiply replacing the (usually) very slow divide.

This section presents an alternate way to achieve the same results, by treating division by an integer constant as an application of fixed-point multiplication. Again, the integer constant divisor is converted into its reciprocal, but this time the process is considered from the viewpoint of fixed-point mathematics. Both methods will achieve exactly the same results, but some people tend to grasp the fixed-point approach better than the purely integer approach.

When writing code to divide by a constant, the programmer must strive to achieve the largest number of significant bits possible, while using the shortest (and most efficient) representation possible. On modern computers, this usually means using 32-bit integers and integer multiply operations which produce 64-bit results. That would be extremely tedious to show in a textbook, so the principals will be demonstrated here using 8-bit integers and an integer multiply which produces a 16-bit result.

Division by constant 23
Suppose we want efficient code to calculate $x \div 23$ using only 8-bit signed integer multiplication. The reciprocal of 23, in binary, is

$$R = \frac{1}{23} = 0.0000101100100001011\ldots_2 .$$

If we store R as an $S(1, 11)$, it would look like this:

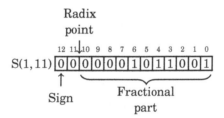

Note that in this format, the reciprocal of 23 has five leading zeros. We can store R in eight bits by shifting it left to remove some of the leading zeros. Each shift to the left changes the format of R. After removing the first leading zero bit, we have:

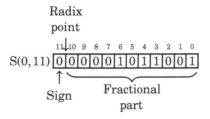

After removing the second leading zero bit, we have:

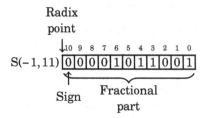

After removing the third leading zero bit, we have:

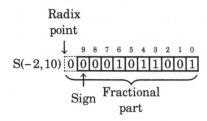

Note that the number in the previous format has a "hidden" bit between the radix point and the sign bit. That bit is not actually stored, but is assumed to be identical to the sign bit. Removing the fourth leading zero produces:

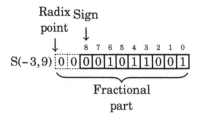

The number in the previous format has two "hidden" bits between the radix point and the sign bit. Those bits are not actually stored, but are assumed to be identical to the sign bit. Removing the fifth leading zero produces:

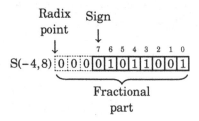

We can only remove five leading zero bits, because removing one more would change the sign bit from 0 to 1, resulting in a completely different number. Note that the final format has three "hidden" bits between the radix point and the sign bit. These bits are all copies of the sign bit. It is an $S(-4, 8)$ number because the sign is four bits to the right of the radix point (resulting in the three "hidden" bits). According to the rules of fixed-point multiplication given earlier, an $S(7, 0)$ number x multiplied by an $S(-4, 8)$ number R will yield an $S(4, 8)$ number y. The value y will be $2^3 \times \frac{x}{23}$ because we have three "hidden" bits to the right of the radix point. Therefore,

$$\frac{x}{23} = R \times x \times 2^{-3},$$

indicating that after the multiplication, we must shift the result right by three bits to restore the radix. Since $\frac{1}{23}$ is positive, the number R must be increased by one to avoid round-off error. Therefore, we will use $R + 1 = 01011010 = 90_{10}$ in our multiply operation. To calculate $y = 101_{10} \div 23_{10}$, we can multiply and perform a shift as follows:

```
                        .  0  1  1  0  0  1  0  1
               ×     0  1  0  1  1  0  1  0
          _____
                  0  .  1  1  0  0  1  0  1  0
            0  1  1  .  0  0  1  0  1
         0  1  1  0  .  0  1  0  1
      0  1  1  0  0  1  .  1  1
   _____
   0  0  1  0  0  1  0  0  .  0  0  0  0  0  0  1  0
```

Because our task is to implement *integer* division, everything to the right of the radix point can be immediately discarded, keeping only the upper eight bits as the integer portion of the result. The integer portion, 1000112, shifted right three bits, is $100_2 = 4_{10}$. If the modulus is required, it can be calculated as: $101 - (4 \times 23) = 9$. Some processors, such as the Motorola HC11, have a special multiply instruction which keeps only the upper half of the result. This method would be especially efficient on that processor. Listing 8.2 shows how the 8-bit division code would be implemented in ARM assembly. Listing 8.3 shows an alternate implementation which uses shift and add operations rather than a multiply.

```
1      @ Assume that r0 already contains x, where -129 < x < 128
2      @ and r1 is available to hold 1/23 * 2^3
3      ldr    r1,=0b01011010   @ Load 1/23 * 2^3 into r1
4      mul    r1,r0,r1         @ Perform multiply
5      asrs   r1,r1,#11        @ shift result right by 8+3 bits
6      addmi  r1,r1,#1         @ add one if result is negative
```

Listing 8.2
Dividing x by 23

```
1    @ Assume that r0 already contains x, where -129 < x < 128
2    @ and r1 is available to hold 1/23 * 2^3
3    add     r0,r0,r0,lsl #2 @ r0 <- x + x*4 = 5x
4    add     r0,r0,r0,lsl #3 @ r0 <- 5x + 5x*8 = 45x
5    asrs    r1,r1,#10        @ shift result right by 8+3 bits
6    addmi   r1,r1,#1         @ add one if result is negative
```

Listing 8.3
Dividing x by 23 Using Only Shift and Add

Division by constant −50

The procedure is exactly the same for dividing by a negative constant. Suppose we want efficient code to calculate $\frac{x}{-50}$ using 16-bit signed integers. We first convert $\frac{1}{50}$ into binary:

$$\frac{1}{50} = 0.0000\overline{010100011110}$$

The two's complement of $\frac{1}{50}$ is

$$\frac{1}{-50} = 1.1111\overline{101011100001}$$

We can represent $\frac{1}{-50}$ as the following $S(1, 21)$ fixed-point number:

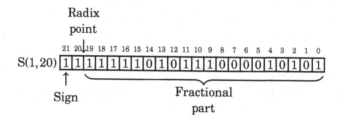

Note that the upper seven bits are all one. We can remove six of those bits and adjust the format as follows. After removing the first leading one, the reciprocal is:

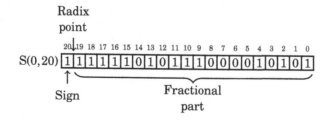

Removing another leading one changes the format to:

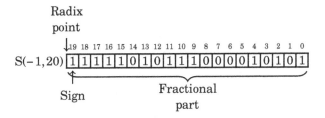

On the next step, the format is:

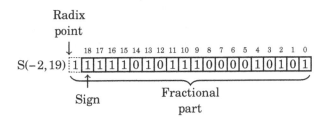

Note that we now have a "hidden" bit between the radix point and the sign bit. The hidden bit is not actually part of the number that we store and use in the computation, but it is assumed to be the same as the sign bit.

After three more leading ones are removed, the format is:

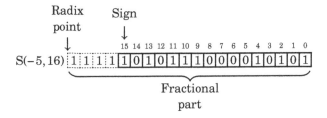

Note that there are four "hidden" bits between the radix point and the sign. Since the reciprocal $\frac{1}{-50}$ is negative, we do not need to round by adding one to the number R. Therefore, we will use $R = 1010111000010101_2 = AE15_{16}$ in our multiply operation.

Since we are using 16-bit integer operations, the dividend, x, will be an $S(15, 0)$. The product of an $S(15, 0)$ and an $S(-5, 16)$ will be an $S(11, 16)$. We will remove the 16 fractional bits by shifting right. The four "hidden" bits indicate that the result must be shifted an additional four bits to the right, resulting in a total shift of 20 bits. Listing 8.4 shows how the 16-bit division code would be implemented in ARM assembly.

```
1    @ Assume that r0 already contains x, where -32769 < x < 32768
2    @ and r1 is available to hold 1/-50 * 2^4
3    ldr     r1,=0xAE15      @ Load 1/-50 * 2^4 into r1
4    mul     r1,r0,r1        @ Perform multiply
5    asrs    r1,r1,#20       @ shift result right by 8+3 bits
6    addmi   r1,r1,#1        @ add one if result is negative
```

Listing 8.4
Dividing x by -50

8.5 Floating Point Numbers

Sometimes we need more range than we can easily get from fixed precision. One approach to solving this problem is to create an aggregate data type that can represent a fractional number by having fields for an exponent, a sign bit, and an integer mantissa. For example, in C, we could represent a fractional number using the data structure shown in Listing 8.5. That data structure, along with some subroutines for addition, subtraction, multiplication and division, would provide the capability to perform arithmetic without explicitly tracking the radix point. The subroutines for the basic arithmetical operations could do that, thereby freeing the programmer to work at a higher level.

The structure shown in Listing 8.5 is a rather inefficient way to represent a fractional number, and may create different data structures on different machines. The sign only requires one bit, and the size of the exponent and mantissa are dependent upon the machine on which the code is compiled. The sign will use one bit, the exponent eight bits, and the mantissa 23 bits.

The C language includes the notion of bit fields. This allows the programmer to specify exactly how many bits are to be used for each field within a struct, Listing 8.6 shows a C data structure that consumes 32 bits on all machines and architectures. It provides the same fields as the structure in Listing 8.5, but specifies exactly how many bits each field consumes.

```
1    typedef struct{
2      int sign;
3      int exponent;
4      int mantissa;
5    } poorfloat;
```

Listing 8.5
Inefficient representation of a binimal.

```
1  typedef struct{
2    int sign:1;
3    int exponent:8;
4    int mantissa: 23;
5  }IEEEsingle;
```

Listing 8.6
Efficient representation of a binimal.

The compiler will compress this data structure into 32 bits, regardless of the natural word size of the machine.

The method of representing fractional numbers as a sign, exponent, and mantissa is very powerful, and IEEE has set standards for various floating point formats. These formats can be described using bit fields in C, as described above. Many processors have hardware that is specifically designed to perform arithmetic using the standard IEEE formatted data. The following sections highlight most of the IEEE defined numerical definitions.

The IEEE standard specifies the bitwise representation for numbers, and specifies parameters for how arithmetic is to be performed. The IEEE standard for numbers includes the possibility of having numbers that cannot be easily represented. For example, any quantity that is greater than the most positive representable value is positive infinity, and any quantity that is less than the most negative representable value is negative infinity. There are special bit patterns to encode these quantities. The programmer or hardware designer is responsible for ensuring that their implementation conforms to the IEEE standards. The following sections describe some of the IEEE standard data formats.

8.5.1 IEEE 754 Half-Precision

The half-precision format gives a 16-bit encoding for fractional numbers with a small range and low precision. There are situations where this format is adequate. If the computation is being performed on a very small machine, then using this format may result in significantly better performance than could be attained using one of the larger IEEE formats. However, in most situations, the programmer can achieve better performance and/or precision by using a fixed-point representation. The format is as follows:

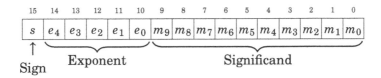

Table 8.1 Format for IEEE 754 half-precision

Exponent	Significand = 0	Significand ≠ 0	Equation
00000	±0	subnormal	$-1^{sign} \times 2^{-14} \times 0.significand$
00001 ... 11110	normalized value		$-1^{sign} \times 2^{exp-15} \times 1.significand$
11111	±∞	NaN	

- The Significand (a.k.a. "Mantissa") is stored using a sign-magnitude coding, with bit 15 being the sign bit.
- The exponent is an excess-15 number. That is, the number stored is 15 greater than the actual exponent.
- There are 10 bits of significand, but there are 11 bits of significand precision. There is a "hidden" bit, m_{10}, between m_9 and e_0. When a number is stored in this format, it is shifted until its leftmost non-zero bit is in the hidden bit position, and the hidden bit is not actually stored. The exception to this rule is when the number is zero or very close to zero. The radix point is assumed to be between the hidden bit and the first bit stored. The radix point is then shifted by the exponent.

Table 8.1 shows how to interpret IEEE 754 Half-Precision numbers. The exponents 00000 and 11111 have special meaning. The value 00000 is used to represent zero and numbers very close to zero, and the exponent value 11111 is used to represent infinity and NaN. NaN, which is the abbreviation for *not a number*, is a value representing an undefined or unrepresentable value. One way to get NaN as a result is to divide infinity by infinity. Another is to divide zero by zero. The NaN value can indicate that there is a bug in the program, or that a calculation must be performed using a different method.

Subnormal means that the value is too close to zero to be completely normalized. The minimum strictly positive (subnormal) value is $2^{-24} \approx 5.96 \times 10^{-8}$. The minimum positive normal value is $2^{-14} \approx 6.10 \times 10^{-5}$. The maximum *exactly* representable value is $(2 - 2^{-10}) \times 2^{15} = 65504$.

Examples
The following bit value:

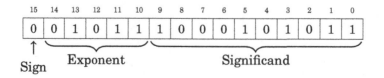

represents

$$+ 1.1000101011 \times 2^{01011-01111} = 1.1000101011 \times 2^{-4} = 0.00011000101011$$
$$\approx 0.09637.$$

The following bit value:

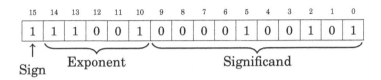

represents

$$- 1.0000100101 \times 2^{11001-01111} = -1.0000100101 \times 2^{10} = -10000100101.0$$
$$= -1061_{10}.$$

8.5.2 IEEE 754 Single-Precision

The single precision format provides a 23-bit mantissa and an 8-bit exponent, which is enough to represent a reasonably large range with reasonable precision. This type can be stored in 32 bits, so it is relatively compact. At the time that the IEEE standards were defined, most machines used a 32-bit word, and were optimized for moving and processing data in 32-bit quantities. For many applications this format represents a good trade-off between performance and precision.

8.5.3 IEEE 754 Double-Precision

The double-precision format was designed to provide enough range and precision for most scientific computing requirements. It provides a 10-bit exponent and a 53-bit mantissa. When the IEEE 754 standard was introduced, this format was not supported by most hardware. That has changed. Most modern floating point hardware is optimized for the IEEE 754 double-precision standard, and most modern processors are designed to move 64-bit or larger quantities. On modern floating-point hardware, this is the most efficient representation.

However, processing large arrays of double-precision data requires twice as much memory, and twice as much memory bandwidth, as single-precision.

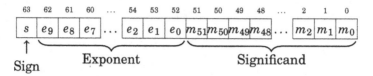

8.5.4 IEEE 754 Quad-Precision

The IEEE 754 Quad-Precision format was designed to provide enough range and precision for very demanding applications. It provides a 14-bit exponent and a 116-bit mantissa. This format is still not supported by most hardware. The first hardware floating point unit to support this format was the SPARC V8 architecture. As of this writing, the popular Intel x86 family, including the 64-bit versions of the processor, do not have hardware support for the IEEE 754 quad-precision format. On modern high-end processors such as the SPARC, this may be an efficient representation. However, for mid-range processors such as the Intel x86 family and the ARM, this format is definitely out of their league.

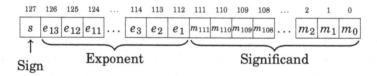

8.6 Floating Point Operations

Many processors do not have hardware support for floating point. On those processors, all floating point must be accomplished through software. Processors that do support floating point in hardware must have quite sophisticated circuitry to manage the basic operations on data in the IEEE 754 standard formats. Regardless of whether the operations are carried out in software or hardware, the basic arithmetic operations require multiple steps.

8.6.1 Floating Point Addition and Subtraction

The steps required for addition and subtraction of floating point numbers is the same, regardless of the specific format. The steps for adding or subtracting to floating point numbers a and b are as follows:

1. Extract the exponents E_a and E_b.
2. Extract the significands M_a and M_b. and convert them into 2's complement numbers, using the signs S_a and S_b.

3. Shift the significand with the smaller exponent right by $|E_a - E_b|$.
4. Perform addition (or subtraction) on the significands to get the significand of the result, M_r. Remember that the result may require one more significant bit to avoid overflow.
5. If M_r is negative, then take the 2's complement and set S_r to 1. Otherwise set S_r to 0.
6. Shift M_r until the leftmost 1 is in the "hidden" bit position, and add the shift amount to the smaller of the two exponents to form the new exponent E_r.
7. Combine the sign S_r, the exponent E_r, and significand M_r to form the result.

The complete algorithm must also provide for correct handling of infinity and NaN.

8.6.2 Floating Point Multiplication and Division

Multiplication and division of floating point numbers also requires several steps. The steps for multiplication and division of two floating point numbers a and b are as follows:

1. Calculate the sign of the result S_r.
2. Extract the exponents E_a and E_b.
3. Extract the significands M_a and M_b.
4. Multiply (or divide) the significands to form M_r.
5. Add (or subtract) the exponents (in excess-N) to get E_r.
6. Shift M_r until the leftmost 1 is in the "hidden" bit position, and add the shift amount to E_r.
7. Combine the sign S, the exponent E_r, and significand M_r to form the result.

The complete algorithm must also provide for correct handling of infinity and NaN.

8.7 Computing Sine and Cosine

It has been said, and is commonly accepted, that "you can't beat the compiler." The meaning of this statement is that using hand-coded assembly language is futile and/or worthless because the compiler is "smarter" than a human. This statement is a myth, as will now be demonstrated.

There are many mathematical functions that are useful in programming. Two of the most useful functions are $\sin x$ and $\cos x$. However, these functions are not always implemented in hardware, particularly for fixed-point representations. If these functions are required for fixed-point computation, then they must be written in software. These two functions have some nice properties that can be exploited. In particular:

- If we have the $\sin x$ function, then we can calculate $\cos x$ using the relationship

$$\cos x = \sin \frac{\pi}{2 - x}. \tag{8.1}$$

Therefore, we only need to get the sine function working, and then we can implement cosine with only a little extra effort.

- $\sin x$ is cyclical, so ... $\sin -2\pi = \sin 0 = \sin 2\pi$ .... This means that we can limit the domain of our function to the range $[-\pi, \pi]$.
- $\sin x$ is symmetric, so that $\sin -x = -\sin x$. This means that we can further restrict the domain to $[0, \pi]$.
- After we restrict the domain to $[0, \pi]$, we notice another symmetry, $\sin x = \sin(\pi - x)$, $\frac{\pi}{2} \leq x \leq \pi$ and we can further restrict the domain to $[0, \frac{\pi}{2}]$.
- The range of both functions, $\sin x$ and $\cos x$, is in the range $[-1, 1]$.

If we exploit all of these properties, then we can write a single shared function to be used by both sine and cosine. We will name this function sinq, and choose the following fixed-point formats:

- sinq will accept x as an $S(1, 30)$, and
- sinq will return an $S(1, 30)$

These formats were chosen because $S(1, 30)$ is a good format for storing a signed number between zero and $\frac{\pi}{2}$, and also the optimal format for storing a signed number between one and negative one.

The sine function will map x into the domain accepted by sinq and then call sinq to do the actual work. If the result should be negative, then the sine function will negate it before returning. The cosine function will use the relationship previously mentioned, and call the sine function.

We have now reduced the problem to one of approximating $\sin x$ within the range $[0, \frac{\pi}{2}]$. An approximation to the function $\sin x$ can be calculated using the Taylor Series:

$$\sin x = \sum_{n=0}^{\infty} (-1)^n \frac{x^{2n+1}}{(2n+1)!}. \tag{8.2}$$

The first few terms of the series should be sufficient to achieve a good approximation. The maximum value possible for the seventh term is $\frac{(0.5 \times \pi)^{13}}{13!} \approx 0.0000000005_{10}$, which indicates that our function should be accurate to at least 25 bits using seven terms. If more accuracy is desired, then additional terms can be added.

8.7.1 Formats for the Powers of x

The numerators in the first nine terms of the Taylor series approximation are: $x, x^3, x^5, x^7, x^9, x^{11}, x^{13}, x^{15}$, and x^{17}. Given an $S(1, 30)$ format for x, we can predict the format for the numerator of each successive term in the Taylor series. If we simply perform successive multiplies, then we would get the following formats for the powers of x:

Term	Format	32-bit
x	$S(1, 30)$	$S(1, 30)$
x^3	$S(3, 90)$	$S(3, 28)$
x^5	$S(5, 150)$	$S(5, 26)$
x^7	$S(7, 210)$	$S(7, 24)$
x^9	$S(9, 270)$	$S(9, 22)$
x^{11}	$S(11, 330)$	$S(11, 20)$
x^{13}	$S(13, 390)$	$S(13, 18)$

The middle column in the table shows that the format for x^{17} would require 528 bits if all of the fractional bits are retained. Dealing with a number at that level of precision would be slow and impractical. We will, of necessity, need to limit the number of bits used. Since the ARM processor provides a multiply instruction involving two 32-bit numbers, we choose to truncate the numerators to 32 bits. The third column in the table indicates the resulting format for each term if precision is limited to 32 bits.

On further consideration of the Taylor series, we notice that each of the above terms will be divided by a constant. Instead of dividing, we can multiply by the reciprocal of the constant. We will create a similar table holding the formats and constants for the factorial terms. With a bit of luck, the division (implemented as multiplication) in each term will result in a reasonable format for each resulting term.

8.7.2 Formats and Constants for the Factorial Terms

The first term of the Taylor series is $\frac{x}{1!}$, so we can simply skip the division. The second term is $-\frac{x^3}{3!} = x^3 \times -\frac{1}{3!}$ and the third term is $\frac{x^5}{5!} = x^5 \times \frac{1}{5!}$ We can convert $-\frac{1}{3!}$ to binary as follows:

Multiplication	Result Integer	Result Fraction
$\frac{1}{6} \times 2 = \frac{2}{6}$	0	$\frac{2}{6}$
$\frac{2}{6} \times 2 = \frac{4}{6}$	0	$\frac{4}{6}$
$\frac{4}{6} \times 2 = \frac{8}{6}$	1	$\frac{2}{6}$
$\frac{2}{6} \times 2 = \frac{4}{6}$	0	$\frac{4}{6}$
$\frac{8}{6} \times 2 = \frac{8}{6}$	1	$\frac{2}{6}$

Since the pattern repeats, we can conclude that $\frac{1}{3!} = 0.0\overline{01}_2$. Since we need a negative number, we take the two's complement, resulting in $-\frac{1}{3!} = \ldots 111.1\overline{10}_2$. Represented as an $S(1, 30)$, this would be

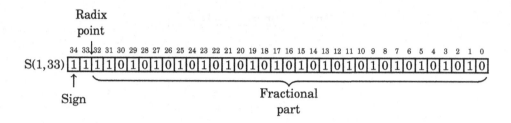

Since the first four bits are one, we can remove three bits and store it as:

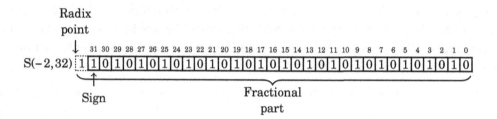

In hexadecimal, this is AAAAAAAA$_{16}$.

Performing the same operations, we find that $\frac{1}{5!}$ can be converted to binary as follows:

Multiplication	Result	
	Integer	Fraction
$\frac{1}{120} \times 2 = \frac{2}{120}$	0	$\frac{2}{120}$
$\frac{2}{120} \times 2 = \frac{4}{120}$	0	$\frac{4}{120}$
$\frac{4}{120} \times 2 = \frac{8}{120}$	0	$\frac{8}{120}$
$\frac{8}{120} \times 2 = \frac{16}{120}$	0	$\frac{16}{120}$
$\frac{16}{120} \times 2 = \frac{32}{120}$	0	$\frac{32}{120}$
$\frac{32}{120} \times 2 = \frac{64}{120}$	0	$\frac{64}{120}$
$\frac{64}{120} \times 2 = \frac{128}{120}$	1	$\frac{8}{120}$

Since the fraction in the seventh row is the same as the fraction in the third row, we know that the table will repeat forever. Therefore, $\frac{1}{5!} = 0.0000\overline{0001}_2$. Since the first six bits to the right of the radix are all zero, we can remove the first five bits. Also adding one to the least significant bit to account for rounding error yields the following S($-6, 32$):

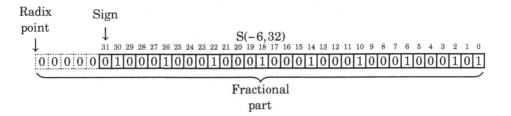

In hexadecimal, the number to be multiplied is 44444445_{16}. Note that since $\frac{1}{5!}$ is a positive number, the reciprocal was incremented by one to avoid round-off errors. We can apply the same procedure to the remaining terms, resulting in the following table:

Term	Reciprocal Format	Reciprocal Value (Hex)
$-\frac{1}{3!}$	$S(-2,32)$	AAAAAAAA
$\frac{1}{5!}$	$S(-6,32)$	44444445
$-\frac{1}{7!}$	$S(-12,32)$	97F97F97
$\frac{1}{9!}$	$S(-18,32)$	5C778E96
$-\frac{1}{11!}$	$S(-25,32)$	9466EA60
$\frac{1}{13!}$	$S(-32,32)$	5849184F

8.7.3 Putting it All Together

We want to keep as much precision as is reasonably possible for our intermediate calculations. Using 64 bits of precision for all intermediate calculations will give a good trade-off between performance and precision. The integer portion should never require more than two bits, so we choose an $S(2,61)$ as our intermediate representation. If we combine the previous two tables, we can determine what the format of each complete term will be. This is shown in Table 8.2.

Note that the formats were truncated to fit in a 64-bit result. We can now see that the formats for the first nine terms of the Taylor series are reasonably similar. They all require exactly 64 bits, and the radix points can be shifted so that they are aligned for addition. In order to make the shifting and adding process easier, we will pre-compute the shift amounts and store them in a look-up table.

Table 8.3 shows the shifts that are necessary to convert each term to an $S(2,61)$ so that it can be added to the running total.

Note that the seventh term contributes very little to the final 32-bit sum which is stored in the upper 32 bits of the running total. We now have all of the information that we need in order to

Table 8.2 Result formats for each term

Term	Numerator Value	Numerator Format	Reciprocal Value	Reciprocal Format	Reciprocal Hex	Result Format
1	x	$S(1,30)$	Extend to 64 bits and shift right			$S(2,61)$
2	x^3	$S(3,28)$	$-\frac{1}{3!}$	$S(-2,32)$	AAAAAAAA	$S(2,61)$
3	x^5	$S(5,26)$	$\frac{1}{5!}$	$S(-6,32)$	44444444	$S(0,63)$
4	x^7	$S(7,24)$	$-\frac{1}{7!}$	$S(-12,32)$	97F97F97	$S(-4,64)$
5	x^9	$S(9,22)$	$\frac{1}{9!}$	$S(-18,32)$	5C778E96	$S(-8,64)$
6	x^{11}	$S(11,20)$	$-\frac{1}{11!}$	$S(-25,32)$	9466EA60	$S(-13,64)$
7	x^{13}	$S(13,18)$	$\frac{1}{13!}$	$S(-32,32)$	5849184F	$S(-18,64)$

Table 8.3 Shifts required for each term

Term Number	Original Format	Shift Amount	Resulting Format
1	$S(1,30)$	1	$S(2,61)$
2	$S(2,61)$	0	$S(2,61)$
3	$S(0,63)$	2	$S(2,61)$
4	$S(-4,64)$	6	$S(2,61)$
5	$S(-8,64)$	10	$S(2,61)$
6	$S(-13,64)$	15	$S(2,61)$
7	$S(-18,64)$	20	$S(2,61)$

implement the function. Listing 8.7 shows how the sine and cosine function can be implemented in ARM assembly using fixed point computation, and Listing 8.8 shows a main program which prints a table of values and their sine and cosines.

```
1   @@**********************************************************
2   @@ Name: sincos.S
3   @@ Author: Larry Pyeatt
4   @@ Date: 2/22/2014
5   @@**********************************************************
6
7   @@ This is a version of the sin/cos functions that uses
8   @@ symmetry to enhance precision.  The actual sin and cos
9   @@ routines convert the input to lie in the range 0 to pi/2,
10  @@ then pass it to the worker routine that computes the
11  @@ result.  The result is then converted back to correspond
12  @@ with the original input.
13
14  @@ We calculate sin(x) using the first seven terms of the
15  @@ Taylor Series: sin(x) = x - x^3/3! + x^5/5! - x^7/7! +
16  @@ x^9/9! - ...  and we calculate cos(x) using the
```

```
17    @@ relationship: cos(x) = sin(pi/2-x)

18

19    @@ We start by defining a helper function, which we call sinq.

20    @@ The sinq function calculates sin(x) for 0<=x<=pi/2.  The

21    @@ input, x, must be an S(1,30) number. The factors of x that

22    @@ sinq will use are: x, x^3, x^5, x^7, x^9, x^11, and x^13.

23

24    @@ Dividing by (2n+1)! is changed to a multiply by a

25    @@ coefficient as we compute each term, we will add it to the

26    @@ sum, stored as an S(2,61). Therefore, we want the product

27    @@ of each power of x and its coefficient to be converted to

28    @@ an S(2,61) for the add.  It turns out that this just

29    @@ requires a small shift.

30

31    @@ We build a table to decide how much to shift each product

32    @@ before adding it to the total.  x^2 will be stored as an

33    @@ S(2,29), and x is given as an S(1,30).  After multiplying

34    @@ x by x^2, we will shift left one bit, so the procedure is:

35    @@ x    will be an  S(1,30) - multiply by x^2 and shift left

36    @@ x^3 will be an  S(3,28) - multiply by x^2 and shift left

37    @@ x^5 will be an  S(5,26) - multiply by x^2 and shift left

38    @@ x^7 will be an  S(7,24) - multiply by x^2 and shift left

39    @@ x^9 will be an  S(9,22) - multiply by x^2 and shift left

40    @@ x^11 will be an S(11,20)- multiply by x^2 and shift left

41    @@ x^13 will be an S(13,18)- multiply by x^2 and shift left

42

43

44    @@ The following table shows the constant coefficients

45    @@ needed for calculating each term.

46

47    @@ -1/3! =  AAAAAAAA as an S(-2,32)

48    @@  1/5! =  44444445 as an S(-6,32)

49    @@ -1/7! =  97F97F97 as an S(-12,32)

50    @@  1/9! =  5C778E96 as an S(-18,32)

51    @@ -1/11! = 9466EA60 as an S(-25,32)

52    @@  1/13! = 5849184F as an S(-32,32)

53

54    @@ Combining the two tables of power and coefficient formats,

55    @@ we can now determine how much shift we need after each

56    @@ step in order to do all sums in S(2,61) format:

57

58    @@ power powerfmt    coef  coeffmt     resultfmt right shift

59    @@ x     S(1,30)  *  1 (skip the multiply)      1 -> S(2,61)

60    @@ x^3  S(3,28)  * -1/3!  S(-2,32)  = S(2,61)    0 -> S(2,61)

61    @@ x^5  S(5,26)  *  1/5!  S(-6,32)  = S(0,63)    2 -> S(2,61)

62    @@ x^7  S(7,24)  * -1/7!  S(-12,32) = S(-4,64)   6 -> S(2,61)
```

```
63      @@  x^9  S(9,22)  *  1/9!  S(-18,32) = S(-8,64)  10-> S(2,61)
64      @@  x^11 S(11,20) * -1/11! S(-25,32) = S(-13,64) 15-> S(2,61)
65      @@  x^13 S(13,18) *  1/13! S(-32,32) = S(-18,64) 20-> S(2,61)
66
67
68          .data
69          .align  2
70          @@ We will define a few constants that may be useful
71          .global pi
72  pi:     .word   0x3243F6A8      @ pi as an S(3,28)
73          .global pi_2
74  pi_2:   .word   0x1921FB54      @ pi/2 as an S(3,28)
75          .global pi_x2
76  pi_x2:  .word   0x6487ED51      @ 2*pi as an S(3,28)
77
78  sintab: @@ This is the table of coefficients and shifts
79          .word 0xAAAAAAAA,  0   @ -1/3!  as an S(-2,32)
80          .word 0x44444445,  2   @  1/5!  as an S(-6,32)
81          .word 0x97F97F97,  6   @ -1/7!  as an S(-12,32)
82          .word 0x5C778E96, 10   @  1/9!  as an S(-18,32)
83          .word 0x9466EA60, 15   @ -1/11! as an S(-25,32)
84          .word 0x5849184F, 20   @  1/13! as an S(-32,32)
85          .equ tablen,(.-sintab)  @ set tablen to the size of table.
86          @@ The '.' refers to the current address counter value.
87          @@ Subtracting the address of sintab from the current
88          @@ address gives the size of the table.
89
90          .text
91          @@---------------------------------------------------------
92          @@ sinq(x)
93          @@ input: x -> S(1,30) s.t. 0 <= x <= pi/2
94          @@ returns sin(x) -> S(1,30)
95  sinq:   stmfd   sp!,{r4-r11,lr}
96          smull   r2,r4,r0,r0     @ r4 will hold x^2.
97          @@ The first term in the Taylor series is simply x, so
98          @@ convert x to an S(2,61) by doing an asr in 64 bits,
99          @@ and use it to initialize the sum.
100         mov     r10, r0,lsl #31 @ low 32 bits of sum
101         mov     r11, r0,asr #1  @ high 32 bits of sum
102         @@ r11:r10 now contains the sum (currently x) as an S(2,61)
103         @@ We are going to convert x^2 to an S(2,28), and round it
104         adds    r2,r2,#0x40000000 @ Round x^2 up by adding 1 to
105                                 @ the first bit that will be lost.
106         adccs   r4,r4,#0        @ Propagate the carry.
107         lsl     r4,r4,#1        @ Make room for one bit in LSB
108         orr     r4,r4,r2,lsr#31 @ Copy least significant bit of x^2
```

```
109        @@ r4 now contains x^2 as an S(2,28)
110        mov     r5,r0          @ r5 will keep x^(2n-1).
111        @@ r5 now contains x as an S(1,30)
112        @@ The multiply will take time, and on some processors,
113        @@ there is an extra clock cycle penalty if the next
114        @@ instruction requires the result, so do the multiply now.
115        smull   r0,r5,r4,r5    @ r5:r0 <- x^(2n+1) as an S(4,59)
116        ldr     r6, =sintab    @ get pointer to beginning of table
117        add     r7, r6, #tablen @ get pointer to end of table
118        @@ We know that we will always execute the loop 6 times,
119        @@ so we use a post-test loop.
120 sloop: ldmia   r6!,{r8,r9}    @ Load two values from the table
121        @@ r8 now has 1/(2n+1)!
122        @@ r9 contains the correcting shift
123        @@ the previous smull r0,r5,r4,r5 should be complete soon
124        lsl     r5,r5,#1       @ Shift and copy the MSB of the
125        orr     r5,r5,r0,lsr#31 @ LSW to LSB of MSW -> S(3,60)
126        @@ r5 now contains x^(2n+1) as an S(3,60)
127        @@ Start next multiply now
128        smull   r0,r1,r5,r8    @ multiply by reciprocal that we
129                               @ loaded earlier (5 cycles)
130        rsb     r2,r9,#32      @ calculate inverse shift amount
131        @@ Apply correcting right shift to make an S(2,61).
132        @@ Note: r9 was loaded from the table earlier.
133        lsr     r0,r0,r9       @ Make room in low word for bits
134        orr     r0,r0,r1,lsl r2 @ paste bits into low word
135        asr     r1,r1,r9       @ shift upper word right
136        @@ accumulate result in r10:r11
137        adds    r10,r10,r0
138        adc     r11,r11,r1
139        @@ check to see if there is another term to compute
140        cmp     r6, r7
141        @@ Start next multiply now
142        smulllt r0,r5,r4,r5    @ r5:r0 <- x^(2n+1) as an S(4,59)
143        @@  The multiply will take three cycles, so start it now
144        blt     sloop          @ Repeat for every table entry
145        @@ shift result left 1 bit and move to r0
146        lsl     r11,r11,#1
147        orr     r0,r11,r10,lsr #31
148        @@ return the result
149        ldmfd   sp!,{r4-r11,pc}
150        @@----------------------------------------------------
151        @@ cos(x)  NOTE: The cos(x) function does not return.
152        @@               It is an alternate entry point to sin(x).
153        @@ input: x -> S(3,28)
154        @@ returns cos(x) -> S(3,28)
```

```
155          .global fixed_cos
156  fixed_cos:
157          ldr     r1,=pi_x2       @ load pointer to 2*pi
158          ldr     r1,[r1]         @ load 2*pi
159          cmp     r0,#0           @ Add 2*pi to x if needed, to make
160          addle   r0,r0,r1        @ sure x does not become too small
161  cosgood:ldr     r1,=pi_2        @ load pointer to pi/2
162          ldr     r1,[r1]         @ load pi/2
163          sub     r0,r1,r0        @ cos(x) = sin(pi/2-x)
164          @@ now we just fall through into the sin function
165
166          @@-------------------------------------------------------
167          @@ sin(x)
168          @@ input: x -> S(3,28)
169          @@ returns sin(x) -> S(3,28)
170          .global fixed_sin
171  fixed_sin:
172          stmfd   sp!,{lr}
173          ldr     r1,=pi_2        @ r1 has pointer to pi/2
174          ldr     r2,=pi          @ r2 has pointer to pi
175          ldr     r3,=pi_x2       @ r3 has pointer to pi*2
176          ldr     r1,[r1]         @ r1 has pi/2
177          ldr     r2,[r2]         @ r2 has pi
178          ldr     r3,[r3]         @ r3 has pi*2
179
180          @@ step 1: make sure x>=0.0 and x<=2pi
181  neg1:   cmp     r0,#0           @ while(x < 0)
182          addlt   r0,r0,r3        @   x = x + 2 * pi
183          blt     neg1            @ end while
184  nonneg: cmp     r0,r3           @ while(x > pi/2)
185          subgt   r0,r0,r3        @   x = x - 2 * pi
186          bgt     nonneg          @ end while
187
188          @@ step 2: find the quadrant and call sinq appropriately
189  inrange:cmp     r0,r1
190          bgt     chkq2
191          @@ it is in the first quadrant... just shift and call sinq
192          lsl     r0,r0,#2
193          bl      sinq
194          b       sin_done
195  chkq2:  cmp     r0,r2
196          bgt     chkq3
197          @@ it is in the second quadrant... mirror, shift, and call
198          @@ sinq
199          sub     r0,r2,r0
200          lsl     r0,r0,#2
```

```
201         bl      sinq
202         b       sin_done
203  chkq3: add     r1,r1,r2        @ we will not need pi/2 again
204         cmp     r0,r1           @ so use r1 to calculate 3pi/2
205         bgt     chkq4
206         @@ it is in the third quadrant... rotate, shift, call sinq,
207         @@ then complement the result
208         sub     r0,r0,r2
209         lsl     r0,r0,#2
210         bl      sinq
211         rsb     r0,r0,#0
212         b       sin_done
213         @@ it is in the fourth quadrant... rotate, mirror, shift,
214         @@ call sinq, then complement the result
215  chkq4: sub     r0,r0,r2
216         sub     r0,r2,r0
217         lsl     r0,r0,#2
218         bl      sinq
219         rsb     r0,r0,#0
220  sin_done:
221         @@ shift result right 2 bits
222         asr     r0,r0,#2
223         @@ return the result
224         ldmfd   sp!,{pc}
225         @@------------------------------------------------------
```

Listing 8.7

ARM assembly implementation of $\sin x$ and $\cos x$ using fixed-point calculations.

```
1          @@ ************************************************************
2          @@ Name: sincosmain.S
3          @@ Author: Larry Pyeatt
4          @@ Date: 2/22/2014
5          @@ ************************************************************
6          @@ This is a short program to print a table of sine and
7          @@ cosine values using the fixed-point sin/cos functions.
8          @@ Compile with:
9          @@ gcc -o sincos sincos.S sincosmain.S fixedfuncs.c
10         .data
11  fmta:   .asciz "%14.6f "
12  head:   .asciz "     x                sin(x)          cos(x)\n"
13  line:   .asciz "     --------------------------------------\n"
14  newline:.asciz "\n"
15  tab:    .asciz "\t"
16
17         .text
18         .global main
```

```
19  main:    stmfd    sp!,{r4-r11,lr}
20           ldr      r0,=head
21           bl       printf
22           ldr      r0,=line
23           bl       printf
24           mov      r4,#0
25  mloop:
26           @@ load count to r0 and convert it to a number x
27           @@ between 0.0 and pi/2
28           mov      r0,r4
29           @@ multiply it by pi
30           ldr      r1,=pi
31           ldr      r1,[r1]
32           smull    r1,r0,r0,r1
33           lsl      r0,r0,#28
34           orr      r0,r0,r1,lsr#4
35           mov      r5,r0            @ save it in r5 for later
36           @@ print x
37           mov      r1,#28
38           ldr      r2,=fmta
39           bl       printS
40           ldr      r0,=tab
41           bl       printf
42           @@ calculate and print sin(x)
43           mov      r0,r5            @ retrieve x
44           bl       sin
45           mov      r1,#28
46           ldr      r2,=fmta
47           bl       printS
48           ldr      r0,=tab
49           bl       printf
50           @@ calculate and print cos(x)
51           mov      r0,r5            @ retrieve x
52           bl       cos
53           mov      r1,#28
54           ldr      r2,=fmta
55           bl       printS
56           ldr      r0,=newline
57           bl       printf
58           add      r4,r4,#1
59           cmp      r4,#33
60           blt      mloop
61           ldmfd    sp!,{r4-r11,pc}
```

Listing 8.8
Example showing how the sin x and cos x functions can be used to print a table.

8.7.4 Performance Comparison

In some situations it can be very advantageous to use fixed-point math. For example, when using an ARMv6 or older processor, there may not be a hardware floating point unit available. Table 8.4 shows the CPU time required for running a program to compute the sine function on 10,000,000 random values, using various implementations of the sine function. In each case, the program `main()` function was written in C. The only difference in the six implementations was the data type (which could be fixed-point, IEEE single precision, or IEEE double precision), and the sine function that was used. The times shown in the table include only the amount of CPU time actually used in the sine function, and do not include the time required for program startup, storage allocation, random number generation, printing results, or program exit. The six implementations are as follows:

32-bit Fixed Point Assembly The sine function is computed using the code shown in Listing 8.7.

32-bit Fixed Point C The sine function is computed using exactly the same algorithm as in Listing 8.7, but it is implemented in C rather than Assembly.

Single Precision Software Float C Sine is computed using the floating point sine function which is provided by the GCC C compiler. The code is compiled for an ARMv6 or earlier processor without hardware floating point support. The C code is written to use IEEE single precision floating point numbers.

Double Precision Software Float C Exactly the same as the previous method, but using IEEE double precision instead of single precision.

Single Precision VFP C Sine is computed using the floating point sine function which is provided by the GCC C compiler. The code is compiled for the ARMv6 or later processor using hardware floating point support. The C code is written to use IEEE single precision floating point numbers.

Table 8.4 Performance of sine function with various implementations

Optimization	Implementation	CPU seconds
None	32-bit Fixed Point Assembly	3.85
	32-bit Fixed Point C	18.99
	Single Precision Software Float C	56.69
	Double Precision Software Float C	55.95
	Single Precision VFP C	11.60
	Double Precision VFP C	11.48
Full	32-bit Fixed Point Assembly	3.22
	32-bit Fixed Point C	5.02
	Single Precision Software Float C	20.53
	Double Precision Software Float C	54.51
	Single Precision VFP C	3.70
	Double Precision VFP C	11.08

Double Precision VFP C Same as the previous method, but using IEEE double precision instead of single precision.

Each of the six implementations was compiled both with and without compiler optimizations, resulting in a total of 12 test cases. All cases were run on a standard Raspberry Pi model B with the default CPU clock rate.

From Table 8.4, it is clear that the fixed-point implementation written in assembly beats the code generated by the compiler in every case. The closest that the compiler can get is when it can use the VFP hardware floating point unit and the compiler is run with full optimization. Even in that case the fixed-point assembly implementation is almost 15% faster than the single precision floating point implementation, and has 33% more precision (32 bits versus 24 bits). In the worst case, when a VFP hardware unit is not available, the assembly code beats the compiler by a whopping 638% in speed and 33% in precision for single precision floats, and is 1692% faster than double precision floating point at a cost of 41% in precision. Note that even with floating point hardware support, fixed point in assembly is still 3.44 times as fast as the C compiler code.

Similar results could be obtained on any processor architecture, and any reasonably complex mathematical problem. When developing software for small systems, the developer must weigh the costs and benefits of alternative implementations. For battery powered systems, it is important to realize that choices of hardware and software can affect power consumption even more strongly than computing performance. First, the power used by a system which includes a hardware floating point processor will be consistently higher than that of a system without one. Second, the reduction in processing time required for the job is closely related to the reduction in power required. Therefore, for battery operated systems, A fixed-point implementation could greatly extend battery life. The following statements summarize the results from the experiment in this section:

1. A competent assembly programmer can beat the assembler, in some cases by a very large margin.
2. If computational performance is critical, then a well-designed fixed-point implementation will usually outperform even a hardware-accelerated floating point implementation.
3. If there is no hardware support for floating point, then floating point performance is extremely poor, and fixed point will always provide the best performance.
4. If battery life is a consideration, then a fixed-point implementation can have an enormous advantage.

Note also from the table that the assembly language version of the fixed-point sine function beats the identical C version by a wide margin. Section 9.8.2 will demonstrate that a good assembly language programmer who is familiar with the floating point hardware can beat the compiler by an even wider performance margin.

8.8 Ethics Case Study: Patriot Missile Failure

Fixed-point arithmetic is very efficient on modern computers. However it is incumbent upon the programmer to track the radix point at all stages of the computation, and to ensure that a sufficient number of bits are provided on both sides of the radix point. The programmer must ensure that all computations are carried out with the desired level of precision, resolution, accuracy, range, and dynamic range. Failure to do so can have serious consequences.

On February 25, 1991, during the Gulf War, an American Patriot Missile battery in Dharan, Saudi Arabia, failed to intercept an incoming Iraqi SCUD missile. The SCUD struck an American army barracks, killing 28 soldiers and injuring around 98 other people. The cause was an inaccurate calculation of the time elapsed since the system was last booted.

The hardware clock on the system counted the time in tenths of a second since the last reboot. Current time, in seconds, was calculated by multiplying that number by $\frac{1}{10}$. For this calculation, $\frac{1}{10}$ was represented as a U(1,23) fixed-point number. Since $\frac{1}{10}$ cannot be represented precisely in a fixed number of bits, there was round-off error in the calculations. The small imprecision, when multiplied by a large number, resulted in significant error. The longer the system ran after boot, the larger the error became.

The system determined whether or not it should fire by predicting where the incoming missile would be at a specific time in the future. The time and predicted location were then fed to a second system which was responsible for locking onto the target and firing the Patriot missile. The system would only fire when the missile was at the proper location at the specified time. If the radar did not detect the incoming missile at the correct time and location, then the system would not fire.

At the time of the failure, the Patriot battery had been up for around 100 h. We can estimate the error in the timing calculations by considering how the binary number was stored. The binary representation of $\frac{1}{10}$ is $0.0\overline{0011}$. Note that it is a non-terminating, repeating binimal. The 24-bit register in the Patriot could only hold the following set of bits:

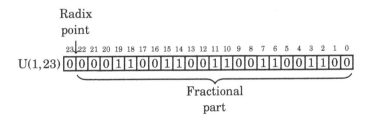

This resulted in an error of $0.00000000000000000000000001\overline{1100}_2$. The error can be computed in base 10 as:

$$e = 2^{-24} + 2^{-25} + 2^{-28} + 2^{-29} + 2^{-32} + 2^{-33} + \cdots \tag{8.3}$$

$$= \sum_{i=0}^{\infty} 2^{-(4i+24)} + 2^{-(4i+25)} \tag{8.4}$$

$$\approx 9.5 \times 10^{-8}. \tag{8.5}$$

To find out how much error was in the total time calculation, we multiply e by the number of tenths of a second in 100 h. This gives $9.5 \times 10^{-8} \times 100 \times 60 \times 60 \times 10 = 0.34$ s. A SCUD missile travels at about 1,676 m/s. Therefore it travels about 570 m in 0.34 s. Because of this, the targeting and firing system was expecting to find the SCUD at a location that was over half a kilometer from where it really was. This was far enough that the incoming SCUD was outside the "range gate" that the Patriot tracked. It did not detect the SCUD at its predicted location, so it could not lock on and fire the Patriot.

This is an example of how a seemingly insignificant error can lead to a major failure. In this case, it led to loss of life and serious injury. Ironically, one factor that contributed to the problem was that part of the code had been modified to provide more accurate timing calculations, while another part had not. This meant that the inaccuracies did not cancel each other. Had both sections of code been re-written, or neither section changed, then the issue probably would not have surfaced.

The Patriot system was originally designed in 1974 to be mobile and to defend against aircraft that move much more slowly than ballistic missiles. It was expected that the system would be moved often, and therefore the computer would be rebooted frequently. Also, the slow-moving aircraft would be much easier to track, and the error in predicting where it is expected to be would not be significant. The system was modified in 1986 to be capable of shooting down Soviet ballistic missiles. A SCUD missile travels at about twice the speed of the Soviet missiles that the system was re-designed for.

The system was deployed to Iraq in 1990, and successfully shot down a SCUD missile in January of 1991. In mid-February of 1991, Israeli troops discovered that the system became inaccurate if it was allowed to run for long periods of time. They claimed that the system would become unreliable after 20 hours of operation. U.S. military did not think the discovery was significant, but on February 16th, a software update was released. Unfortunately, the update could not immediately reach all units because of wartime difficulties in transportation. The Army released a memo on February 21st, stating that the system was not to be run for "very long times," but did not specify how long a "very long time" would be. The software update reached Dhahran one day after the Patriot Missile system failed to intercept a SCUD missile, resulting in the death of 28 Americans and many more injuries.

Part of the reason this error was not found sooner was that the program was written in assembly language, and had been patched several times in its 15-year life. The code was difficult to understand and maintain, and did not conform to good programming practices. The people who worked to modify the code to handle the SCUD missiles were not as familiar with the code as they would have been if it were written more recently, and time was a critical factor. Prolonged testing could have caused a disaster by keeping the system out of the hands of soldiers in a time of war. The people at Raytheon Labs had some tough decisions to make. It cannot be said that Raytheon was guilty of negligence or malpractice. The problem with the system was not necessarily the developers, but that the system was modified often and in inconsistent ways, without complete understanding.

8.9 Chapter Summary

Sometimes it is desirable to perform calculations involving non-integral numbers. The two common ways to represent non-integral numbers in a computer are fixed point and floating point. A fixed point representation allows the programmer to perform calculations with non-integral numbers using only integer operations. With fixed point, the programmer must track the radix point throughout the computation. Floating point representations allow the radix point to be tracked automatically, but require much more complex software and/or hardware. Fixed point will usually provide better performance than floating point, but requires more programming skill.

Fractional numbers in radix notation may not terminate in all bases. Numbers which terminate in base two will also terminate in base ten, but the converse is not true. Programmers should avoid counting using fractions which do not terminate in base two, because it leads to the accumulation of round-off errors.

Exercises

8.1 Perform the following base conversions:
   (a) Convert 10110.001_2 to base ten.
   (b) Convert 11000.0101_2 to base ten.
   (c) Convert 10.125_{10} to binary.
8.2 Complete the following table (assume all values represent positive fixed-point numbers):

Base 10	Base 2	Base 16	Base 13
49.125			
	101011.011		
		AF.3	
			12

8.3 You are working on a problem involving real numbers between -2 and 2 on a computer that has 16-bit integer registers and no hardware floating point support. You decide to use 16-bit fixed-point arithmetic.

(a) What fixed-point format should you use?

(b) Draw a diagram showing the sign, if any, radix point, integer part, and fractional part.

(c) What is the precision, resolution, accuracy, and range of your format?

8.4 What is the resulting type of each of the following fixed-point operations?

(a) $S(24, 7) \times S(27, 15)$

(b) $S(3, 4) \div U(4, 20)$

8.5 Convert 26.640625_{10} to a binary $U(18,14)$ representation. Show the ARM assembly code necessary to load that value into register r4.

8.6 For each of the following fractions, indicate whether or not it will terminate in bases 2, 5, 7, and 10.

(a) $\frac{13}{64}$

(b) $\frac{37}{60}$

(c) $\frac{25}{74}$

(d) $\frac{39}{1250}$

(e) $\frac{17}{343}$

8.7 What is the exact value of the binary number 0011011100011010 when interpreted as an IEEE half-precision number? Give your answer in base ten.

8.8 The "Software Engineering Code of Ethics And Professional Practice" states that a responsible software engineer should "Approve software only if they have well-founded belief that it is safe, meets specifications, passes appropriate tests. . ." (sub-principle 1.03) and "Ensure adequate testing, debugging, and review of software. . .on which they work" (sub-principle 3.10).

The software engineering code of ethics also states that a responsible software engineer should "Treat all forms of software maintenance with the same professionalism as new development."

(a) Explain how the Software Engineering Code of Ethics And Professional Practice were violated by the Patriot Missile system developers.

(b) How should the engineers and managers at Raytheon have responded when they were asked to modify the Patriot Missile System to work outside of its original design parameters?

(c) What other ethical and non-ethical considerations may have contributed to the disaster?

CHAPTER 9

The ARM Vector Floating Point Coprocessor

Chapter Outline

Some ARM processors have dedicated hardware to support floating point operations. For ARMv7 and previous architectures, floating point is provided by an optional Vector Floating Point (VFP) *coprocessor*. Many newer processors also support the NEON extensions, which are covered in Chapter 10. The remainder of this chapter will explain the VFP coprocessor.

9.1 Vector Floating Point Overview

There are four major revisions of the VFP coprocessor:

VFPv1: Obsolete

VFPv2: An optional extension to the ARMv5 and ARMv6 processors. VFPv2 has 16 64-bit FPU registers.

VFPv3: An optional extension to the ARMv7 processors. It is backwards compatible with VFPv2, except that it cannot trap floating-point exceptions. VFPv3-D32 has 32 64-bit FPU registers. Some processors have VFPv3-D16, which supports only 16 64-bit FPU registers. VFPv3 adds several new instructions to the VFP instruction set.

VFPv4: Implemented on some Cortex ARMv7 processors. VFPv4 has 32 64-bit FPU registers. It adds both half-precision extensions and multiply-accumulate instructions to the features of VFPv3. Some processors have VFPv4-D16, which supports only 16 64-bit FPU registers.

Fig. 9.1 shows the 16 ARM integer registers, and the additional registers provided by the VFP coprocessor. Banks four through seven are only present on the VFPv3-D32 and VFPv4-D32 versions of the coprocessor. Note that each register in Banks zero through three can be used to store either one 64-bit number or two 32-bit numbers. For example, double precision register d0 may also be referred to as single precision registers s0 and s1. Each 32-bit VFP register can hold an integer or a single precision floating point number. Registers in Banks four through seven cannot be used as single precision registers.

The VFP adds about 23 new instructions to the ARM instruction set. The exact number of VFP instructions depends on the specific version of the VFP coprocessor. Instructions are provided to:

- transfer floating point values between VFP registers,
- transfer floating-point values between the VFP coprocessor registers and main memory,
- transfer 32-bit values between the VFP coprocessor registers and the ARM integer registers,
- perform addition, subtraction, multiplication, and division, involving two source registers and a destination register,
- compute the square root of a value,
- perform combined multiply-accumulate operations,
- perform conversions between various integer, fixed point, and floating point representations, and
- compare floating-point values.

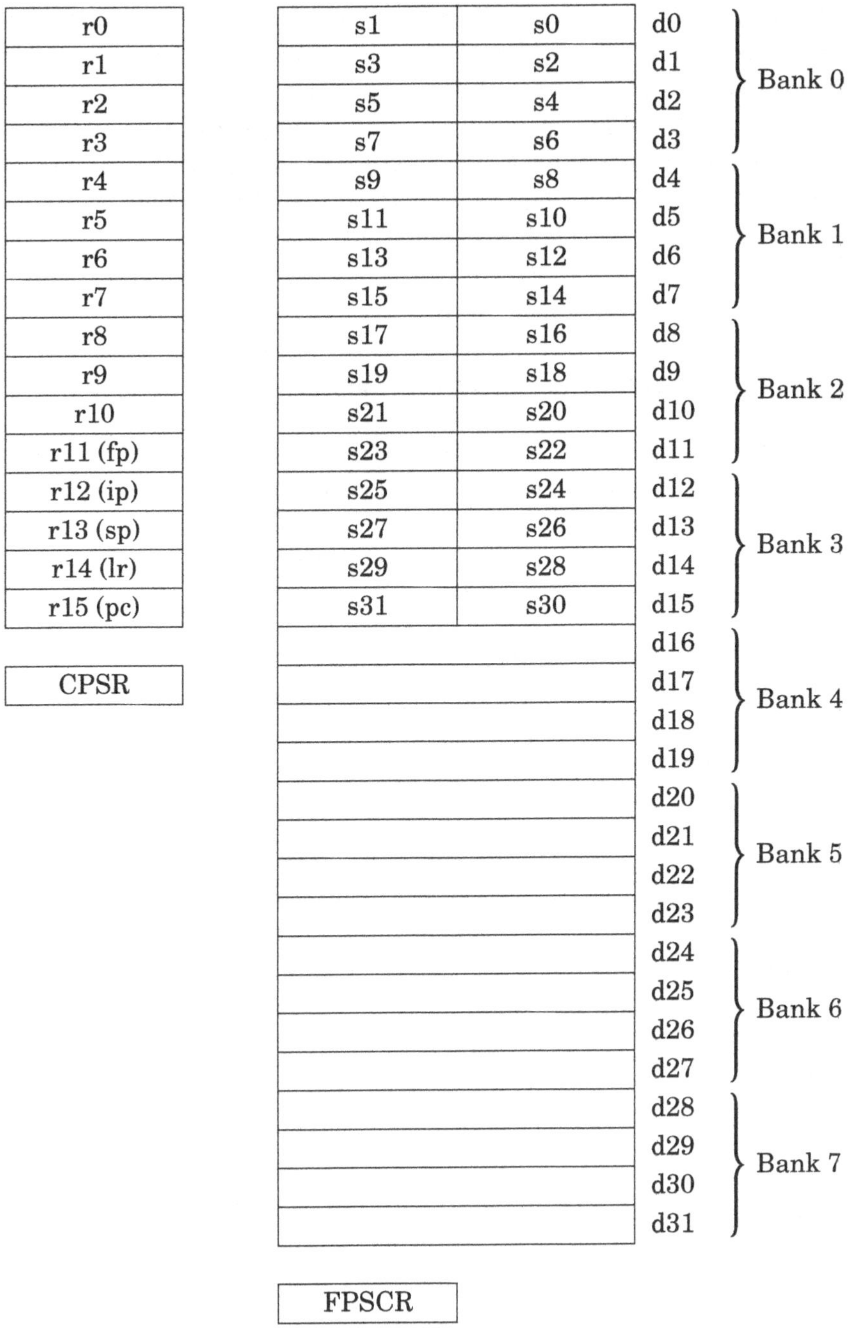

r0	s1	s0	d0
r1	s3	s2	d1
r2	s5	s4	d2
r3	s7	s6	d3
r4	s9	s8	d4
r5	s11	s10	d5
r6	s13	s12	d6
r7	s15	s14	d7
r8	s17	s16	d8
r9	s19	s18	d9
r10	s21	s20	d10
r11 (fp)	s23	s22	d11
r12 (ip)	s25	s24	d12
r13 (sp)	s27	s26	d13
r14 (lr)	s29	s28	d14
r15 (pc)	s31	s30	d15

Bank 0 → d0–d3
Bank 1 → d4–d7
Bank 2 → d8–d11
Bank 3 → d12–d15
Bank 4 → d16–d19
Bank 5 → d20–d23
Bank 6 → d24–d27
Bank 7 → d28–d31

CPSR

d16
d17
d18
d19
d20
d21
d22
d23
d24
d25
d26
d27
d28
d29
d30
d31

FPSCR

Figure 9.1
ARM integer and vector floating point user program registers.

In addition to performing basic operations involving two source registers and one destination register, VFP instructions can also perform operations involving registers arranged as short *vectors* (arrays) of up to eight single-precision values or four double-precision values. A single instruction can be used to perform operations on all of the elements of such vectors. This feature can substantially accelerate computation on arrays and matrices of floating point data. This type of data is common in graphics and signal processing applications. Vector mode can reduce code size and increase speed of execution by supporting parallel operations and multiple transfers.

9.2 Floating Point Status and Control Register

The Floating Point Status and Control Register (FPSCR) is similar to the CPSR register. The FPSCR stores status bits from floating point operations in much the same way as the CPSR stores status bits from integer operations. The programmer can also write to certain bits in the FPSCR to control the behavior of the VFP coprocessor. The layout of the FPSCR is shown in Fig. 9.2. The meaning of each field is as follows:

N The Negative flag is set to one by `vcmp` if Fd < Fm.
Z The Zero flag is Set to one by `vcmp` if Fd = Fm.
C The Carry flag is set to one by `vcmp` if Fd = Fm, or Fd > Fm, or Fd and Fm are unordered.
V The oVerflow flag is set to one by `vcmp` if Fd and Fm are unordered.

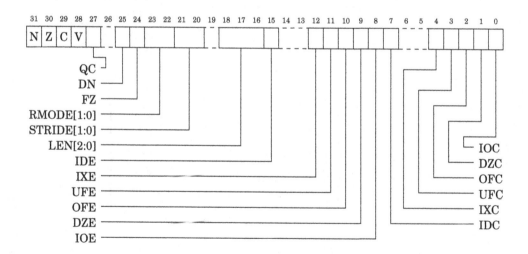

Figure 9.2
Bits in the FPSCR.

QC NEON only. The saturation cumulative flag is set to one by saturating instructions if saturation has occurred.

DN Default NaN enable:

0: Disable Default NaN mode. NaN operands propagate through to the output of a floating-point operation.

1: Enable Default NaN mode. Any operation involving one or more NaNs returns the default NaN.

The default single precision NaN is $7FC00000_{16}$ and the default double-precision NaN is $7FF8000000000000_{16}$. Default NaN mode does not comply with IEEE 754 standard, but may increase performance. NEON instructions ignore this bit and always use Default NaN mode.

FZ Flush-to-Zero enable:

0: Disable Flush-to-Zero mode.

1: Enable Flush-to-Zero mode.

Flush-to-Zero mode replaces subnormal numbers with 0. This does not comply with IEEE 754 standard, but may increase performance. NEON instructions ignore this bit and always use flush-to-Zero mode.

RMODE Rounding mode:

00   Round to Nearest (RN).

01   Round towards Plus infinity (RP).

10   Round towards Minus infinity (RM).

11   Round towards Zero (RZ).

NEON instructions ignore these bits and always use Round to Nearest mode.

STRIDE Sets the stride (distance between items) for vector operations:

00   Stride is 1.

01   Reserved.

10   Reserved.

11   Stride is 2.

LEN Sets the vector length for vector operations:

000   Vector length is 1 (scalar mode).

001   Vector length is 2.

010   Vector length is 3.

011   Vector length is 4.

100   Vector length is 5.

101   Vector length is 6.

110   Vector length is 7.

111   Vector length is 8.

IDE Input Denormal (subnormal) exception Enable:

    0: Exception disabled.

    1: An exception is generated when one or more operand is subnormal.

IXE IneXact exception Enable:

    0: Exception disabled.

    1: An exception is generated when the result contains more significand bits than the destination format can contain, and must be rounded.

UFE UnderFlow exception Enable:

    0: Exception disabled.

    1: An exception is generated when the result is closer to zero than can be represented by the destination format.

OFE OverFlow exception Enable:

    0: Exception disabled.

    1: An exception is generated when the result is farther from zero than can be represented by the destination format.

DZE Division by Zero exception Enable:

    0: Exception disabled.

    1: An exception is generated by divide instructions when the divisor is zero or subnormal.

IOE Invalid Operation exception Enable:

    0: Exception disabled.

    1: An exception is generated when the result is not defined, or cannot be represented. For example, adding positive and negative infinity gives an invalid result.

IDC The Input Subnormal Cumulative flag is set to one when an IDE condition has occurred.

IXC The IneXact Cumulative flag is set to one when an IXE condition has occurred.

UFC The UnderFlow Cumulative flag is set to one when a UFE condition has occurred.

OFC The OverFlow Cumulative flag is set to one when an OFE condition has occurred.

DZC The Division by Zero Cumulative flag is set to one when a DZE condition has occurred.

IOC The Invalid Operation Cumulative flag is set to one when an OFE condition has occurred.

The only VFP instruction that can be used to update the status flags in the FPSCR is `fcmp`, which is similar to the integer `cmp` instruction. To use the FPSCR flags to control conditional instructions, including conditional VFP instructions, they must first be moved into the CPSR register. Table 9.1 shows the meanings of the FPSCR flags when they are transferred to the

Table 9.1 Condition code meanings for ARM and VFP

<cond>	ARM Data Processing Instruction	VFP fcmp Instruction
AL	Always	Always
EQ	Equal	Equal
NE	Not Equal	Not equal, or unordered
GE	Signed greater than or equal	Greater than or equal
LT	Signed less than	Less than, or unordered
GT	Signed greater than	Greater than
LE	Signed less than or equal	Less than or equal, or unordered
HI	Unsigned higher	Greater than, or unordered
LS	Unsigned lower or same	Less than or equal
HS	Carry set/unsigned higher or same	Greater than or equal, or unordered
CS	Same as HS	Same as HS
LO	Carry clear/ unsigned lower	less than
CC	Same as LO	Same as LO
MI	Negative	Less than
PL	Positive or zero	Greater than or equal, or unordered
VS	Overflow	Unordered (at least one NaN operand)
VC	No overflow	Not unordered

CPSR and used for conditional execution on following instructions. The following rules govern how the bits in the FPSCR may be changed by subroutines:

1. Bits 27-31, 0-4, and 7 do not need to be preserved.
2. Subroutines *may* modify bits 8-12, 15, and 22-25 but the practice is discouraged. These bits should only be changed by specific support subroutines which change the global state of the program. If they are modified within a subroutine, then their original value must be restored before the function returns or calls another function.
3. Bits 16–18 and bits 20–21 may be changed by a subroutine, but must be set to zero before the function returns or calls another function.
4. All other bits are reserved for future use and must not be modified.

9.2.1 Performance Versus Compliance

Floating point operations are complex, and there are many special cases, such as dealing with NaNs, infinities, and subnormals. These special cases are a normal part of performing floating point math, but they are relatively infrequent. In order to simplify the hardware, many special situations which occur infrequently are handled by software. When one of these exceptional situations occurs, the VFP hardware sets the appropriate flags in the FPSCR and generates an interrupt. The ARM CPU then executes an interrupt handler to deal with the exceptional situation. When the routine finishes, it returns to the point where the exception occurred and execution resumes just as if the situation had been dealt with by the hardware. This approach is taken by many processor architectures to reduce the complexity, cost, and/or power consumption of the floating point hardware, This approach also allows the programmer to make a trade-off between performance and strict IEEE 754 compliance.

Full-compliance mode

The support code for dealing with VFP exceptions is included in most ARM-based operating systems. Even bare-metal embedded systems can include the VFP support service routines. With the support code enabled, the VFP coprocessor is fully compliant with the IEEE 754 standard. However, using the fully compliant mode does increase the average run-time for floating point code, and increases the size of the operating system kernel or embedded system code.

RunFast mode

When all of the VFP exceptions are disabled, Default NaN mode is enabled, and Flush-to-Zero is enabled, the VFP is not fully compliant with the IEEE 754 standard. However, floating point code runs significantly faster. For that reason, the state when bits 8–12 and bit 15 are set to zero while bits 24 and 25 are set to one is referred to as RunFast mode. There is some loss of accuracy for very small values, but the hardware no longer has to check for many of the conditions that may stall the floating point pipeline. This results in fewer stalls and much higher throughput in the hardware, as well as eliminating the necessity to handle exceptions in software. Many other floating point architectures have similar modes, so the GCC developers have found it worthwhile to provide programmers with the option of using them. User applications can be compiled to use this mode with GCC by using the `-ffast-math` and/or `-Ofast` options during compilation and linking. The startup code in the C standard library will then set the VFP to RunFast mode before calling the `main` function.

9.2.2 Vector Mode

A VFP vector consists of up to eight single-precision registers, or up to four double-precision registers. All of the registers in a vector must be in the same bank. Also, vectors cannot be stored in Bank 0 or Bank 4. For example, registers `s8` through `s10` could be treated as a vector of three single-precision values. Registers `s14` through `s17` cannot be treated as a vector because some of those registers are in Bank 1 and others are in Bank 2. Registers `d0` through `d3` cannot be treated as a vector because they are in Bank 0.

The LEN field in the FPSCR controls the length of vectors that are used for vector operations. In vector operations, the first register in the vector is given as the operand, and the remaining registers are inferred from the settings of LEN and STRIDE. The STRIDE field allows data to be interleaved. For example, if the stride is set to two, and length is set to four, then the vector starting at s8 would consist of registers s8, s10, s12, and s14, while the vector starting at s9 would consist of registers s9, s11, s13, and s15. If a vector runs off the end of a bank, then the address wraps around to the first register in the bank. For example, if length is set to six and stride is set to one, then the vector starting at s13 would consist of s13, s14, s15, s8, s9, and s10, in that order.

The vector-capable data-processing instructions have one of the following two forms:

```
1   Op  Fd,Fn,Fm
2   Op  Fd,Fm
```

where Op is the VFP instruction, Fd is the destination register (or the first register in a vector), Fn is an operand register (or the first register in a vector), and Fm is an operand register (or the first register in a vector). Most data-processing instructions can operate in scalar mode, mixed mode, or vector mode. The mode depends on the LEN bits in the FPSCR, as well as on which register banks contain the destination and operand(s).

- The operation is *scalar* if the LEN field is set to zero (scalar mode) or the destination operand, Fd, is in Bank 0 or Bank 4. The operation acts on Fm (and Fn if the operation uses two operands) and places the result in Fd.
- The operation is *mixed* if the LEN field is not set to zero and Fm is in Bank 0 or Bank 4 but Fd is not. If the operation has only one operand, then the operation is applied to Fm and copies of the result are stored into each register in the destination vector. If the operation has two operands, then it is applied with the scalar Fm and each element in the vector starting at Fn, and the result is stored in the vector beginning at Fd.
- The operation is *vector* if the LEN field is not set to zero and neither Fd nor Fm is in Bank 0 or Bank 4. If the operation has only one operand, then the operation is applied to the vector starting at Fm and the results are placed in the vector starting at Fd. If the operation has two operands, then it is applied with corresponding elements from the vectors starting at Fm and Fn, and the result is stored in the vector beginning at Fd.

9.3 Register Usage Rules

As with the integer registers, there are rules for using the VFP registers. These rules are a convention, and following the convention ensures interoperability between code written by different programmers and compilers. Registers s16 through s31 are *non-volatile*. This implies that d8 through d15 are also non-volatile, since they are really the same registers. The contents of these registers must be preserved across subroutine calls. The remaining registers (s0 through s15, also known as d0 through d7) are *volatile*. They are used for passing arguments, returning results, and for holding local variables. They do not need to be preserved by subroutines. If registers d16 through d31 are present, then they are also considered *volatile*.

In addition to the FPSCR, all VFP implementations contain at least two additional system registers. The Floating-point System ID register (FPSID) is a read-only register whose value indicates which VFP implementation is being provided. The contents of the FPSID can be transferred to an ARM integer register, then examined to determine which VFP version is available. There is also a Floating-point Exception register (FPEXC). Two bits of the FPEXC register provide system-level status and control. The remaining bits of this register are defined by the sub-architecture. These additional system registers should not be accessed by user applications.

9.4 Load/Store Instructions

The VFP provides several instructions for moving data between memory and the VFP registers. There are instructions for loading and storing single and double precision registers, and for moving multiple registers to or from memory.. All of the load and store instructions require a memory address to be in one of the ARM integer registers.

9.4.1 Load/Store Single Register

The following instructions are used to load or store a single VFP register:

vldr   Load VFP Register, and
vstr   Store VFP Register.

Syntax

```
v<op>r{<cond>}{.<prec>}  Fd, [Rn{,#offset}]
v<op>r{<cond>}{.<prec>}  Fd, =label
```

-     <op> may be either ld or st.
-     Fd may be any single or double precision register.
-     Rn may be any ARM integer register.
-     <cond> is an optional condition code.
-     <prec> may be either f32 or f64.

Operations

Name	Effect	Description
vldr	$Fd \leftarrow Mem[Rn + offset]$	Load Fd using Rn as a pointer
vstr	$Mem[Rn + offset] \leftarrow Fd$	Store Fd using Rn as a pointer

Examples

```
1    vldr       s5,[r0]     @ load s5 from address in r0
2    vstr.f64   d4,[r2]     @ store d4 using address in r2
3    vstreq.f32 s0,[r1]     @ if eq condition is true,
4                           @ store s0 using address in r1
```

9.4.2 Load/Store Multiple Registers

These instructions load or store multiple floating-point registers:

vldm   Load Multiple VFP Registers, and
vstm   Store Multiple VFP Registers.

As with the integer ldm and stm instructions, there are multiple versions for use in moving data and accessing stacks.

Syntax

```
v<op>m<mode>{<cond>}{.<prec>}  Rn{!},<list>
vpush{<cond>}{.<prec>}         <list>
vpop{<cond>}{.<prec>}          <list>
```

- <op> may be either ld or st.
- <mode> is one of
  ia   Increment address after each transfer.
  db   Decrement address before each transfer.
- Rn may be any ARM integer register.
- <cond> is an optional condition code.
- <prec> may be either f32 or f64.
- <list> may be any set of *contiguous* single precision registers, or any set of *contiguous* double precision registers.
- If mode is db then the ! is required.
- vpop <list> is equivalent to vldmia sp!,<list>.
- vpush <list> is equivalent to vstmdb sp!,<list>.

Operations

Name	Effect	Description

vldmia	*addr ← Rd*
	for *i* ∈ *register_list* **do**
	  i ← Mem[addr]
	  if single **then**
	    addr ← addr + 4
	  else
	    addr ← addr + 8
	  end if
	end for
	if ! is present **then**
	  Rd ← addr
	end if
		Load multiple registers from memory starting at the address in Rd. Increment address after each load.

vstmia	*addr ← Rd*
	for *i* ∈ *register_list* **do**
	  Mem[addr] ← i
	  if single **then**
	    addr ← addr + 4
	  else
	    addr ← addr + 8
	  end if
	end for
	if ! is present **then**
	  Rd ← addr
	end if
		Store multiple registers in memory starting at the address in Rd. Increment address after each store.

vldmdb	*addr ← Rd*
	for *i* ∈ *register_list* **do**
	  if single **then**
	    addr ← addr − 4
	  else
	    addr ← addr − 8
	  end if
	  i ← Mem[addr]
	end for
	Rd ← addr
		Load multiple registers from memory starting at the address in Rd. Decrement address before each load.

vstmdb	*addr ← Rd*
	for *i* ∈ *register_list* **do**
	  if single **then**
	    addr ← addr − 4
	  else
	    addr ← addr − 8
	  end if
	  Mem[addr] ← i
	end for
	Rd ← addr
		Store multiple registers in memory starting at the address in Rd. Decrement address before each store.

Examples

```
1    vstmdb    sp!,{s0-s3}      @ Store s0 through s3 on stack
2    vstmia    r1,{s0-s31}      @ Store all fp registers
3                               @ at address in r1
4    vldmia    sp!,{d4-d7}      @ Pop four doubles from the stack
5    vldmiaeq  sp!,{d4-d7}      @ If eq, then pop four doubles
6                               @ from the stack
```

9.5 Data Processing Instructions

These operations are vector-capable. For details on how to use vector mode, refer to Section 9.2.2. Instructions are provided to perform the four basic arithmetic functions, plus absolute value, negation, and square root. There are also special forms of the multiply instructions that perform multiply-accumulate.

9.5.1 Copy, Absolute Value, Negate, and Square Root

The unary operations require on source operand and a destination register. The source and destination can be the same register. There are four unary operations:

vcpy   Copy VFP Register (equivalent to move),
vabs   Absolute Value,
vneg   Negate, and
vsqrt   Square Root.

Syntax

```
v<op>{<cond>}.<prec>  Fd, Fm
```

- `<op>` is one of `cpy`, `abs`, `neg`, or `sqrt`.
- `<cond>` is an optional condition code.
- `<prec>` may be either `f32` or `f64`.

Operations

Name	Effect	Description
vcpy	$Fd \leftarrow Fn$	Copy
vabs	$Fd \leftarrow |Fn|$	Absolute Value
vneg	$Fd \leftarrow -Fn$	Negate
vsqrt	$Fd \leftarrow \sqrt{Fn}$	Square Root

Examples

```
vabs    d3, d5    @ Store absolute value of d1 in d3
vnegmi  s15, s15  @ if mi, then negate s15
```

9.5.2 Add, Subtract, Multiply, and Divide

The basic mathematical operations require two source operands and one destination. There are five basic mathematical operations:

vadd   Add,
vsub   Subtract,
vmul   Multiply,
vnmul   Negate and Multiply, and
vdiv   Divide.

Syntax

```
v<op>{<cond>}.<prec>  Fd, Fn, Fm
```

- <op> is one of add, sub, mul, nmul, or div.
- <cond> is an optional condition code.
- <prec> may be either f32 or f64.

Operations

Name	Effect	Description
vadd	$Fd \leftarrow Fn + Fm$	Add
vsub	$Fd \leftarrow Fn - Fm$	Subtract
vmul	$Fd \leftarrow Fn \times Fm$	Multiply
vnmul	$Fd \leftarrow Fn \times -Fm$	Negate and multiply
vdiv	$Fd \leftarrow Fn \div Fm$	Divide

Examples

```
vadd.f64    d0, d1, d2    @ d0 <- d1 + d2
vaddgt.f32  s0, s1, s2    @ if (gt) then s0 <- s1 + s2
vnmul.f32   s10, s10, s14 @ s10 <- -(s10 * s14)
vdivlt.f64  d0, d7, d8    @ if lt, then d0 <- d7 / d8
```

9.5.3 Compare

The compare instruction subtracts the value in `Fm` from the value in `Fd` and sets the flags in the FPSCR based on the result. The comparison operation will raise an exception if one of the operations is a signalling NaN. There is also a version of the instruction that will raise an exception if either operand is any type of NaN. The two comparison instructions are:

vcmp   Compare, and
vcmpe   Compare with Exception.

Syntax

```
vcmp{e}{<cond>}.<prec>    Fd, Fm
```

- If `e` is present, an exception is raised if either operand is any kind of NaN. Otherwise, an exception is raised only if either operand is a signaling NaN.
- `<cond>` is an optional condition code.
- `<prec>` may be either `f32` or `f64`.

Operations

Name	Effect	Description
fcmp	*FPSCR ← flags(Fd − Fm)*	Compare two registers

Examples

```
1    vcmp.f32    s0, s1    @ Subtract s1 from s0 and set
2    @ FPSCR flags
```

9.6 Data Movement Instructions

With the addition of all of the VFP registers, there many more possibilities for how data can be moved. There are many more registers, and VFP registers may be 32 or 64 bit. This results in several possible combinations for moving data among all of the registers. The VFP instruction set includes instructions for moving data between two VFP registers, between VFP and integer registers, and between the various system registers.

9.6.1 Moving Between Two VFP Registers

The most basic move instruction involving VFP registers simply moves data between two floating point registers. The instruction is:

vmov   Move Between VFP Registers.

Syntax

```
vmov{<cond>}{.<prec>} Fd, Fm
```

- F can be s or d.
- Fd and Fm must be the same size.
- <cond> is an optional condition code.
- <prec> is either f32 or f64.

Operations

Name	Effect	Description
vmov	*Fd ← Fm*	Move Fm to Fd

Examples

```
1   vmov.f64 d3,d4        @ d3 <- d4
2   vmov.f32 s5,s12       @ s5 <- s12
```

9.6.2 Moving Between VFP Register and One Integer Register

This version of the move instruction allows 32 bits of data to be moved between an ARM integer register and a floating point register. The instruction is:

vmov   Move Between VFP and One ARM Integer Register.

Syntax

```
vmov{<cond>}    Rd, Sn
vmov{<cond>}    Sn, Rd
```

- Rd is an ARM integer register.
- Sd is a VFP single precision register.
- <cond> is an optional condition code.

Operations

Name	Effect	Description
vmov Rd,Sm	*Rd ← Sm*	Move Sm to Rd
vmov Sm,Rd	*Sm ← Rd*	Move Rd to Sm

Examples

```
1    vmov r3,s4           @ r2 <= s4
2    vmov s12,r8          @ s12 <- r8
```

9.6.3 Moving Between VFP Register and Two Integer Registers

This version of the move instruction is used to transfer 64 bits of data between ARM integer registers and floating point registers:

vmov   Move Between VFP and Two ARM Integer Registers.

Syntax

```
vmov{<cond>}   destination(s), source(s)
```

- Source and destination must be VFP or integer registers. One of them must be a set of ARM integer registers, and the other must be VFP coprocessor registers. The following table shows the possible choices for sources and destinations.

ARM Integer	Floating Point
Rl,Rh	Dd
	Sd,Sd'

- Sd and Sd' must be adjacent, and Sd' must be the higher-numbered register.
- <cond> is an optional condition code.

Operations

Name	Effect	Description
vmov Dd,Rl,Rh	$Dd \leftarrow Rh : Rl$	Move Rh and Rl to Dd
vmov Rl,Rh,Dm	$Rh : Rl \leftarrow Dm$	Move Dm to Rh and Rl
vmov Sd,Sd',Rl,Rh	$Sd \leftarrow Rh, Sd' \leftarrow Rl$	Move Rh and Rl to Sd and Sd'.
vmov Rl,Rh,Sd,Sd'	$Rh \leftarrow Sd, Rl \leftarrow Sd'$	Move Sd and Sd' to Rh and Rl.

Examples

```
1    vmov d9,r0,r1        @ d9 <- r1:r0
2    vmov r2,r3,d12       @ r3:r2 <- d12
3    vmov s1,s2,r2,r4     @ s1 <- r2, s2 <- r4
4    vmov r5,r7,s0,s1     @ r1 <- s0, r7 <- s1
```

9.6.4 Move Between ARM Register and VFP System Register

There are two instructions which allow the programmer to examine and change bits in the VFP system register(s):

vmrs   Move From VFP System Register to ARM Register, and

vmsr   Move From ARM Register to VFP System Register.

User programs should only access the FPSCR to check the flags and control vector mode.

Syntax

```
vmrs{<cond>} Rd, VFPsysreg
vmsr{<cond>} VFPsysreg, Rd
```

* VFPsysreg can be any of the VFP system registers.
* Rd can be APSR_nzcv or any ARM integer register.,
* <cond> is an optional condition code.

Operations

Name	Effect	Description
mrs	*Rd ← VFP sysreg*	Move data from VFP system register to integer register
msr	*VFP sysreg ← Rd*	Move data from integer register to VFP system register

Examples

```
1    vmrs    APSR_nzcv,fpscr   @ Copy flags from FPSCR to CPSR
2    vmrs    r3, FPSCR         @ Copy FPSCR flags to CPSR
3    vmsr    FPSCR,r5          @ Copy FPSCR flags to CPSR
```

9.7 Data Conversion Instructions

The ARM VFP provides several instructions for converting between various floating point and integer formats. Some VFP versions also have instructions for converting between fixed point and floating point formats.

9.7.1 Convert Between Floating Point and Integer

These instructions are used to convert integers to single or double precision floating point, or for converting single or double precision to integer:

vcvt   Convert Between Floating Point and Integer
vcvtr   Convert Floating Point to Integer with Rounding

These instructions always use a single precision register for the integer, but the floating point argument can be single precision or double precision. Some versions of the VFP do not support the double precision versions.

Syntax

```
vcvt{r}{<cond>}.<type>.f64   Sd, Dm
vcvt{r}{<cond>}.<type>.f32   Sd, Sm
vcvt{<cond>}.f64.<type>      Dd, Sm
vcvt{<cond>}.f32.<type>      Sd, Sm
```

- The optional r makes the operation use the rounding mode specified in the FPSCR. The default is to round toward zero.
- <cond> is an optional condition code.
- The <type> can be either u32 or s32 to specify unsigned or signed integer.
- These instructions can also convert from fixed point to floating point if followed by an appropriate vmul.

Operation

Opcode	Effect	Description
vcvt.f64.s32	$Dd \leftarrow double(Sm)$	Convert signed integer to double
vcvt.f32.s32	$Sd \leftarrow single(Sm)$	Convert signed integer to single
vcvt.f64.u32	$Dd \leftarrow double(Sm)$	Convert unsigned integer to double
vcvt.f32.u32	$Sd \leftarrow single(Sm)$	Convert unsigned integer to single
vcvt.s32.f32	$Sd \leftarrow int(Sm)$	Convert single to signed integer
vcvt.u32.f32	$Sd \leftarrow unsigned(Sm)$	Convert single to unsigned integer
vcvt.s32.f64	$Sd \leftarrow int(Dm)$	Convert double to signed integer
vcvt.u32.f64	$Sd \leftarrow unsigned(Dm)$	Convert double to unsigned integer

Examples

```
vcvt.f64.u32   d5, s7   @ Convert unsigned integer to double
vcvt.f64.f32   d0, s4   @ Convert signed integer to double
vcvt.u32.f64   s0, d7   @ Convert double to unsigned integer
vcvt.s32.f64   s1, d4   @ Convert double to signed integer
@@ Convert s10 to an S(15,16)
consta:    .float  65536.0
    ⋮
```

```
8    vldr.f32     s11,consta  @ Load floating point constant
9    vmul.f32     s10,s10,s11 @ Multiply equates to shift
10   vcvt.s32.f32 s10,s10     @ Convert single to S(15,16)
```

9.7.2 Convert Between Fixed Point and Single Precision

VFPv3 and higher coprocessors have additional instructions used for converting between fixed point and single precision floating point:

vcvt   Convert To or From Fixed Point.

Syntax

```
vcvt{<cond>}.<td>.f32   Sd, Sm, #fbits
vcvt{<cond>}.f32.<td>   Sd, Sm, #fbits
```

* <cond> is an optional condition code.
* <td> specifies the type and size of the fixed point number, and must be one of the following:

  s32   signed 32 bit value,
  u32   unsigned 32 bit value,
  s16   signed 16 bit value, or
  u16   unsigned 16 bit value.

* The #fbits operand specifies the number of fraction bits in the fixed point number, and must be less than or equal to the size of the fixed point number indicated by <td>.

Operations

Name	Effect	Description
vcvt.s32.f32	$Dd \leftarrow fixed32(Sm)$	Convert single precision to 32-bit signed fixed point.
vcvt.u32.f32	$Sd \leftarrow ufixed32(Sm)$	Convert single precision to 32-bit unsigned fixed point.
vcvt.s16.f32	$Dd \leftarrow fixed16(Sm)$	Convert single precision to 16-bit signed fixed point.
vcvt.u16.f32	$Sd \leftarrow ufixed16(Sm)$	Convert single precision to 16-bit unsigned fixed point.
vcvt.f32.s32	$Dd \leftarrow single(Sm)$	Convert signed 32-bit fixed point to single precision
vcvt.f32.u32	$Sd \leftarrow single(Sm)$	Convert unsigned 32-bit fixed point to single precision

| vcvt.f32.s16 | $Dd \leftarrow single(Sm)$ | Convert signed 16-bit fixed point to single precision |
| vcvt.f32.16 | $Sd \leftarrow single(Sm)$ | Convert unsigned 16-bit fixed point to single precision |

Examples

```
1   vcvt.f32.u16  s0,s0,#4 @ Convert from U(12,4) to single
2   vcvt.s32.f32  s1,s1,#8 @ Convert from single to S(23,8)
```

9.8 Floating Point Sine Function

A fixed point implementation of the sine function was discussed in Section 8.7, and shown to be superior to the floating point sine function provided by GCC. Now that we have covered the VFP instructions, we can write an assembly version using floating point which also performs better than the routines provided by GCC.

9.8.1 Sine Function Using Scalar Mode

Listing 9.1 shows a single precision floating point implementation of the sine function, using the ARM VFPv3 instruction set. It works in a similar way to the previous fixed point code. There is a table of constants, each of which is the reciprocal of one of the factorial divisors in the Taylor series for sine. The subroutine calculates the powers of x one-by-one, and multiplies each power by the next constant in the table, summing the results as it goes. Note that the table of constants is shorter than the fixed point version of the code, because there are fewer bits of precision in a single precision floating point number than there are in the fixed point representation that was used previously.

```
1          .data
2          @@ The following is a table of constants used in the
3          @@ Taylor series approximation for sine
4          .align  5               @ Align to cache
5   ctab:  .word   0xBE2AAAAA      @ -1.666666e-01
6          .word   0x3C088889      @  8.333334e-03
7          .word   0xB9500D00      @ -1.984126e-04
8          .word   0x3638EF1D      @  2.755732e-06
9          .word   0xB2D7322A      @ -2.505210e-08
10  @@@ --------------------------------------------------------
11         .text
12         .align  2
```

```
13        @@ sin_a_f implements the sine function using IEEE single
14        @@ precision floating point. It computes sine by summing
15        @@ the first six terms of the Taylor series.
16        .global sin_a_f
17 sin_a_f:
18        @@ set runfast mode and rounding to nearest
19        fmrx    r1, fpscr      @ get FPSCR contents in r1
20        bic     r2, r1, #(0b1111<<23)
21        orr     r2, r2, #(0b1100<<23)
22        fmxr    fpscr, r2 @ store in FPSCR
23        @@ initialize variables
24        vmul.f32        s1,s0,s0 @ s1 <- x^2
25        vmul.f32        s3,s1,s0 @ s3 <- x^3
26        ldr             r0,=ctab @ load pointer to coefficients
27        mov             r3,#5    @ load loop counter
28 loop:  vldr.f32        s4,[r0] @ load coefficient
29        add             r0,r0,#4 @ increment pointer
30        vmul.f32        s4,s3,s4 @ s4 <- next term
31        vadd.f32        s0,s0,s4 @ add term to result
32        subs            r3,r3,#1 @ decrement and test loop count
33        vmulne.f32      s3,s1,s3 @ s4 <- x^2n
34        bne             loop     @ loop five times
35        @@ restore original FPSCR
36        fmxr    fpscr, r1
37        mov     pc,lr
```

Listing 9.1

Simple scalar implementation of the $\sin x$ function using IEEE single precision.

Listing 9.2 shows a double precision floating point implementation of the sine function, using the ARM VFPv3 instruction set. Again, there is a table of constants, each of which is the reciprocal of one of the factorial divisors in the Taylor series for sine. The subroutine calculates the powers of x one-by-one, and multiplies each power by the next constant in the table, summing the results as it goes. Note that the table of constants is longer than the fixed point version of the code, because there are more bits of precision in a double precision floating point number than there are in the fixed point representation that was used previously.

```
1        .data
2        @@ The following is a table of constants used in the
3        @@ Taylor series approximation for sine
4        .align 6        @ Align for efficient caching
5 ctab:  .word 0x55555555, 0xBFC55555    @ -1.666666666666667e-01
6        .word 0x11111111, 0x3F811111    @  8.333333333333333e-03
7        .word 0x1A01A01A, 0xBF2A01A0    @ -1.984126984126984e-04
8        .word 0xA556C734, 0x3EC71DE3    @  2.755731922398589e-06
```

```
 9        .word 0x67F544E4, 0xBE5AE645     @ -2.505210838544172e-08
10        .word 0x13A86D09, 0x3DE61246     @  1.605904383682161e-10
11        .word 0xE733B81F, 0xBD6AE7F3     @ -7.647163731819816e-13
12        .word 0x7030AD4A, 0x3CE952C7     @  2.811457254345521e-15
13        .word 0x46814157, 0xBC62F49B     @ -8.220635246624329e-18
14 @@@ -----------------------------------------------------------
15        .text
16        .align 2
17        @@ sin_a_f_d implements the sine function using IEEE
18        @@ double precision floating point.  It computes sine
19        @@ by summing the first ten terms of the Taylor series.
20        .global sin_a_d
21 sin_a_d:
22        @@ set runfast mode and rounding to nearest
23        fmrx    r1,fpscr       @ get FPSCR contents in r1
24        bic     r2,r1, #(0b1111<<23)
25        orr     r2,r2, #(0b1100<<23)
26        fmxr    fpscr, r2      @ store settings in FPSCR
27        @@ initialize variables
28        vmul.f64 d1,d0,d0       @ d1 <- x^2
29        vmul.f64 d3,d1,d0       @ d3 <- x^3
30        ldr     r0,=ctab       @ load pointer to coefficient table
31        mov     r3,#9          @ load loop counter
32 loop:  vldr.f64 d4,[r0]       @ load coefficient
33        add     r0,r0,#8       @ increment pointer
34        vmul.f64 d4,d3,d4       @ d4 <- next term
35        vadd.f64 d0,d0,d4       @ add term to result
36        subs    r3,r3,#1       @ decrement and test loop counter
37        vmulne.f64 d3,d1,d3     @ d4 <- x^2n
38        bne     loop           @ loop nine times
39        @@ restore original FPSCR
40        fmxr    fpscr, r1
41        mov     pc,lr
```

Listing 9.2
Simple scalar implementation of the $\sin x$ function using IEEE double precision.

9.8.2 Sine Function Using Vector Mode

The previous implementations are already faster than the implementations provided by GCC, However, it may be possible to gain a little more performance by using VFP vector mode. In the single precision code, there are five terms to be added. Since single precision vectors can have up to eight elements, the code should not require any loop at all.

Listing 9.3 shows a single precision floating point implementation of the sine function, using the ARM VFPv3 instruction set in vector mode. It performs the same operations as the previous implementation, but instead of using a loop, all of the data is pre-loaded into vector banks and then a vector multiply operation is performed. The processor is then returned to scalar mode, and the summation is performed. This implementation is slightly faster than the previous version.

```
1          .data
2          @@ The following is a table of constants used in the
3          @@ Taylor series approximation for sine
4          .align  6               @ Align to cache
5   ctab:  .word   0xBE2AAAAB      @ -1.666667e-01
6          .word   0x3C088889      @  8.333334e-03
7          .word   0xB9500D01      @ -1.984127e-04
8          .word   0x3638EF1D      @  2.755732e-06
9          .word   0xB2D7322B      @ -2.505211e-08
10  @@@ --------------------------------------------------------------
11         .text
12         .align 2
13         @@ sin_a_f implements the sine function using IEEE single
14         @@ precision floating point.  It takes advantage of the
15         @@ ARM VFP vector processing instructions.  It computes
16         @@ sine by summing the first six terms of the Taylor
17         @@ series.
18         .global sin_v_f
19  sin_v_f:
20         vmrs    r1, fpscr       @ get FPSCR contents in r1
21         .if SET_RUNFAST
22         @@ set runfast mode and rounding to nearest
23         bic     r2, r1, #(0b1111<<23)
24         orr     r2, r2, #(0b1100<<23)
25         vmsr    fpscr, r2       @ store settings in FPSCR
26         .endif
27         @@ Put x^2 in s1
28         vmul.f32 s1,s0,s0       @ s1 = x^2
29         @@ load vector of coefficients into Bank 2
30         ldr     r0,=ctab        @ get address of coefficients
31         vldmia  r0!,{s16-s20}
32         @@ Set up vector containing powers of x in Bank 1
33         vmul.f32 s8,s0,s1       @ s8 = x^3
34         vmul.f32 s9,s8,s1       @ s9 = x^5
35         vmul.f32 s10,s9,s1      @ s10 = x^7
36         vmul.f32 s11,s10,s1     @ s11 = x^9
37         vmul.f32 s12,s11,s1     @ s11 = x^11
```

```
38        @@ Set VFP for vector mode
39        @@ set rounding, stride to 1, and vector length to 5
40        .if SET_RUNFAST
41        bic    r2, r2, #(0b11111<<16)
42        .else
43        bic    r2, r1, #(0b11111<<16)
44        .endif
45        orr    r2, r2, #(0b00100<<16)
46        vmsr   fpscr, r2      @ store settings in FPSCR
47        @@ Multiply powers by coefficients. Put results in Bank 3
48        vmul.f32 s24,s8,s16     @ VECTOR operation
49        @@ restore original FPSCR
50        vmsr   fpscr, r1
51        @@ Add terms in Bank 3 to the result in s0
52        vadd.f32 s24,s24,s25
53        vadd.f32 s26,s26,s27
54        vadd.f32 s0,s0,s24
55        vadd.f32 s26,s26,s28
56        vadd.f32 s0,s0,s26
57        mov    pc,lr
```

Listing 9.3
Vector implementation of the sin x function using IEEE single precision.

```
1         .data
2         @@ The following is a table of constants used in the
3         @@ Taylor series approximation for sine
4         .align 7        @ Align for efficient caching
5  ctab:  .word 0x55555555, 0xBFC55555   @ -1.666666666666667e-01
6         .word 0x11111111, 0x3F811111   @  8.333333333333333e-03
7         .word 0x1A01A01A, 0xBF2A01A0   @ -1.984126984126984e-04
8         .word 0xA556C734, 0x3EC71DE3   @  2.755731922398589e-06
9         .word 0x67F544E4, 0xBE5AE645   @ -2.505210838544172e-08
10        .word 0x13A86D09, 0x3DE61246   @  1.605904383682161e-10
11        .word 0xE733B81F, 0xBD6AE7F3   @ -7.647163731819816e-13
12        .word 0x7030AD4A, 0x3CE952C7   @  2.811457254345521e-15
13        .word 0x46814157, 0xBC62F49B   @ -8.220635246624329e-18
14
15 @@@ -------------------------------------------------------------
16        .text
17        .align 2
18        @@ sin_a_d implements the sine function using IEEE
19        @@ double precision floating point.  It takes advantage
20        @@ of the ARM VFP vector processing instructions and
21        @@ computes sine by summing the first ten terms of the
```

```
22          @@ Taylor series.
23          .global sin_v_d
24   sin_v_d:
25          vmul.f64 d1,d0,d0        @ d1 <- x^2
26          vmrs    r1, fpscr       @ get FPSCR contents in r1
27          .if SET_RUNFAST
28          @@ set runfast mode and rounding to nearest
29          bic     r2, r1, #(0b1111<<23)
30          orr     r2, r2, #(0b1100<<23)
31          vmsr    fpscr, r2 @ store settings in FPSCR
32          .endif
33          @@ Set up vector of the initial powers of x in Bank 1
34          @@      vmul.f64 d4,d0,d1        @ d8 <- x^3
35          @@      vmul.f64 d5,d4,d1        @ d9 <- x^5
36          @@      vmul.f64 d6,d5,d1        @ d10 <- x^7
37          @@ (The second and third multiply each require the result
38          @@  from the previous multiply, so the instructions are
39          @@  spread out for better scheduling to get 5% better
40          @@  performance overall.)
41          vmul.f64 d4,d0,d1        @ d8 <- x^3
42          @@ load vector of coefficients into Bank 2
43          ldr     r0,=ctab        @ get address of coefficient table
44          vmul.f64 d5,d4,d1        @ d9 <- x^5
45          vldmia  r0!,{d8-d10}    @ load first three coefficients
46          @@ Make three copies of x^6 in Bank 3
47          vmul.f64 d12,d5,d0       @ d12 <- x^6
48          vmul.f64 d6,d5,d1        @ d10 <- x^7
49          vmov.f64 d13,d12         @ d13 <- x^6
50          vmov.f64 d14,d12         @ d14 <- x^6
51          @@ Set VFP for vector mode (stride = 1, vector length = 3)
52          .if SET_RUNFAST
53          bic     r2, r2, #(0b11111<<16)
54          .else
55          bic     r2, r1, #(0b11111<<16)
56          .endif
57          orr     r2, r2, #(0b00010<<16)
58          vmsr    fpscr, r2
59          @@ Multiply powers by coefficients. Put results in Bank 3
60          vmul.f64 d8,d8,d4        @ VECTOR operation
61          @@ Add terms in Bank 3 to the result in d0
62          vadd.f64 d3,d8,d9
63          vadd.f64 d0,d0,d10
64          mov     r3,#2           @ load loop counter
65          vadd.f64 d0,d0,d3
66   loop:  @@ load vector of next three coefficients into Bank 2
67          vldmia  r0!,{d8-d10}
```

```
68      @@ Set up vector of the required powers of x in Bank 1
69      vmul.f64 d4,d4,d12      @ VECTOR operation
70      @@ Multiply powers by coefficients Put results in Bank 2
71      vmul.f64 d8,d8,d4       @ VECTOR operation
72      @@ Add terms in Bank 2 to the result in d0
73      vadd.f64 d3,d8,d9
74      vadd.f64 d0,d0,d10
75      subs     r3,r3,#1       @ decrement and perform loop test
76      vadd.f64 d0,d0,d3       @ placed here for performance
77      bne      loop           @ perform loop twice
78      @@ restore original FPSCR
79      vmsr     fpscr, r1
80      mov      pc,lr
```

Listing 9.4
Vector implementation of the $\sin x$ function using IEEE double precision.

Listing 9.4 shows a double precision floating point implementation of the sine function, using the ARM VFPv3 instruction set in vector mode. It performs the same operations as the previous implementation, but performs the nine multiplications in three groups of three, using vector operations. Also, computing the powers of x is done within the loop, using a vector multiply. In this case, the vector code is significantly faster than the scalar version.

9.8.3 Performance Comparison

Table 9.2 shows the performance of various implementations of the sine function, with and without compiler optimization. The Single Precision C and Double Precision C implementations are the standard implementations provided by GCC.

When compiler optimization is not used, the single precision scalar VFP implementation achieves a speedup of about 2.96, and the vector implementation achieves a speedup of about 3.33 compared to the GCC implementation. The double precision scalar VFP implementation achieves a speedup of about 2.01, and the vector implementation achieves a speedup of about 2.46 compared to the GCC implementation.

When the best possible compiler optimization is used (`-Ofast`), the single precision scalar VFP implementation achieves a speedup of about 1.20, and the vector implementation achieves a speedup of about 1.26 compared to the GCC implementation. The double precision scalar VFP implementation achieves a speedup of about 2.19, and the vector implementation achieves a speedup of about 2.69 compared to the GCC implementation.

In most cases, the assembly versions were significantly faster than the functions provided by GCC. GCC with full optimization using single-precision numbers was competitive, but the assembly language vector implementation still beat it by over 25%. It is clear that writing some functions in assembly can result in large performance gains.

Table 9.2 Performance of sine function with various implementations

Optimization	Implementation	CPU seconds
None	Single Precision Scalar Assembly	2.96
	Single Precision Vector Assembly	2.63
	Single Precision C	8.75
	Double Precision Scalar Assembly	4.59
	Double Precision Vector Assembly	3.75
	Double Precision C	9.21
Full	Single Precision Scalar Assembly	2.16
	Single Precision Vector Assembly	2.06
	Single Precision C	2.59
	Double Precision Scalar Assembly	3.88
	Double Precision Vector Assembly	3.16
	Double Precision C	8.49

9.9 Alphabetized List of VFP Instructions

Name	Page	Operation
vabs	277	Absolute Value
vadd	278	Add
vcmp	279	Compare
vcmpe	279	Compare with Exception
vcpy	277	Copy VFP Register
vcvt	283	Convert Between Floating Point and Integer
vcvt	284	Convert To or From Fixed Point
vcvtr	283	Convert Floating Point to Integer with Rounding
vdiv	278	Divide
vldm	275	Load Multiple VFP Registers
vldr	274	Load VFP Register
vmov	280	Move Between VFP and One ARM Integer Register
vmov	281	Move Between VFP and Two ARM Integer Registers
vmov	279	Move Between VFP Registers
vmrs	282	Move From VFP System Register to ARM Register
vmsr	282	Move From ARM Register to VFP System Register
vmul	278	Multiply
vneg	277	Negate
vnmul	278	Negate and Multiply
vsqrt	277	Square Root
vstm	275	Store Multiple VFP Registers
vstr	274	Store VFP Register
vsub	278	Subtract

9.10 Chapter Summary

The ARM VFP coprocessor adds a great deal of power to the ARM architecture. The register set is expanded to hold up to four times the amount of data that can be held in the ARM integer registers. The additional instructions allow the programmer to deal directly with the most common IEEE 754 formats for floating point numbers. The ability to treat groups of registers as vectors adds a significant performance improvement. Access to the vector features is only possible through assembly language. The GCC compiler is not capable of using these advanced features, which gives the assembly programmer a big advantage when high-performance code is needed.

Exercises

9.1   How many registers does the VFP coprocessor add to the ARM architecture?

9.2   What is the purpose of the FZ, DN, and IDE, IXE, UFE, OFE, DZE, and IOE bits in the FPSCR? What is it called when FZ and DN are set to one and all of the others are set to zero?

9.3   If a VFP coprocessor is present, how are floating point parameters passed to subroutines? How is a *pointer* to a floating point value (or array of values) passed to a subroutine?

9.4   Write the following C code in ARM assembly:

```
1    for ( x = 0.0; x != 10.0; x += 0.1)
2        {
3            .
4            .
5            .
6        }
```

9.5   In the previous exercise, the C code contains a subtle bug.
   a.   What is the bug?
   b.   Show two ways to fix the code in ARM assembly. Hint: One way is to change the amount of the increment, which will change the number of times that the loop executes.

9.6   The fixed point sine function from the previous chapter was not compared directly to the hand-coded VFP implementation. Based on the information in Tables 9.2 and 8.4, would you expect the fixed point sine function from the previous chapter to beat the hand-coded assembly VFP sine function in this chapter? Why or why not?

9.7   3-D objects are often stored as an array of points, where each point is a vector (array) consisting of four values, x, y, z, and the constant 1.0. Rotation, translation, scaling and other operations are accomplished by multiplying each point by a 4 × 4 transformation matrix. The following C code shows the data types and the transform operation:

```
1   typedef float[4] point;     // Point is an array of floats
2   typedef float[4][4] matrix; // Matrix is a 2-D array of floats
3   .
4   .
5   .
6   void xform(matrix *m, point* p)
7     {
8       int i,j;
9       point result;
10      for(i=0;i<4;i++)
11      {
12        result[i] = 0.0;
13        for(j=0;j<4;j++)
14          result[i] += m[j][i] * p[j];
15      }
16      *p = result;
17    }
```

Write the equivalent ARM assembly code.

9.8   Optimize the ARM assembly code you wrote in the previous exercise. Use vector mode if possible.

9.9   Since the fourth element of the point is always 1.0, there is no need to actually store it. This will reduce memory requirements by about 25%, and require one fewer multiply. The C code would look something like this:

```
1   typedef float[3] point;     // Point is an array of floats
2   typedef float[4][4] matrix; // Matrix is a 2-D array of floats
3   .
4   .
5   .
6   void xform(matrix *m, point* p)
7     {
8       int i,j;
9       point result;
10      for(i=0;i<4;i++)
11      {
12        result[i] = m[3][i];
13        for(j=0;j<3;j++)
14          result[i] += m[j][i] * p[j];
15      }
16      *p = result;
17    }
```

Write optimal ARM VFP code to implement this function.

9.10 The function in the previous problem would typically be called multiple times to process an array of points, as in the following function:

```
void xformall(matrix *m, point* p, int num_points)
  {
    int i;
    for(i=0;i<num_points;i++)
      xform(m,p+i);
  }
```

This could be somewhat inefficient. Re-write this function in assembly so that the transformation of each point is done without resorting to a function call. Make your code as efficient as possible.

The ARM NEON Extensions

Chapter Outline

Modern Assembly Language Programming with the ARM Processor. http://dx.doi.org/10.1016/B978-0-12-803698-3.00010-3

The ARM VFP coprocessor has been replaced or augmented by the NEON architecture on ARMv7 and higher systems. NEON extends the VFP instruction set with about 125 instructions and pseudo-instructions to support not only floating point, but also integer and fixed point. NEON also supports Single Instruction, Multiple Data (SIMD) operations. All NEON processors have the full set of 32 double precision VFP registers, but NEON adds the ability to view the register set as 16 128-bit (quadruple-word) registers, named q0 through q15.

A single NEON instruction can operate on up to 128 bits, which may represent multiple integer, fixed point, or floating point numbers. For example, if two of the 128-bit registers each contain eight 16-bit integers, then a single NEON instruction can add all eight integers from one register to the corresponding integers in the other register, resulting in eight simultaneous additions. For certain applications, this SIMD architecture can result in extremely fast and efficient implementations. NEON is particularly useful at handling streaming video and audio, but also can give very good performance on floating point intensive tasks. NEON instructions perform parallel operations on vectors. NEON deprecates the use of VFP vector mode covered in Section 9.2.2. On most NEON systems, using the VFP vector mode will result in an exception, which transfers control to the support code which emulates vector mode in software. This causes a severe performance penalty, so VFP vector mode should not be used on NEON systems.

Fig. 10.1 shows the ARM integer, VFP, and NEON register set. NEON views each register as containing a vector of 1, 2, 4, 8, or 16 elements, all of the same size and type. Individual elements of each vector can also be accessed as scalars. A scalar can be 8 bits, 16 bits, 32 bits, or 64 bits. The instruction syntax is extended to refer to scalars using an index, x, in a doubleword register. Dm[x] is element x in register Dm. The size of the elements is given as part of the instruction. Instructions that access scalars can access any element in the register bank.

10.1 NEON Intrinsics

The GCC compiler gives C (and C++) programs direct access to the NEON instructions through the NEON intrinsics. The intrinsics are a large set of functions that are built into the compiler. Most of the intrinsics functions map to one NEON instruction. There are additional functions provided for typecasting (reinterpreting) NEON vectors, so that the C compiler does not complain about mismatched types. It is usually shorter and more efficient to write the NEON code directly as assembly language functions and link them to the C code. However only those who know assembly language are capable of doing that.

10.2 Instruction Syntax

Some instructions require specific register types. Other instructions allow the programmer to choose single word, double word, or quad word registers. If the instruction requires single precision registers, then the registers are specified as Sd for the destination register, Sn for the first operand register, and Sm for the second operand register. If the instruction requires only two registers, then Sn is not used. The lower-case letter is replaced with a valid register number. The register name is not case sensitive, so S10 and s10 are both valid names for single precision register 10.

The syntax of the NEON instructions can be described using a relatively simple notation. The notation consists of the following elements:

{item}	Braces around an item indicate that the item is optional. For example, many operations have an optional condition, which is written as {<cond>}.
Ry	An ARM integer register. y can be any number in the range 0–15.
Sy	A 32-bit or single precision register. y can be any number in the range 0–31.
Dy	A 64-bit or double precision register. y can be any number in the range 0–31.

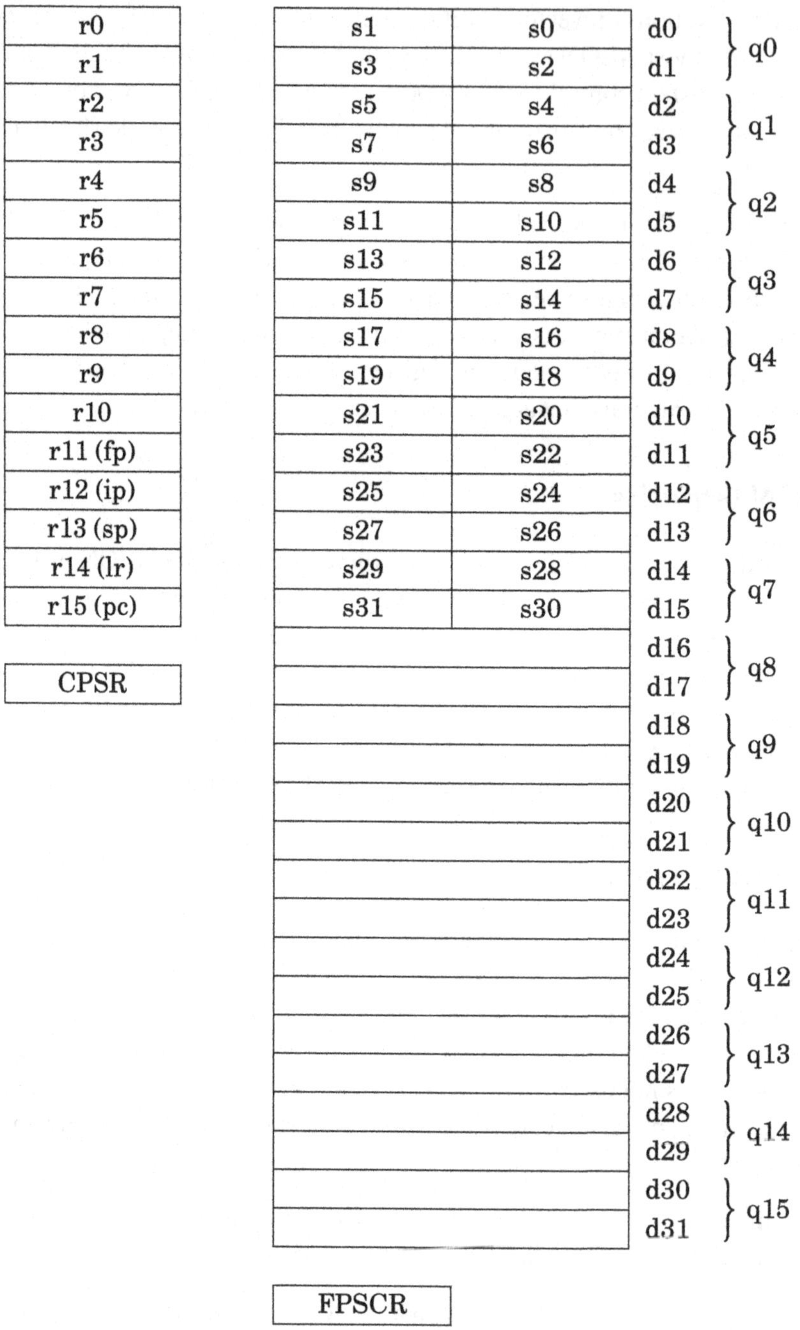

r0		s1	s0	d0 } q0
r1		s3	s2	d1
r2		s5	s4	d2 } q1
r3		s7	s6	d3
r4		s9	s8	d4 } q2
r5		s11	s10	d5
r6		s13	s12	d6 } q3
r7		s15	s14	d7
r8		s17	s16	d8 } q4
r9		s19	s18	d9
r10		s21	s20	d10 } q5
r11 (fp)		s23	s22	d11
r12 (ip)		s25	s24	d12 } q6
r13 (sp)		s27	s26	d13
r14 (lr)		s29	s28	d14 } q7
r15 (pc)		s31	s30	d15
				d16 } q8
				d17
CPSR				d18 } q9
				d19
				d20 } q10
				d21
				d22 } q11
				d23
				d24 } q12
				d25
				d26 } q13
				d27
				d28 } q14
				d29
				d30 } q15
				d31

FPSCR

Figure 10.1
ARM integer and NEON user program registers.

Qy	A quad word register. y can be any number in the range 0–15.
Fy	A VFP register. F must be either s for a single word register, or d for a double word register. y can be any valid register number.
Ny	A NEON or VFP register. N must be either s for a single word register, d for a double word register, or q for a quad word register. y can be any valid register number.
Vy	A NEON vector register. V must be replaced with d for a double word register, or q for a quad word register. y can be any valid register number.
Vy[x]	A NEON scalar (vector element). The size of the scalar is defined as part of the instruction. V must be replaced with d for a double word register, or q for a quad word register. y can be any valid register number. x specifies which scalar element of Vy is to be used. Valid values for x can be deduced by the size of Vy and the size of the scalars that the instruction uses.
<op>	Operation specific part of a general instruction format
<n>	An integer usually indicating a specific instruction version
<size>	An integer indicating the number of bits used
<cond>	ARM condition code from Table 3.2
<type>	Many instructions operate on one or more of the following specific data types:

 i8 Untyped 8 bits
 i16 Untyped 16 bits
 i32 Untyped 32 bits
 i64 Untyped 64 bits
 s8 Signed 8-bit integer
 s16 Signed 16-bit integer
 s32 Signed 32-bit integer
 s64 Signed 64-bit integer
 u8 Unsigned 8-bit integer
 u16 Unsigned 16-bit integer
 u32 Unsigned 32-bit integer
 u64 Unsigned 64-bit integer
 f16 IEEE 754 half precision floating point
 f32 IEEE 754 single precision floating point
 f64 IEEE 754 double precision floating point

<list>	A brace-delimited list of up to four NEON registers, vectors, or scalars. The general form is {Dn,D(n+a),D(n+2a),D(n+3a)} where a is either 1 or 2.
<align>	Specifies the memory alignment of structured data for certain load and store operations.
<imm>	An immediate value. The required format for immediate values depends on the instruction.
<fbits>	Specifies the number of fraction bits in fixed point numbers.

The following function definitions are used in describing the effects of many of the instructions:

$\lfloor x \rfloor$ The *floor* function maps a real number, x, to the next smallest integer.

$\lceil x \rceil$ The *saturate* function limits the value of x to the highest or lowest value that can be stored in the destination register.

$\|x\|$ The *round* function maps a real number, x, to the nearest integer.

$\succ x \prec$ The *narrow* function reduces a $2n$ bit number to an n bit number, by taking the n least significant bits.

$\prec x \succ$ The *extend* function converts an n bit number to a $2n$ bit number, performing zero extension if the number is unsigned, or sign extension if the number is signed.

10.3 Load and Store Instructions

These instructions can be used to perform interleaving of data when structured data is loaded or stored. The data should be properly aligned for best performance. These instructions are very useful for common multimedia data types.

For example, image data is typically stored in arrays of pixels, where each pixel is a small data structure such as the `pixel` struct shown in Listing 5.37. Since each pixel is three bytes, and a d register is 8 bytes, loading a single pixel into one register would be inefficient. It would be much better to load multiple pixels at once, but an even number of pixels will not fit in a register. It will take three doubleword or quadword registers to hold an even number of pixels without wasting space, as shown in Fig. 10.2. This is the way data would be loaded using a VFP `vldr` or `vldm` instruction. Many image processing operations work best if each color "channel" is processed separately. The NEON load and store vector instructions can be used to split the image data into color channels, where each channel is stored in a different register, as shown in Fig. 10.3.

Other examples of interleaved data include stereo audio, which is two interleaved channels, and surround sound, which may have up to nine interleaved channels. In all of these cases, most processing operations are simplified when the data is separated into non-interleaved channels.

$green_2$	red_2	$blue_1$	$green_1$	red_1	$blue_0$	$green_0$	red_0	d0
red_5	$blue_4$	$green_4$	red_4	$blue_3$	$green_3$	red_3	$blue_2$	d1
$blue_7$	$green_7$	red_7	$blue_6$	$green_6$	red_6	$blue_5$	$green_5$	d2

Figure 10.2
Pixel data interleaved in three doubleword registers.

red_7	red_6	red_5	red_4	red_3	red_2	red_1	red_0	d0
$green_7$	$green_6$	$green_5$	$green_4$	$green_3$	$green_2$	$green_1$	$green_0$	d1
$blue_7$	$blue_6$	$blue_5$	$blue_4$	$blue_3$	$blue_2$	$blue_1$	$blue_0$	d2

Figure 10.3
Pixel data de-interleaved in three doubleword registers.

10.3.1 Load or Store Single Structure Using One Lane

These instructions are used to load and store structured data across multiple registers:

vld<n>   Load Structured Data, and
vst<n>   Store Structured Data.

They can be used for interleaving or deinterleaving the data as it is loaded or stored, as shown in Fig. 10.3.

Syntax

```
v<op><n>.<size> <list>,[Rn{:<align>}]{!}
v<op><n>.<size> <list>,[Rn{:<align>}],Rm
```

- <op> must be either ld or st.
- <n> must be one of 1, 2, 3, or 4.
- <size> must be one of 8, 16, or 32.
- <list> specifies the list of registers. There are four list formats:
  1. {Dd[x]}
  2. {Dd[x], D(d+a)[x]}
  3. {Dd[x], D(d+a)[x], D(d+2a)[x]}
  4. {Dd[x], D(d+a)[x], D(d+2a)[x], D(d+3a)[x]}
  where a can be either 1 or 2. Every register in the list must be in the range d0-d31.
- Rn is the ARM register containing the base address. Rn cannot be pc.
- <align> specifies an optional alignment. If <align> is not specified, then standard alignment rules apply.
- The optional ! indicates that Rn is updated after the data is transferred. This is similar to the ldm and stm instructions.
- Rm is an ARM register containing an offset from the base address. If Rm is present, Rn is updated to Rn + Rm after the address is used to access memory. Rm cannot be sp or pc.

Table 10.1 shows all valid combinations of parameters for these instructions. Note that the same vector element (scalar) x must be used in each register. Up to four registers can be specified. If the structure has more than four fields, then these instructions can be used repeatedly to load or store all of the fields.

Table 10.1 Parameter combinations for loading and storing a single structure

\<n\>	\<size\>	\<list\>	\<align\>	Alignment
1	8	Dd[x]		Standard only
	16	Dd[x]	16	2 byte
	32	Dd[x]	32	4 byte
2	8	Dd[x], D(d+1)[x]	16	2 byte
	16	Dd[x], D(d+1)[x]	32	4 byte
		Dd[x], D(d+2)[x]	32	4 byte
	32	Dd[x], D(d+1)[x]	64	8 byte
		Dd[x], D(d+2)[x]	64	8 byte
3	8	Dd[x], D(d+1)[x], D(d+2)[x]		Standard only
	16 or 32	Dd[x], D(d+1)[x], D(d+2)[x]		Standard only
		Dd[x], D(d+2)[x], D(d+4)[x]		Standard only
4	8	Dd[x], D(d+1)[x], D(d+2)[x], D(d+3)[x]	32	4 byte
	16	Dd[x], D(d+1)[x], D(d+2)[x], D(d+3)[x]	64	8 byte
		Dd[x], D(d+2)[x], D(d+4)[x], D(d+6)[x]	64	8 byte
	32	Dd[x], D(d+1)[x], D(d+2)[x], D(d+3)[x]	64 or 128	(\<align\>÷8) bytes
		Dd[x], D(d+2)[x], D(d+4)[x], D(d+6)[x]	64 or 128	(\<align\>÷8) bytes

Operations

Name	Effect	Description

vld\<n\>	$tmp \leftarrow Rn$
	$incr \leftarrow (\text{\<size\>}÷8)$
	for $D \in regs(\text{\<list\>})$ **do**
	　$D[x] \leftarrow Mem[tmp]$
	　$tmp \leftarrow tmp + incr$
	end for
	if ! is present **then**
	　$Rn \leftarrow tmp$
	else
	　if Rm is specified **then**
	　　$Rn \leftarrow Rm$
	　end if
	end if
		Load one or more data items into a single lane of one or more registers

vst\<n\>	$tmp \leftarrow Rn$
	$incr \leftarrow (\text{\<size\>}÷8)$
	for $D \in regs(\text{\<list\>})$ **do**
	　$Mem[tmp] \leftarrow D[x]$
	　$tmp \leftarrow tmp + incr$
	end for
	if ! is present **then**
	　$Rn \leftarrow tmp$
	else
	　if Rm is specified **then**
	　　$Rn \leftarrow Rm$
	　end if
	end if
		Store one or more data items from a single lane of one or more registers

Examples

```
1    vld3.8    {d0[0],d1[0],d2[0]},[r0]! @ load first pixel
2    vld3.8    {d0[1],d1[1],d2[1]},[r0]!
3    vld3.8    {d0[2],d1[2],d2[2]},[r0]!
4    vld3.8    {d0[3],d1[3],d2[3]},[r0]!
5    vld3.8    {d0[4],d1[4],d2[4]},[r0]!
6    vld3.8    {d0[5],d1[5],d2[5]},[r0]!
7    vld3.8    {d0[6],d1[6],d2[6]},[r0]!
8    vld3.8    {d0[7],d1[7],d2[7]},[r0]! @ load eighth pixel
```

10.3.2 Load Copies of a Structure to All Lanes

This instruction is used to load multiple copies of structured data across multiple registers:

vld<n>   Load Copies of Structured Data.

The data is copied to all lanes. This instruction is useful for initializing vectors for use in later instructions.

Syntax

```
vld<n>.<size> <list>,[Rn{:<align>}]]{!}
vld<n>.<size> <list>,[Rn{:<align>}],Rm
```

- <n> must be one of 1, 2, 3, or 4.
- <size> must be one of 8, 16, or 32.
- <list> specifies the list of registers. There are four list formats:
    1. {Dd[]}
    2. {Dd[], D(d+a)[]}
    3. {Dd[], D(d+a)[], D(d+2a)[]}
    4. {Dd[], D(d+a)[], D(d+2a)[], D(d+3a)[]}
    where a can be either 1 or 2. Every register in the list must be in the range d0-d31.
- Rn is the ARM register containing the base address. Rn cannot be pc.
- <align> specifies an optional alignment. If <align> is not specified, then standard alignment rules apply.
- The optional ! indicates that Rn is updated after the data is transferred. This is similar to the ldm and stm instructions.
- Rm is an ARM register containing an offset from the base address. If Rm is present, Rn is updated to Rn + Rm after the address is used to access memory. Rm cannot be sp or pc.

Table 10.2 shows all valid combinations of parameters for this instruction. Note that the vector element number is not specified, but the brackets [] must be present. Up to four registers can be specified. If the structure has more than four fields, then this instruction can be repeated to load or store all of the fields.

Table 10.2 Parameter combinations for loading multiple structures

`<n>`	`<size>`	`<list>`	`<align>`	Alignment
1	8	`Dd[]`		Standard only
		`Dd[], D(d+1)[]`		Standard only
	16	`Dd[]`	16	2 byte
		`Dd[], D(d+1)[]`	16	2 byte
	32	`Dd[]`	32	4 byte
		`Dd[], D(d+1)[]`	32	4 byte
2	8	`Dd[], D(d+1)[]`	8	1 byte
	8	`Dd[], D(d+2)[]`	8	1 byte
	16	`Dd[], D(d+1)[]`	16	2 byte
		`Dd[], D(d+2)[]`	16	2 byte
	32	`Dd[], D(d+1)[]`	32	4 byte
		`Dd[], D(d+2)[]`	32	4 byte
3	8, 16, or 32	`Dd[], D(d+1)[], D(d+2)[]`		Standard only
		`Dd[], D(d+2)[], D(d+4)[]`		Standard only
4	8	`Dd[], D(d+1)[], D(d+2)[], D(d+3)[]`	32	4 byte
		`Dd[], D(d+2)[], D(d+4)[], D(d+6)[]`	32	4 byte
	16	`Dd[], D(d+1)[], D(d+2)[], D(d+3)[]`	64	8 byte
		`Dd[], D(d+2)[], D(d+4)[], D(d+6)[]`	64	8 byte
	32	`Dd[], D(d+1)[], D(d+2)[], D(d+3)[]`	64 or 128	(`<align>`÷8) bytes
		`Dd[], D(d+2)[], D(d+4)[], D(d+6)[]`	64 or 128	(`<align>`÷8) bytes

Operations

Name	Effect	Description

`vld<n>`	$tmp \leftarrow Rn$
	$incr \leftarrow (\texttt{<size>} \div 8)$
	$nlanes \leftarrow (64 \div \texttt{<size>})$
	for $D \in regs(\texttt{<list>})$ **do**
	  for $0 \le x < nlanes$ **do**
	  $D[x] \leftarrow Mem[tmp]$
	  $tmp \leftarrow tmp + incr$
	  end for
	end for
	if ! is present **then**
	  $Rn \leftarrow tmp$
	else
	  if `Rm` is specified **then**
	    $Rn \leftarrow Rm$
	  end if
	end if
		Load one or more data items into all lanes of one or more registers.

Examples

```
1   @ Load multiple copies of an rgb struct into
2   @ d0(red),d1(green),and d2(blue)
3   vld3.8    {d0[],d1[],d2[]},[r0]! @ load first pixel
```

10.3.3 Load or Store Multiple Structures

These instructions are used to load and store multiple data structures across multiple registers with interleaving or deinterleaving:

vld<n>   Load Multiple Structured Data, and

vst<n>   Store Multiple Structured Data.

Syntax

```
v<op><n>.<size> <list>,[Rn{:<align>}]{!}
v<op><n>.<size> <list>,[Rn{:<align>}],Rm
```

- <op> must be either ld or st.
- <n> must be one of 1, 2, 3, or 4.
- <size> must be one of 8, 16, or 32.
- <list> specifies the list of registers. There are four list formats:
  1. {Dd}
  2. {Dd, D(d+a)}
  3. {Dd, D(d+a), D(d+2a)}
  4. {Dd, D(d+a), D(d+2a), D(d+3a)}

  where a can be either 1 or 2. Every register in the list must be in the range d0-d31.
- Rn is the ARM register containing the base address. Rn cannot be pc.
- <align> specifies an optional alignment. If <align> is not specified, then standard alignment rules apply.
- The options ! indicates that Rn is updated after the data is transferred, similar to the ldm and stm instructions.
- Rm is an ARM register containing an offset from the base address. If Rm is present, Rn is updated to Rn + Rm after the address is used to access memory. Rm cannot be sp or pc.

Table 10.3 shows all valid combinations of parameters for this instruction. Note that the scalar is not specified and the instructions work on all multiple vector elements. Up to four registers can be specified. If the structure has more than four fields, then this instruction can be repeated to load or store all of the fields.

Table 10.3 Parameter combinations for loading copies of a structure

\<n\>	\<size\>	\<list\>	\<align\>	Alignment
1	8, 16, 32, or 64	Dd	64	8 bytes
		Dd, D(d+1)	64 or 128	(\<align\>÷8) bytes
		Dd, D(d+1), D(d+2)	64	8 bytes
		Dd, D(d+1), D(d+2), D(d+3)	64, 128, or 256	(\<align\>÷8) bytes
2	8, 16, or 32	Dd, D(d+1)	64 or 128	(\<align\>÷8) bytes
		Dd, D(d+2)	64 or 128	(\<align\>÷8) bytes
		Dd, D(d+1), D(d+2), D(d+3)	64, 128, or 256	(\<align\>÷8) bytes
3	8, 16, or 32	Dd, D(d+1), D(d+2)	64	8 bytes
		Dd, D(d+2), D(d+3)	64	8 bytes
4	8, 16, or 32	Dd, D(d+1), D(d+2), D(d+3)	64, 128, or 256	(\<align\>÷8) bytes
		Dd, D(d+2), D(d+4), D(d+6)	64, 128, or 256	(\<align\>÷8) bytes

Operations

Name	Effect	Description

vld\<n\>	$tmp \leftarrow Rn$
	$incr \leftarrow (\text{\<size\>}÷8)$
	$nlanes \leftarrow (64÷\text{\<size\>})$
	for $0 \leq x < nlanes$ **do**
	　for $D \in \text{\<list\>}$ **do**
	　　$D[x] \leftarrow Mem[tmp]$
	　　$tmp \leftarrow tmp + incr$
	　end for
	end for
	if ! is present **then**
	　$Rn \leftarrow tmp$
	else
	　if Rm is specified **then**
	　　$Rn \leftarrow Rm$
	　end if
	end if
		Load one or more data items into all lanes of one or more registers.

vst\<n\>	$tmp \leftarrow Rn$
	$incr \leftarrow (\text{\<size\>}÷8)$
	$nlanes \leftarrow (64÷\text{\<size\>})$
	for $0 \leq x < nlanes$ **do**
	　for $D \in \text{\<list\>}$ **do**
	　　$Mem[tmp] \leftarrow D[x]$
	　　$tmp \leftarrow tmp + incr$
	　end for
	end for
	if ! is present **then**
	　$Rn \leftarrow tmp$
	else
	　if Rm is specified **then**
	　　$Rn \leftarrow Rm$
	　end if
	end if
		Load one or more data items into all lanes of one or more registers.

Examples

```
1    @ Load multiple copies of an rgb struct into
2    @ d0(red),d1(green),and d2(blue)
3    vld3.8    {d0,d1,d2},[r0]! @ load 8 pixels, deinterlaced
```

10.4 Data Movement Instructions

Because they use the same set of registers, VFP and NEON share some instructions for loading, storing, and moving registers. The shared instructions are vldr, vstr, vldm, vstm, vpop, vpush, vmov, vmrs, and vmsr. These were explained in Chapter 9. NEON extends the vmov instructions to allow specification of NEON scalars and quadwords, and adds the ability to perform one's complement during a move.

10.4.1 Moving Between NEON Scalar and Integer Register

This version of the move instruction allows data to be moved between the NEON registers and the ARM integer registers as 8-bit, 16-bit, or 32-bit NEON scalars:

vmov   Move Between NEON and ARM.

Syntax

```
vmov{<cond>}.<size>    Dn[x],Rd
vmov{<cond>}.<type>    Rd,Dn[x]
```

- <cond> is an optional condition code.
- <size> must be 8, 16, or 32, and specifies the number of bits that are to be moved.
- The <type> must be u8, u16, u32, s8, s16, s32, or f32, and specifies the number of bits that are to be moved and whether or not the result should be sign-extended in the ARM integer destination register.

Operations

Name	Effect	Description
vmov Dd[x],Rm	$Dn[x] \leftarrow Rd$	Move least significant size bits of Rd to NEON scalar Dn[x].
vmov Rd,Dn[x]	$Rd \leftarrow Dn[x]$	Move NEON scalar Dn[x] to Rd, storing as specified type

Examples

```
 1    @ 32 bit moves ( x can be 0 or 1 )
 2    vmov.32  d0[0],r6     @ d0[0] <- r6
 3    vmov.f32 r7,d1[1]     @ r7 <- d1[1]
 4    vmov.u32 r8,d2[0]     @ r8 <- d2[0]
 5    vmoveq.s32 r9,d2[1]   @ if eq, r9 <- d2[1]
 6
 7    @ 16 bit moves ( x can be 0, 1, 2, or 3)
 8    vmov.16  d0[1],r8     @ d0[1] <- r8
 9    @ (least significant 16 bits)
10    vmov.s16 r7,d1[2]     @ r7 <- d1[2] (sign extend)
11
12    @ 8 bit moves ( x can be 0, 1, 2, 3, 4, 5, 6, or 7)
13    vmov.8   d0[5],r6     @ d0[5] <- r6
14    @ (least significant 8 bits)
15    vmov.u8 r5,d1[5]      @ r5 <- d1[5] (no sign extend)
```

10.4.2 Move Immediate Data

NEON extends the VFP `vmov` instruction to include the ability to move an immediate value, or the one's complement of an immediate value, to every element of a register. The instructions are:

`vmov`   Move Immediate, and

`vmvn`   Move Immediate NOT.

Syntax

```
v<op>.<type>  Vd, #<imm>
```

- `<op>` must be either `<mov>` or `<mvn>` .
- `<type>` must be `i8`, `i16`, `i32`, `f32`, or `i64`, and specifies the size of items in the vector.
- `V` can be `s`, `d`, or `q`.
- `<imm>` is an immediate value that matches `<type>`, and is copied to every element in the vector. The following table shows valid formats for `imm`:

<type>	vmov	vmvn
i8	0xXY	0xXY
i16	0x00XY	0xFFXY
	0xXY00	0xXYFF
i32	0x000000XY	0xFFFFFFXY
	0x0000XY00	0xFFFFXYFF
	0x00XY0000	0xFFXYFFFF
	0xXY000000	0xXYFFFFFF
i64	0xABCDEFGH	0xABCDEFGH
	Each letter represents a byte, and must be either FF or 00	
f32	Any number that can be written as $\pm n \times (2 - r)$, where n and r are integers, such that $16 \leq n \leq 31$ and $0 \leq r \leq 7$	

Operations

Name	Effect	Description
vmov	Vd[] ← *immed*	Copy immediate value to all elements of Vd.
vmvn	Vd[] ← ¬*immed*	Copy one's complement of immediate value to all elements of Vd.

Examples

```
1   vmov.i8   d3,#0x0A    @ d3[7..0] <— 10
2   vmvn.i16  q0,#0xFFF5  @ q0[7..0] <— 10
```

10.4.3 Change Size of Elements in a Vector

It is sometimes useful to increase or decrease the number of bits per element in a vector. NEON provides these instructions to convert a doubleword vector with elements of size y to a quadword vector with size $2y$, or to perform the inverse operation:

vmovl   Move and Lengthen,
vmovn   Move and Narrow,
vqmovn   Saturating Move and Narrow, and
vqmovun   Saturating Move and Narrow Unsigned.

Syntax

```
vmovl.<type>      Qd, Dm
v{q}movn.<type>   Dd, Qm
vqmovun.<type>    Dd, Qm
```

- The valid choices for `<type>` are given in the following table:

Opcode	Valid Types
vmovl	s8, s16, s32, u8, u16, or u32
vmovn	i8, i16, or i32
vqmovn	s8, s16, s32, u8, u16, or u32
vqmovun	s8, s16, or s32

- q indicates that the results are saturated.

Operations

Name	Effect	Description

vmovl	**for** $0 \leq i < (64 \div size)$ **do**
	    $Qd[i] \leftarrow \prec Dm[i] \succ$
	end for
		Sign or zero extends (depending on `<type>`) each element of a doubleword vector to twice their length

v{q}movn	**for** $0 \leq i < (64 \div size)$ **do**
	    if q is present **then**
	        $Dd[i] \leftrightarrow \lceil Qm[i] \rceil \prec$
	    else
	        $Dd[i] \leftrightarrow \lfloor Qm[i]) \rfloor \prec$
	    end if
	end for
		Copy the least significant half of each element of a quadword vector to the corresponding elements of a doubleword vector. If q is present, then the value is saturated

vqmovun	**for** $0 \leq i < (64 \div size)$ **do**
	    $Dd[i] \leftrightarrow \lceil Qm[i] \rceil \prec$
	end for
		Copy each element of the operand vector to the corresponding element of the destination vector. The destination element is unsigned, and the value is saturated

Examples

```
1    vmovl.s16   q3,d2    @ Convert vector elements
2                         @ from 16 to 32 bits
3    vqmovn.u16  d0,q4    @ Convert vector elements
4                         @ from 16 to 8 bits
```

10.4.4 Duplicate Scalar

The duplicate instruction copies a scalar into every element of the destination vector. The scalar can be in a NEON register or an ARM integer register. The instruction is:

vdup   Duplicate Scalar.

Syntax

```
vdup.<size>    Vd, Rm
vdup.<size>    Vd, Dm[x]
```

- `<size>` must be one of 8, 16 or 32.
- V can be d or q.
- Rm cannot be r15.

Operations

Name	Effect	Description
vdup.<size>	Vd[] < −Rm	Copy <size> least significant bits of Rm to all elements of Vd
vdup.<size>	Vd[] < −Dm[x]	Copy element x of Dm to all elements of Vd

Examples

```
1    vdup.8     d0,r1     @ copy 8 bits from r1 to
2                         @ 8 8-bit elements of d0
3    vdup.32    q3,d2[1]  @ copy top 32 bits to from d2
4                         @ to four 32-bit elements of q3
```

10.4.5 Extract Elements

This instruction extracts 8-bit elements from two vectors and concatenates them. Fig. 10.4 gives an example of what this instruction does. The instruction is:

vext   Extract Elements.

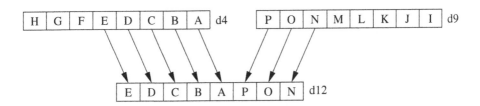

Figure 10.4
Example of vext.8 d12,d4,d9,#5.

Syntax

```
vext.<size>    Vd, Vn, Vm, #<imm>
```

- `<size>` must be one of 8, 16, 32, or 64.
- `V` can be d or q.
- `<imm>` is the number of elements to extract from the bottom of Vm. The remaining elements required to fill Vd are taken from the top of Vn.

Operation

Name	Effect	Description
vext	**if** V is double **then** $\quad size \leftarrow 8$ **else** $\quad size \leftarrow 16$ **end if** **for** $imm > i \geq 0$ **do** $\quad Vd[i + size - imm] \leftarrow Vm[i]$ **end for** **for** $size > i \geq imm$ **do** $\quad Vd[i - imm] \leftarrow Vm[i]$ **end for**	Concatenate the top of first operand to the bottom of the second operand.

Examples

```
1   vext.8    d0,d0,d1,#3    @ d0[7..0] <- d1[2..0]:d0[7..3]
2   vext.16   d5,d6,d3,#2    @ d5[3..0] <- d6[1..0]:d3[3..2]
```

10.4.6 Reverse Elements

This instruction reverses the order of data in a register:

vrev   Reverse Elements.

One use of this instruction is for converting data from big-endian to little-endian order, or from little-endian to big-endian order. It could also be useful for swapping data and transforming matrices. Fig. 10.5 shows three examples.

Syntax

```
vrev<n>.<size>    Vd, Vm
```

- `<n>` can be 16, 32, or 64.
- `<size>` is either 8, 16, or 32 and indicates the size of the elements to be reversed. `<size>` must be less than `<n>`.
- `V` can be q or d.

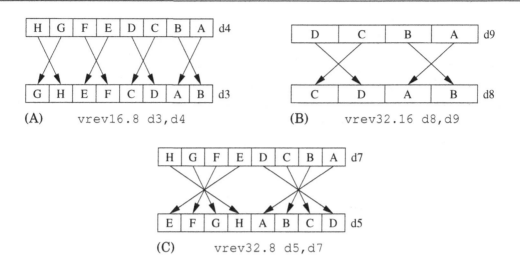

Figure 10.5

Examples of the `vrev` instruction. (A) `vrev16.8 d3,d4`; (B) `vrev32.16 d8,d9`; (C) `vrev32.8 d5,d7`.

Operation

Name	Effect	Description

vrev	$n \leftarrow$ # of groups
	$g \leftarrow$ size of group
	for $0 \leq i < n$ **do**
	  for $0 \leq j < g$ **do**
	    $Vd[i \times g + j] \leftarrow Vm[i \times g + (g - j - 1)]$
	  end for
	end for
		Reverse the order of elements of `<size>` bits within every element of `<n>` bits.

Examples

```
1    vrev64.32    d3,d4    @ s6:s7 <— s9:s8
2    vrev64.8     d5,d6    @ reverse bytes in d6
```

10.4.7 Swap Vectors

This instruction simply swaps two NEON registers:

`vswp`   Swap Vectors.

Syntax

```
vswp{.<type>}    Vd, Vm
```

- `<type>` can be any NEON data type. The assembler ignores the type, but it can be useful to the programmer as extra documentation.
- V can be q or d.

Operation

Name	Effect	Description
vswp	Vd ← Vm; Vm ← Vd	Swap registers

Examples

```
1   vswp.i64    d3,d4        @ swap d3 and d4
2   vswp        q3,q4        @ swap q3 and q4
```

10.4.8 Transpose Matrix

This instruction transposes 2 × 2 matrices:

vtrn   Transpose Matrix.

Fig. 10.6 shows two examples of this instruction. Larger matrices can be transposed using a divide-and-conquer approach.

Syntax

```
vtrn.<size>    Vd, Vm
```

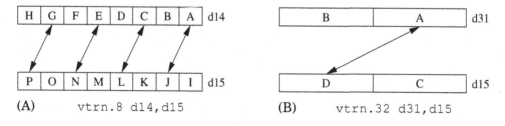

Figure 10.6
Examples of the vtrn instruction. (A) vtrn.8 d14,d15; (B) vtrn.32 d31,d15.

- `<size>` is either 8, 16, or 32 and indicates the size of the elements in the matrix (or matrices).
- V can be q or d.

Operation

Name	Effect	Description

vtrn	$n \leftarrow$ # of elements
	for $0 \leq i < n$ by 2 **do**
	  $tmp \leftarrow Vm[i]$
	  $Vm[i] \leftarrow Vd[i+1]$
	  $Vd[i+1] \leftarrow tmp$
	end for
		Treat two vectors as an array of 2×2 matrices and transpose them.

Examples

```
1    vtrn.32   d3,d4          @ Transpose a 2x2 matrix
2                             @  of 32-bit elements
3    vtrn.16   q8,q10         @ Transpose four 2x2 matrices
4                             @  of 16-bit elements
```

Fig. 10.7 shows how the vtrn instruction can be used to transpose a 3×3 matrix. Transposing a 4×4 matrix requires the transposition of 13 2×2 matrices. However, this instruction can operate on multiple 2×2 sub-matrices in parallel, and can group elements into different sized sub-matrices. There is also a very useful swap instruction that can exchange the rows of a matrix. Using the swap and transpose instructions, transposing a 4×4 matrix of 16-bit elements can be done with only four instructions, as shown in Fig. 10.8.

10.4.9 Table Lookup

The table lookup instructions use indices held in one vector to lookup values from a table held in one or more other vectors. The resulting values are stored in the destination vector. The table lookup instructions are:

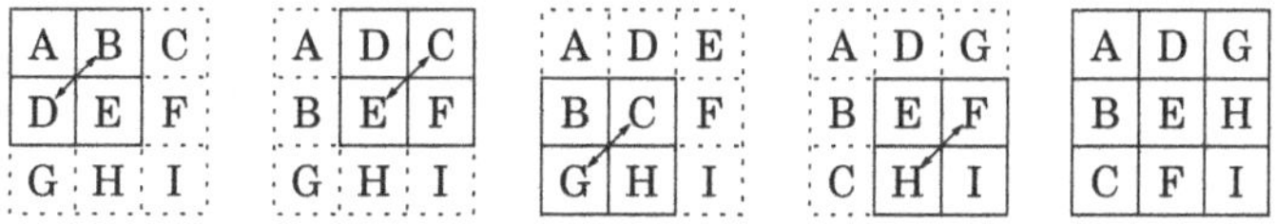

Figure 10.7
Transpose of a 3×3 matrix.

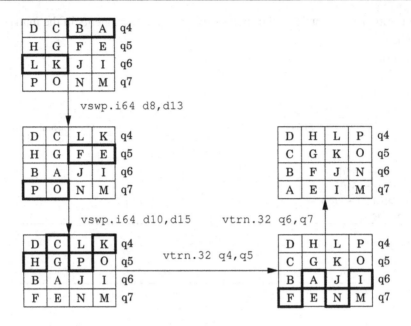

Figure 10.8
Transpose of a 4 × 4 matrix of 32-bit numbers.

vtbl Table Lookup, and
vtbx Table Lookup with Extend.

Syntax

```
v<op>.8    Dd, <list>, Dm
```

- <op> is one of tbl or tbx
- <list> specifies the list of registers. There are five list formats:
  1. {Dn},
  2. {Dn, D(n+1)},
  3. {Dn, D(n+1), D(n+2)},
  4. {Dn, D(n+1), D(n+2), D(n+3)}, or
  5. {Qn, Q(n+1)}.
- Dm is the register holding the indices.
- The table can contain up to 32 bytes.

Operations

Name	Effect	Description

vtbl	$Minr \leftarrow$ first register
	$Maxr \leftarrow$ last register
	for $0 \leq i < 8$ **do**
	  $r \leftarrow Minr + (Dm[i] \div 8)$
	  if $r > Maxr$ **then**
	    $Dd[i] \leftarrow 0$
	  else
	    $e \leftarrow Dm[i] \mod 8$
	    $Dd[i] \leftarrow Dr[e]$
	  end if
	end for
		Use indices Dm to look up values in a table and store them in Dd. If the index is out of range, zero is stored in the corresponding destination.

vtbx	$Minr \leftarrow$ first register
	$Maxr \leftarrow$ last register
	for $0 \leq i < 8$ **do**
	  $r \leftarrow Minr + (Dm[i] \div 8)$
	  if $r \leq Maxr$ **then**
	    $e \leftarrow Dm[i] \mod 8$
	    $Dd[i] \leftarrow Dr[e]$
	  end if
	end for
		Use indices Dm to look up values in a table and store them in Dd. If the index is out of range, the corresponding destination is unchanged.

Examples

```
1    vtbl.8      d0,{d4,d5,d6,d7},d1 @ do table lookup
2    vtbx.8      d4,{d8,d9},d1       @ do table lookup
```

10.4.10 Zip or Unzip Vectors

These instructions are used to interleave or deinterleave the data from two vectors:

vzip   Zip Vectors, and
vuzp   Unzip Vectors.

Fig. 10.9 gives an example of the vzip instruction. The vuzp instruction performs the inverse operation.

Syntax

```
v<op>.<size>    Vd, Vm
```

- <op> is either zip or uzp.
- <size> is either 8, 16, or 32 and indicates the size of the elements in the matrix (or matrices).
- V can be q or d.

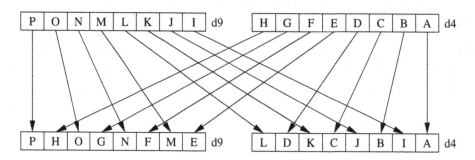

Figure 10.9
Example of vzip.8 d9,d4.

Operations

Name	Effect	Description

vzip	$n \leftarrow$ # of elements
	for $0 \leq i < (n \div 2)$ by 2 **do**
	$\quad tmp1[2 \times i] \leftarrow Vm[i]$
	$\quad tmp1[2 \times i + 1] \leftarrow Vd[i]$
	end for
	for $(n \div 2) \leq i < n$ by 2 **do**
	$\quad tmp2[2 \times i] \leftarrow Vm[i]$
	$\quad tmp2[2 \times i + 1] \leftarrow Vd[i]$
	end for
	$Vm \leftarrow tmp1$
	$Vd \leftarrow tmp2$
		Interleave data from two vectors. *tmp* is a vector of suitable size.

vuzp	$n \leftarrow$ # of elements
	for $0 \leq i < (n \div 2)$ by 2 **do**
	$\quad tmp1[i] \leftarrow Vm[2 \times i]$
	$\quad tmp2[i] \leftarrow Vm[2 \times i + 1]$
	end for
	for $(n \div 2) \leq i < n$ by 2 **do**
	$\quad tmp1[i] \leftarrow Vd[2 \times i]$
	$\quad tmp2[i] \leftarrow Vd[2 \times i + 1]$
	end for
	$Vm \leftarrow tmp1$
	$Vd \leftarrow tmp2$
		Interleave data from two vectors. *tmp* is a vector of suitable size.

Examples

```
1    vzip.i32   d3,d4   @ interleave d3:d4
2    vuzp.i16   q8,q10  @ de-interleave q8:q10
```

10.5 Data Conversion

When high precision is not required, The IEEE half-precision format can be used to store floating point numbers in memory. This can reduce memory requirements by up to 50%. This can also result in a significant performance improvement, since only half as much data needs to be moved between the CPU and main memory. However, on most processors half-precision data must be converted to single precision before it is used in calculations. NEON provides enhanced versions of the `vcvt` instruction which support conversion to and from IEEE half precision. There are also versions of `vcvt` which operate on vectors, and perform integer or fixed-point to floating-point conversions.

10.5.1 Convert Between Fixed Point and Single-Precision

This instruction can be used to perform a data conversion between single precision and fixed point on each element in a vector:

`vcvt`   Convert Data Format.

The elements in the vector must be a 32-bit single precision floating point or a 32-bit integer. Fixed point (or integer) arithmetic operations are up to twice as fast as floating point operations. In some cases it is much more efficient to make this conversion, perform the calculations, then convert the results back to floating point.

Syntax

```
vcvt{<cond>}.<type>.f32   Sd, Sm{, #<fbits>}
vcvt{<cond>}.f32.<type>   Sd, Sm{, #<fbits>}
```

- `<cond>` is an optional condition code.
- `<type>` must be either `s32` or `u32`.
- The optional `<fbits>` operand specifies the number of fraction bits for a fixed point number, and must be between 0 and 32. If it is omitted, then it is assumed to be zero.

Operations

Name	Effect	Description
`vcvt.s32.f32`	$Fd[] \leftarrow fixed(Fm[])$	Convert single precision to 32-bit signed fixed point or integer.
`vcvt.u32.f32`	$Fd[] \leftarrow ufixed(Fm[])$	Convert single precision to 32-bit unsigned fixed point or integer.
`vcvt.f32.s32`	$Fd[] \leftarrow single(Fm[])$	Convert signed 32-bit fixed point or integer to single precision
`vcvt.f32.u32`	$Fd[] \leftarrow single(Fm[])$	Convert unsigned 32-bit fixed point or integer to single precision

Examples

```
1    vcvt.f32.u32   s0,s0,#4 @ Convert from U(28,4) to float
2    vcvteq.s32.f32 s1,s1    @ Convert from float to integer
```

10.5.2 Convert Between Half-Precision and Single-Precision

NEON systems with the half-precision extension provide the following instruction to perform conversion between single precision and half precision floating point formats:

vcvt   Convert Between Half and Single.

Syntax

```
vcvt<op>{<cond>}.f16.f32   Sd, Sm
vcvt<op>{<cond>}.f32.f16   Sd, Sm
```

- The <op> must be either b or t and specifies whether the top or bottom half of the register should be used for the half-precision number.
- <cond> is an optional condition code.

Operations

Name	Effect	Description
vcvtb.f16.f32	$Sd \leftarrow half(Sm)$	Convert single precision to half precision and store in bottom half of destination
vcvtt.f16.f32	$Sd \leftarrow half(Sm)$	Convert single precision to half precision and store in top half of destination
vcvtb.f32.f16	$Sd \leftarrow single(Sm)$	Convert half precision number from bottom half of source to single precision
vcvtt.f32.f16	$Sd \leftarrow single(Sm)$	Convert half precision number from top half of source to single precision

Examples

```
1    vcvtb.f32.f16 s0,s1 @ convert bottom of s1 to single
2    vcvtt.f16.f32 s3,s4 @ convert s4 to half precision
```

10.6 Comparison Operations

NEON adds the ability to perform integer comparisons between vectors. Since there are multiple pairs of items to be compared, the comparison instructions set one element in a result vector for each pair of items. After the comparison operation, each element of the result vector will have every bit set to zero (for false) or one (for true). Note that if the elements of the result vector are interpreted as signed two's-complement numbers, then the value 0 represents false and the value −1 represents true.

10.6.1 Vector Compare

The following instructions perform comparisons of all of the corresponding elements of two vectors in parallel:

vceq   Compare Equal,
vcge   Compare Greater Than or Equal,
vcgt   Compare Greater Than,
vcle   Compare Less Than or Equal, and
vclt   Compare Less Than.

The vector compare instructions compare each element of a vector with the corresponding element in a second vector, and sets an element in the destination vector for each comparison. If the comparison is true, then all bits in the result element are set to one. Otherwise, all bits in the result element are set to zero. Note that summing the elements of the result vector (as signed integers) will give the two's complement of the number of comparisons which were true.

Note: vcle and vclt are actually pseudo-instructions. They are equivalent to vcgt and vcge with the operands reversed.

Syntax

```
vc<op>.<type> Vd, Vn, Vm
vc<op>.<type> Vd, Vn, #0
```

-     <op> must be one of eq, ge, gt, le, or lt.
-     If <op> is eq, then <type> must be i8, i16, i32, or f32.
-     If <op> is not eq and Rop is #0, then <type> must be s8, s16, s32, or f32.
-     If <op> is not eq and the third operand is a register, then <type> must be s8, s16, s32, u8, u16, u32, or f32.
-     The result data type is determined from the following table:

Operand Type	Result Type
i32, s32, u32, or f32	i32
i16, s16, or u16	i16
i8, s8, or u8	i8

- If the third operand is #0, then it is taken to be a vector of the correct size in which every element is zero.
- V can be d or q.

Operations

Name	Effect	Description

vc\<op>	**for** $i \in vector\_length$ **do**
	  if $Fm[i]$\<op>$Rop[i]$
	  then
	    $Fd[i] \leftarrow 111\ldots$
	  else
	    $Fd[i] \leftarrow 000\ldots$
	  end if
	end for
		Compare each scalar in Fn to the corresponding scalar in Fm. Set the corresponding scalar in Fd to all ones if \<op> is true, and all zeros if \<op> is not true.

Examples

```
1   vceq.i8   d0,d1,d2   @ 8 8-bit comparisons
2   vcge.s16  d0,d1,d2   @ 4 16-bit signed comparisons
3   vcgt.u16  q0,q1,q2   @ 8 16-bit unsigned comparisons
4   vcle.f32  d0,d1,d2   @ 2 single precision comparisons
5   vclt.f32  q0,q1,q2   @ 4 single precision comparisons
6   vceq.i8   q0,q1,#0   @ 16 8-bit comparisons
7   vcge.s16  d0,d1,#0   @ 8 8-bit signed comparisons
8   vcgt.f32  d0,d1,#0   @ 2 single precision comparisons
```

10.6.2 Vector Absolute Compare

The following instructions perform comparisons between the absolute values of all of the corresponding elements of two vectors in parallel:

vacgt   Absolute Compare Greater Than, and

vacge   Absolute Compare Greater Than or Equal.

The vector absolute compare instruction compares the absolute value of each element of a vector with the absolute value of the corresponding element in a second vector, and sets an element in the destination vector for each comparison. If the comparison is true, then all bits in the result element are set to one. Otherwise, all bits in the result element are set to zero. Note that summing the elements of the result vector (as signed integers) will give the two's complement of the number of comparisons which were true.

Syntax

```
vac<op>.f32    Vd, Vn, Vm
```

- <op> must be either ge or gt.
- V can be d or q.
- The operand element type must be f32.
- The result element type is i32.

Operations

Name	Effect	Description

vac<op>	**for** *i* ∈ *vector_length* **do**
	  if |*Fm*[*i*]|<op>|*Fn*[*i*]|
	  then
	    Fd[*i*] ← 111...
	  else
	    Fd[*i*] ← 000...
	  end if
	end for
		Compare each scalar in Fn to the corresponding scalar in Fm. If the comparison is true, then set all bits in the corresponding scalar in Fd to one. Otherwise set all bits in the corresponding scalar in Fd to zero.

Examples

```
  vacgt.f32  d0,d1,d2  @ 2 single precision comparisons
  vacge.f32  q0,q1,q2  @ 4 single precision comparisons
```

10.6.3 Vector Test Bits

NEON provides the following vector version of the ARM tst instruction:

vtst   Test Bits.

The vector test bits instruction performs a logical AND operation between each element of a vector and the corresponding element in a second vector. If the result is not zero, then every bit in the corresponding element of the result vector is set to one. Otherwise, every bit in the corresponding element of the result vector is set to zero.

Syntax

```
vtst.<size>  Vd, Vn, Vm
```

- V can be d or q.
- <size> must be one of 8, 16 or 32
- The result element type is defined by the following table:

`<size>`	Result Type
32	i32
16	i16
8	i8

Operations

Name	Effect	Description

vtst	**for** *i* ∈ *vector_length* **do**
	    if (*Fm[i]* ∧ *Fn[i]*) ≠ 0 **then**
	        Fd[i] ← 111...
	    else
	        Fd[i] ← 000...
	    end if
	end for
		Perform logical AND between each scalar in Fn and the corresponding scalar in Fm. Set the corresponding scalar in Fd to all ones if the result is not zero, and all zeros otherwise

Examples

```
1    vtst.8   d0,d1,d2  @ Test bits in d1, using d2
2    vtst.16  q0,q1,q2
```

10.7 Bitwise Logical Operations

NEON adds the ability to perform integer and bitwise logical operations on the VFP register set. Recall that integer operations can also be used on fixed-point data. These operations add a great deal of power to the ARM processor.

10.7.1 Bitwise Logical Operations

NEON includes vector versions of the following five basic logical operations:

vand   Bitwise AND,
veor   Bitwise Exclusive-OR,
vorr   Bitwise OR,
vorn   Bitwise Complement and OR, and
vbic   Bit Clear.

All of them involve two source operands and a destination register.

Syntax

```
v<op>{.<type>}  Vd, Vn, Vm
```

- `<op>` must be one of and, eor, orr, orn, or bic.
- V must be either q or d.
- type must be i8, i16, i32, or i64. For these bitwise logical operations, type does not matter.

Operations

Name	Effect	Description
vand	$Vd \leftarrow Vn \wedge Vm$	Logical AND
veor	$Vd \leftarrow Vn \oplus Vm$	Exclusive OR
vorr	$Vd \leftarrow Vn \vee Vm$	Logical OR
vorn	$Vd \leftarrow \neg(Vn \vee Vm)$	Complement of Logical OR
vbic	$Vd \leftarrow Vn \wedge \neg Vm$	Bit Clear

Examples

```
1    vand.i64   q0,q1,q2   @ q0=q1 & q2
2    vbic.i32   d3,d3,d5   @ if (eq) then d3=d3 & !d4
3    vorr.i8    q0,q1,q2   @ q0=q1 | q2
4    vorr.i64   q0,q1,q2   @ q0=q1 | q2
```

10.7.2 Bitwise Logical Operations with Immediate Data

It is often useful to clear and/or set specific bits in a register. The NEON instruction set provides the following vector versions of the logical OR and bit clear instructions:

vorr   Bitwise OR Immediate, and
vbic   Bit Clear Immediate.

Syntax

```
v<op>.<type>   Vd, #<imm>
```

- `<op>` must be either orr, or bic.
- V must be either q or d to specify whether the operation involves quadwords or doublewords.
- `<type>` must be i16 or i32.
- `<imm>` is a 16-bit or 32-bit immediate value, which is interpreted as a pattern for filling the immediate operand. The following table shows acceptable patterns for `<imm>`, based on what was chosen for `<type>`:

<type>	
i16	i32
0x00XY	0x000000XY
0xXY00	0x0000XY00
	0x00XY0000
	0xXY000000

Operations

Name	Effect	Description
vorr	$Vd \leftarrow Vd \vee imm : imm \ldots$	Logical OR
vbic	$Vd \leftarrow Vd \wedge imm : imm \ldots$	Bit Clear

Examples

```
1    vbic.i16    q0, #0x00FF     @ q0=q0 & !0x00FF00FF...
2    vbic.i16    q0, #0xFF00     @ q0=q0 & !0xFF00FF00...
3    vbic.i32    q0, #0x000000FF @ q0=q0 &
4                                @ !0x000000FF000000FF...
5    vorr.i16    q0, #0x0001     @ q0=q0 | 0x00010001...
```

10.7.3 Bitwise Insertion and Selection

NEON provides three instructions which can be used to combine the bits in two registers or to extract specific bits from a register, according to a pattern:

vbit   Bitwise Insert,

vbif   Bitwise Insert if False, and

vbsl   Bitwise Select.

Syntax

```
v<op>{.<type>} Vd, Vn, Vm
```

- <op> can be bif, bit, or bsl.
- V can be d or q.
- The <type> must be i8, i16, i32, or i64, and specifies the size of items in the vectors. Note that for these bitwise logical operations, the type does not matter. so the assembler ignores it. However, it can be useful to the programmer as extra documentation.

Operations

Name	Effect	Description
vbit	$Fd \leftarrow (Fd \wedge \neg Fm) \vee (Fn \wedge Fm)$	Insert each bit from the first operand into the destination if the corresponding bit of the second operand is 1
vbif	$Fd \leftarrow (Fd \wedge Fm) \vee (Fn \wedge \neg Fm)$	Insert each bit from the first operand into the destination if the corresponding bit of the second operand is 0
vbsl	$Fd \leftarrow (Fd \wedge Fn) \vee (\neg Fd \wedge Fm)$	Select each bit for the destination from the first operand if the corresponding bit of the destination is 1, or from the second operand if the corresponding bit of the destination is 0

Examples

```
1    vbit.i8    d3,d2,d1
2    vbsl.i16   q0,q8,q9
```

10.8 Shift Instructions

The NEON shift instructions operate on vectors. Shifts are often used for multiplication and division by powers of two. The results of a left shift may be larger than the destination register, resulting in overflow. A shift right is equivalent to division. In some cases, it may be useful to round the result of a division, rather than truncating. NEON provides versions of the shift instruction which perform saturation and/or rounding of the result.

10.8.1 Shift Left by Immediate

These instructions shift each element in a vector left by an immediate value:

vshl   Shift Left Immediate,
vqshl   Saturating Shift Left Immediate,
vqshlu   Saturating Shift Left Immediate Unsigned, and
vshll   Shift Left Immediate Long.

Overflow conditions can be avoided by using the saturating version, or by using the long version, in which case the destination is twice the size of the source.

Syntax

```
    vshl.<type>        Vd, Vm, #<imm>
    vqshl{u}.<type>    Vd, Vm, #<imm>
    vshll.<type>       Qd, Dm, #<imm>
```

- If u is present, then the results are unsigned.
- The valid choices for <type> are given in the following table:

Opcode	Valid Types
vshl	i8, i16, i32, i64, s8, s16, or s32
vqshl	s8, s16, s32, s64, u8, u16, u32, or u64
vqshlu	s8, s16, s32, or s64
vshll	u8, u16, u32, u64, s8, s16, or s32

Operations

Name	Effect	Description
vshl	$Vd[] \leftarrow Vm[] \ll imm$	Each element of Vm is shifted left by the immediate value and stored in the corresponding element of Vd. Bits shifted past the end of an element are lost.
vshll	$Qd[] \leftarrow Dm[] \ll imm$	Each element of Vm is shifted left by the immediate value and stored in the corresponding element of Vd. The values are sign or zero extended, depending on <type>
vqshl{u}	$Vd[] \leftarrow \lfloor Vm[] \ll imm \rceil$	Each element of Vm is shifted left by the immediate value and stored in the corresponding element of Vd. If the result of the shift is outside the range of the destination element, then the value is saturated. If u was specified, then the destination is unsigned. Otherwise, it is signed

Examples

```
1   vshl.s16   q1,q6,#4  @ shift each 16-bit word left
2   vqshl.u8   d1,d6,#1  @ Multiply each byte by two
```

10.8.2 Shift Left or Right by Variable

These instructions shift each element in a vector, using the least significant byte of the corresponding element of a second vector as the shift amount:

vshl	Shift Left or Right by Variable,
vrshl	Shift Left or Right by Variable and Round,
vqshl	Saturating Shift Left or Right by Variable, and
vqrshl	Saturating Shift Left or Right by Variable and Round.

If the shift value is positive, the operation is a left shift. If the shift value is negative, then it is a right shift. A shift value of zero is equivalent to a move. If the operation is a right shift, and r is specified, then the result is rounded rather than truncated. Results are saturated if q is specified.

Syntax

```
v{q}{r}shl.<type>  Vd, Vn, Vm
```

- If q is present, then the results are saturated.
- If r is present, then right shifted values are rounded rather than truncated.
- V can be d or q.
- \<type\> must be one of s8, s16, s32, s64, s8, s16, s32, or s64.

Operations

Name	Effect	Description

vshl	if q is present **then**
	  if r is present **then**
	    $Vd[] \leftarrow \lfloor \|Vn[] \ll Vm[]\| \rceil$
	  else
	    $Vd[] \leftarrow \lfloor Vn[] \ll Vm[] \rceil$
	  end if
	else
	  if r is present **then**
	    $Vd[] \leftarrow \|Vn[] \ll Vm[]\|$
	  else
	    $Vd[] \leftarrow Vn[] \ll Vm[]$
	  end if
	end if
		Each element of Vm is shifted left by the immediate value and stored in the corresponding element of Vd. Bits shifted past the end of an element are lost.

Examples

```
1    vshl    q0,q1,q3  @ use elements in q3 to shift
2                      @ elements of q1
3    vrshl   q0,q1,q3  @ use elements in q3 to shift
4                      @ elements of q1 with rounding
```

10.8.3 Shift Right by Immediate

These instructions shift each element in a vector right by an immediate value:

vshr   Shift Right Immediate,
vrshr   Shift Right Immediate and Round,
vshrn   Shift Right Immediate and Narrow,
vrshrn Shift Right Immediate Round and Narrow,
vsra   Shift Right and Accumulate Immediate, and
vrsra   Shift Right Round and Accumulate Immediate.

Syntax

```
v{r}shr{<cond>}.<type>        Vd, Vm, #<imm>
v{r}shrn{<cond>}.<type>       Vd, Vm, #<imm>
v{r}sra{<cond>}.<type>        Vd, Vm, #<imm>
```

- V can be d or q.
- If r is present, then right shifted values are rounded rather than truncated.
- <cond> is an optional condition code.
- The valid choices for <type> are given in the following table:

Opcode	Valid Types
v{r}shr	u8, u16, u32, u64, s8, s16, s32, or s64,
v{r}shrn	i16, i32, or i64
v{r}sra	u8, u16, u32, u64, s8, s16, s32, or s64,

Operations

Name	Effect	Description

v{r}shr	**if** r is present **then**
	 $\quad Vd[] \leftarrow \|Vm[] \gg imm\|$
	 else
	 $\quad Vd[] \leftarrow Vm[] \gg imm$
	 end if
		Each element of Vm is shifted right with zero extension by the immediate value and stored in the corresponding element of Vd. Results can be rounded both.

v{r}shrn	**if** r is present **then**
	 $\quad Vd[] \leftarrow\succ \|Vm[] \gg imm\| \prec$
	 else
	 $\quad Vd[] \leftarrow\succ Vm[] \gg imm \prec$
	 end if
		Each element of Vm is shifted right with zero extension by the immediate value, optionally rounded, then narrowed and stored in the corresponding element of Vd.

v{r}sra	**if** r is present **then**
	 $\quad Vd[] \leftarrow Vd[] + \|Vm[] \gg imm\|$
	 else
	 $\quad Vd[] \leftarrow Vd[] + Vm[] \gg imm$
	 end if
		Each element of Vm is shifted right with sign or zero extension by the immediate value and accumulated in the corresponding element of Vd. Results can be rounded.

Examples

```
1    vsra.S32      q1,q6,#4  @ shift each 32-bit integer
2    vrsra.u16     d1,d6,#1  @ Divide by 2 with rounding
```

10.8.4 Saturating Shift Right by Immediate

These instructions shift each element in a quad word vector right by an immediate value:

vqshrn   Saturating Shift Right Immediate,
vqrshrn   Saturating Shift Right Immediate Round,

vqshrun   Saturating Shift Right Immediate Unsigned, and
vqrshrun Saturating Shift Right Immediate Round Unsigned.

The result is optionally rounded, then saturated, narrowed, and stored in a double word vector.

Syntax

```
vq{r}shr{u}n.<type>    Dd, Qm, #<imm>
```

- If r is present, then right shifted values are rounded rather than truncated.
- If u is present, then the results are unsigned, regardless of the type of elements in Qm.
- The valid choices for <type> are given in the following table:

Opcode	Valid Types
vq{r}shrn	u16, u32, u64, s16, s32, or s64,
vq{r}shrun	s16, s32, or s64,

- <imm> Is the amount that elements are to be shifted, and must be between zero and one less than the number of bits in <type>.

Operations

Name	Effect	Description

vq{r}shrn	**if** r is present **then**
	 $Vd[] \leftarrow \lfloor \lVert Vm[] \gg imm \rVert \rceil$
	 else
	 $Vd[] \leftarrow \lfloor Vm[] \gg imm \rceil$
	 end if
		Each element of Vm is shifted right with sign extension by the immediate value, optionally rounded, then saturated and narrowed, and stored in the corresponding element of Vd.

vq{r}shrun	**if** r is present **then**
	 $Vd[] \leftarrow \lfloor \lVert Vm[] \gg imm \rVert \rceil$
	 else
	 $Vd[] \leftarrow \lfloor Vm[] \gg imm \rceil$
	 end if
		Each element of Vm is shifted right with zero extension by the immediate value, optionally rounded, then saturated and narrowed, and stored in the corresponding element of Vd.

Examples

```
1   vqshrn.S32    d1,q6,#4  @ shift, saturate and narrow
2   vqrshrn.S32   d1,q6,#4  @ shift, round,
3   @ saturate and narrow
```

10.8.5 Shift and Insert

These instructions perform bitwise shifting of each element in a vector, then combine the results with the contents of the destination register:

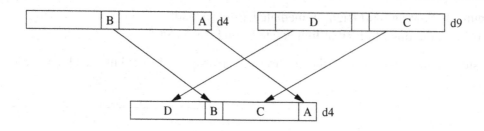

Figure 10.10

Effects of `vsli.32 d4,d9,#6`.

vsli   Shift Left and Insert,
vsri   Shift Right and Insert.

Fig. 10.10 provides an example.

Syntax

```
vs<dir>i.<size>     Vd, Vm, #<imm>
```

- `<dir>` must be l for a left shift, or r for a right shift.
- `<size>` must be 8, 16, 32, or 64.
- `<imm>` is the amount that elements are to be shifted, and must be between zero and `<size>` − 1 for `vsli`, or between one and `<size>` for `vsri`.

Operations

Name	Effect	Description
vsli	$mask \leftarrow (1 \ll imm + 1) - 1$ $Vd[] \leftarrow (mask \wedge Vd[]) \vee (Vm[] \ll imm)$	Each element of Vm is shifted left and combined with lower `<imm>` bits of the corresponding element of Vd.
vsri	$mask \leftarrow \neg(1 \ll size - imm + 1) - 1$ $Vd[] \leftarrow (mask \wedge Vd[]) \vee (Vm[] \gg imm)$	Each element of Vm is shifted right and combined with upper `<imm>` bits of the corresponding element of Vd.

Examples

```
1    vsli.32  q1,q6,#4  @ shift each element left
2    vsri.8   d0,d4,#4  @ vector shift right
```

10.9 Arithmetic Instructions

NEON provides several instructions for addition, subtraction, and multiplication, but does not provide a divide instruction. Whenever possible, division should be performed by multiplying the reciprocal. When dividing by constants, the reciprocal can be calculated in advance, as shown in Chapter 8. For dividing by variables, NEON provides instructions for quickly calculating the reciprocals for all elements in a vector. In most cases, this is faster than using a divide instruction. When division is absolutely unavoidable, the VFP divide instructions can be used.

10.9.1 Vector Add and Subtract

The following eight instructions perform vector addition and subtraction:

vadd	Add
vqadd	Saturating Add
vaddl	Add Long
vaddw	Add Wide
vsub	Subtract
vqsub	Saturating Subtract
vsubl	Subtract Long
vsubw	Subtract Wide

The Vector Add (vadd) instruction adds corresponding elements in two vectors and stores the results in the corresponding elements of the destination register. The Vector Subtract (vsub) instruction subtracts elements in one vector from corresponding elements in another vector and stores the results in the corresponding elements of the destination register. Other versions allow mismatched operand and destination sizes, and the saturating versions prevent overflow by limiting the range of the results.

Syntax

```
v{q}<op>.<type>    Vd, Vn, Vm
v<op>l.<type>      Qd, Dn, Dm
v<op>w.<type>      Qd, Qn, Dm
```

- <op> is either add or sub .
- The valid choices for <type> are given in the following table:

Opcode	Valid Types
v<op>	i8, i16, i32, i64, or f32
vq<op>	s8, s16, s32, s64, u8, u16, u32, or u64
v<op>l	s8, s16, s32, u8, u16, or u32
v<op>w	s8, s16, s32, u8, u16, or u32

Operations

Name	Effect	Description
v<op>	$Vd[] \leftarrow Vn[]$<op>$Vm[]$	The operation is applied to corresponding elements of Vn and Vm. The results are stored in the corresponding elements of Vd.
vq<op>	$Vd[] \leftarrow \lfloor Vn[]$<op>$Vm[] \rfloor$	The operation is applied to corresponding elements of Vn and Vm. The results are saturated then stored in the corresponding elements of Vd.
v<op>l	$Qd[] \leftarrow \prec Dn[]$<op>$Dm[] \succ$	The operation is applied to corresponding elements of Dn and Dm. The results are zero or sign extended then stored in the corresponding elements of Qd.
v<op>w	$Qd[] \leftarrow Qn[]$<op>$\prec Dm[] \succ$	The elements of Vm are sign or zero extended, then the operation is applied with corresponding elements of Vn. The results are stored in the corresponding elements of Vd.

Examples

```
1    vadd.s8    q1,q6,q8  @ Add elements
2    vqadd.s8   q1,q6,q8  @ Add elements and saturate
```

10.9.2 Vector Add and Subtract with Narrowing

These instructions add or subtract the corresponding elements of two vectors, and narrow by taking the most significant half of the result:

vaddhn   Add and Narrow
vraddhn   Add, Round, and Narrow
vsubhn   Subtract and Narrow
vrsubhn   Subtract, Round, and Narrow

The results are stored in the corresponding elements of the destination register. Results can be optionally rounded instead of truncated.

Syntax

```
v{r}<op>hn.<type> Dd, Qn, Qm
```

- <op> is either add or sub.
- If <r> is specified, then the result is rounded instead of truncated.
- <type> must be either i16, i32, or i64.

Operations

Name	Effect	Description

v<op>hn	*shift ← size ÷ 2*
	if r is present **then**
	  x ← ‖Vn[]<op>Vm[]‖
	  Vd[] ↔ x ≫ shift ≺
	else
	  x ← Vn[]<op>Vm[]
	  Vd[] ↔ x ≫ shift ≺
	end if
		The operation is applied to corresponding elements of Vn and Vm. The results are optionally rounded, then narrowed by taking the most significant half, and stored in the corresponding elements of Vd.

Examples

```
1    vaddhn.i32    d1,q6,q8  @ Add and narrow
2    vrsubhn.i16   d4,q5,q3  @ Subtract round and narrow
```

10.9.3 Add or Subtract and Divide by Two

These instructions add or subtract corresponding elements from two vectors then shift the result right by one bit:

vhadd   Halving Add
vrhadd   Halving Add and Round
vhsub   Halving Subtract

The results are stored in corresponding elements of the destination vector. If the operation is addition, then the results can be optionally rounded.

Syntax

```
v{r}hadd.<type> Vd, Vn, Vm
vhsub.<type>    Vd, Vn, Vm
```

- If <r> is specified, then the result is rounded instead of truncated.
- <type> must be either s8, s16, s32, u8, u16, ar u32.

Operations

Name	Effect	Description

v{r}hadd	**if** r is present **then**
	  Vd[] ← ‖Vn[] + Vm[]‖ ≫ 1
	else
	  Vd[] ← Vn[] + Vm[] ≫ 1
	end if
		The corresponding elements of Vn and Vm are added together, optionally rounded, then shifted right one bit. Results are stored in the corresponding elements of Vd.

vhsub	*Vd[] ← Vn[] − Vm[] ≫ 1*	The elements of Vn are subtracted from the corresponding elements of Vm. Results are shifted right one bit and stored in the corresponding elements of Vd.

Examples

```
1   vrhadd.s8   q1,q6,q8  @ Add elements and divide by 2
2   vhsub.s16   q1,q6,q8  @ Subtract and divide by 2
```

10.9.4 Add Elements Pairwise

These instructions add vector elements pairwise:

vpadd   Add Pairwise
vpaddl   Add Pairwise Long
vpadal   Add Pairwises and Accumulate Long

The long versions can be used to prevent overflow.

Syntax

```
vpadd.<type>     Dd, Dn, Dm
vp<op>l.<type>   Vd, Vm
```

- `<op>` must be either add or ada.
- The valid choices for `<type>` are given in the following table:

Opcode	Valid Types
vpadd	i8, i16, i32, or f32
vp<op>l	s8, s16, s32, u8, u16, or u32

Operations

Name	Effect	Description

vpadd	$n \leftarrow$ # of elements
	for $0 \leq i < (n \div 2)$ **do**
	$\quad Dd[i] \leftarrow Dm[i] + Dm[i+1]$
	end for
	for $(n \div 2) \leq i < n$ **do**
	$\quad j \leftarrow i - (n \div 2)$
	$\quad Dd[i] \leftarrow Dn[j] + Dn[j+1]$
	end for
		Add elements of two vectors pairwise and store the results in another vector.

vpaddl	$n \leftarrow$ # of elements
	for $0 \leq i < (n \div 2)$ by 2 **do**
	$\quad Vd[i] \leftarrow \prec Vm[i] \succ + \prec Vm[i+1] \succ$
	end for
		Add elements of a vector pairwise and store the results in another vector.

vpadal	$n \leftarrow$ # of elements
	for $0 \leq i < (n \div 2)$ by 2 **do**
	$\quad Vd[i] \leftarrow Vd[i] + \prec Vm[i] \succ + \prec Vm[i+1] \succ$
	end for
		Add elements of a vector pairwise and accumulate the results in another vector.

Examples

```
1    vpadd.s16    d1,d6,d8    @ Add pairwise
2    vpadal.s8    q1,q6       @ Extend, add pairwise, and
3                             @ accumulate
4    vpadal.s32   q0,q8       @ Extend and add pairwise
```

10.9.5 Absolute Difference

These instructions subtract the elements of one vector from another and store or accumulate the absolute value of the results:

vaba   Absolute Difference and Accumulate
vabal   Absolute Difference and Accumulate Long
vabd   Absolute Difference
vabdl   Absolute Difference Long

The long versions can be used to prevent overflow.

Syntax

```
v<op>.<type>     Vd, Vn, Vm
v<op>l.<type>    Qd, Dn, Dm
```

- <op> is either aba or abd .
- The valid choices for <type> are given in the following table:

Opcode	Valid Types
vabd	s8, s16, s32, u8, u16, u32, or f32
vaba	s8, s16, s32, u8, u16, or u32
vabdl	s8, s16, s32, u8, u16, or u32
vabal	s8, s16, s32, u8, u16, or u32

Operations

Name	Effect	Description
vabd	$Vd[] \leftarrow |Vn[] - Vm[]|$	Subtract corresponding elements and take the absolute value
vaba	$Vd[] \leftarrow Vd[] + |Vn[] - Vm[]|$	Subtract corresponding elements and take the absolute value. Accumulate the results
vabdl	$Qd[] \leftarrow |\prec Dn[] \succ - \prec Dm[] \succ|$	Extend and subtract corresponding elements, then take the absolute value
v<op>w	$Qd[] \leftarrow Qd[] + |\prec Dn[] \succ - \prec Dm[] \succ|$	Extend and subtract corresponding elements, then take the absolute value. Accumulate the results

Examples

```
1    vaba.s8     q1,q6,q8  @ Accumulate absolute difference
2    vabd.s16    q3,q4,q5  @ Absolute value of differences
```

10.9.6 Absolute Value and Negate

These operations compute the absolute value or negate each element in a vector:

vabs   Absolute Value
vneg   Negate
vqabs   Saturating Absolute Value
vqneg   Saturating Negate

The saturating versions can be used to prevent overflow.

Syntax

```
v{q}<op>.<type>        Vd, Vm
```

- If q is present then results are saturated.
- <op> is either abs or neg .
- The valid choices for <type> are given in the following table:

Opcode	Valid Types
vabs	s8, s16, s32, or f32
vneg	s8, s16, s32, or f32
vqabs	s8, s16, or s32
vqneg	s8, s16, or s32

Operations

Name	Effect	Description
v{q}abs	**if** q is present **then** $Vd[] \leftarrow \lfloor \lvert Vm[] \rvert \rfloor$ **else** $Vd[] \leftarrow \lvert Vm[] \rvert$ **end if**	Copy absolute value of each element of Vm to the corresponding element of Vd, optionally saturating the result
v{q}neg	**if** q is present **then** $Vd[] \leftarrow \lfloor -Vm[] \rfloor$ **else** $Vd[] \leftarrow -Vm[]$ **end if**	Copy absolute value of each element of Vm to the corresponding element of Vd, optionally saturating the result

Examples

```
1    vabs.f32   q1,q6  @ Get absolute values
2    vqneg.s16  q3,q4  @ Negate and saturate
```

10.9.7 Get Maximum or Minimum Elements

The following four instructions select the maximum or minimum elements and store the results in the destination vector:

vmax   Maximum
vmin   Minimum
vpmax   Pairwise Maximum
vpmin   Pairwise Minimum

Syntax

```
v<op>.<type>      Vd, Vn, Vm
vp<op>.<type>     Dd, Dn, Dm
```

- `<op>` is either max or min.
- `<type>` must be one of s8, s16, s32, u8, u16, u32, or f32.

Operations

Name	Effect	Description

vmax	$n \leftarrow \#$ of elements
	for $0 \le i < n$ **do**
	  if $Vn[i] > Vm[i]$ **then**
	    $Vd[i] \leftarrow Vn[i]$
	  else
	    $Vd[i] \leftarrow Vm[i]$
	  end if
	end for
		Compare corresponding elements and copy the greater of each pair into the corresponding element in the destination vector

vpmax	$n \leftarrow \#$ of elements
	for $0 \le i < (n \div 2)$ **do**
	  if $Dm[i] > Dm[i+1]$ **then**
	    $Dd[i] \leftarrow Dm[i]$
	  else
	    $Dd[i] \leftarrow Dm[i+1]$
	  end if
	end for
	for $(n \div 2) \le i < n$ **do**
	  if $Dn[i] > Dn[i+1]$ **then**
	    $Dd[i+(n \div 2)] \leftarrow Dn[i]$
	  else
	    $Dd[i+(n \div 2)] \leftarrow Dn[i+1]$
	  end if
	end for
		Compare elements pairwise and copy the greater of each pair into an element in the destination vector, another vector

vmin	$n \leftarrow$ # of elements
	for $0 \leq i < n$ **do**
	  if $Vn[i] < Vm[i]$ **then**
	    $Vd[i] \leftarrow Vn[i]$
	  else
	    $Vd[i] \leftarrow Vm[i]$
	  end if
	end for
		Compare corresponding elements and copy the lesser of each pair into the corresponding element in the destination vector

vpmin	$n \leftarrow$ # of elements
	for $0 \leq i < (n \div 2)$ **do**
	  if $Dm[i] < Dm[i+1]$ **then**
	    $Dd[i] \leftarrow Dm[i]$
	  else
	    $Dd[i] \leftarrow Dm[i+1]$
	  end if
	end for
	for $(n \div 2) \leq i < n$ **do**
	  if $Dn[i] < Dn[i+1]$ **then**
	    $Dd[i+(n \div 2)] \leftarrow Dn[i]$
	  else
	    $Dd[i+(n \div 2)] \leftarrow Dn[i+1]$
	  end if
	end for
		Compare elements pairwise and copy the lesser of each pair into an element in the destination vector, another vector

Examples

```
1   vmin.u8    q1,q6,q7  @ Get minimum values
2   vpmax.f32  d0,d4,d5  @ Get maximum values
```

10.9.8 Count Bits

These instructions can be used to count leading sign bits or zeros, or to count the number of bits that are set for each element in a vector:

vcls   Count Leading Sign Bits
vclz   Count Leading Zero Bits
vcnt   Count Set Bits

Syntax

```
v<op>.<type>      Vd, Vm
```

- `<op>` is either `cls`, `clz` or `cnt`.
- The valid choices for `<type>` are given in the following table:

Opcode	Valid Types
vcls	s8, s16, or s32
vclz	u8, u16, or u32
vcnt	i8

Operations

Name	Effect	Description

vcls	$n \leftarrow$ # of elements
	for $0 \leq i < n$ **do**
	    $Vd[i] \leftarrow leading\_sign\_bits(Vm[i])$
	end for
		Count the number of consecutive bits that are the same as the sign bit for each element in Fm, and store the counts in the corresponding elements of Fd

vcls	$n \leftarrow$ # of elements
	for $0 \leq i < n$ **do**
	    $Vd[i] \leftarrow leading\_zero\_bits(Vm[i])$
	end for
		Count the number of leading zero bits for each element in Fm, and store the counts in the corresponding elements of Fd.

vcnt	$n \leftarrow$ # of elements
	for $0 \leq i < n$ **do**
	    $Vd[i] \leftarrow count\_one\_bits(Vm[i])$
	end for
		Count the number of bits in Fm that are set to one, and store the counts in the corresponding elements of Fd

Examples

```
1    vcls.s8   q1,q6  @ Count leading sign bits
2    vcnt.i8   d0,d4  @ Count bits that are 1
```

10.10 Multiplication and Division

There is no vector divide instruction in NEON. Division is accomplished with multiplication by the reciprocals of the divisors. The reciprocals are found by making an initial estimate, then using the Newton-Raphson method to improve the approximation. This can actually be faster than using a hardware divider. NEON supports single precision floating point and unsigned fixed point reciprocal calculation. Fixed point reciprocals provide higher precision. Division using the NEON reciprocal method may not provide the best precision possible. If the best possible precision is required, then the VFP divide instruction should be used.

10.10.1 Multiply

These instructions are used to multiply the corresponding elements from two vectors:

vmul	Multiply
vmla	Multiply Accumulate
vmls	Multiply Subtract
vmull	Multiply Long

`vmlal`   Multiply Accumulate Long
`vmlsl`   Multiply Subtract Long

The long versions can be used to avoid overflow.

Syntax

```
v<op>.<type>    Vd, Vn, Vm
v<op>l.<type>   Qd, Dn, Dm
```

- `<op>` is either `mul`, `mla`. or `mls`.
- The valid choices for `<type>` are given in the following table:

Opcode	Valid Types
`vmul`	p8, i8, i16, or i32
`vmla`	i8, i16, or i32
`vmls`	i8, i16, or i32
`vmull`	p8, s8, s16, s32, u8, u16, or u32
`vmlal`	s8, s16, s32, u8, u16, or u32
`vmlsl`	s8, s16, s32, u8, u16, or u32

Operations

Name	Effect	Description
`vmul`	$Vd[] \leftarrow Vn[] \times Vm[]$	Multiply corresponding elements from two vectors and store the results in a third vector
`vmla`	$Vd[] \leftarrow Vd[] + (Vn[] \times Vm[])$	Multiply corresponding elements from two vectors and add the results in a third vector
`vmul`	$Vd[] \leftarrow Vd[] - (Vn[] \times Vm[])$	Multiply corresponding elements from two vectors and subtract the results from a third vector
`vmull`	$Qd[] \leftarrow Dn[] \times Dm[]$	Multiply corresponding elements from two vectors and store the results in a third vector
`vmlal`	$Qd[] \leftarrow Qd[] + (Dn[] \times Dm[])$	Multiply corresponding elements from two vectors and add the results in a third vector
`vmul`	$Qd[] \leftarrow Qd[] - (Dn[] \times Dm[])$	Multiply corresponding elements from two vectors and subtract the results from a third vector

Examples

```
1    vmul.i8     q1,q6,q8  @ Multiply elements
2    vmlal.s8    q0,d4,d5  @ Multiply-accumulate long
```

10.10.2 Multiply by Scalar

These instructions are used to multiply each element in a vector by a scalar:

vmul   Multiply by Scalar
vmla   Multiply Accumulate by Scalar
vmls   Multiply Subtract by Scalar
vmull   Multiply Long by Scalar
vmlal   Multiply Accumulate Long by Scalar
vmlsl   Multiply Subtract Long by Scalar

The long versions can be used to avoid overflow.

Syntax

```
v<op>.<type>    Vd, Vn, Dm[x]
v<op>l.<type>   Qd, Dn, Dm[x]
```

- <op> is either mul, mla. or mls.
- The valid choices for <type> are given in the following table:

Opcode	Valid Types
vmul	i16, i32, or f32
vmla	i16, i32, or f32
vmls	i16, i32, or f32
vmull	s16, s32, u16, or u32
vmlal	s16, s32, u16, or u32
vmlsl	s16, s32, u16, or u32

- x must be valid for the chosen <type>.

Operations

Name	Effect	Description
vmul	$Vd[] \leftarrow Vn[] \times Dm[x]$	Multiply corresponding elements from two vectors and store the results in a third vector
vmla	$Vd[] \leftarrow Vd[] + (Vn[] \times Dm[x])$	Multiply corresponding elements from two vectors and add the results in a third vector
vmul	$Vd[] \leftarrow Vd[] - (Vn[] \times Dm[x])$	Multiply corresponding elements from two vectors and subtract the results from a third vector
vmull	$Qd[] \leftarrow Dn[] \times Dm[x]$	Multiply corresponding elements from two vectors and store the results in a third vector
vmlal	$Qd[] \leftarrow Qd[] + (Dn[] \times Dm[x])$	Multiply corresponding elements from two vectors and add the results in a third vector
vmul	$Qd[] \leftarrow Qd[] - (Dn[] \times Dm[x])$	Multiply corresponding elements from two vectors and subtract the results from a third vector

Examples

```
1    vmul.s16    q1,q6,d7[1]  @ Multiply elements
2    vmlal.u32   q0,d4,d5[0]  @ Multiply-accumulate long
```

10.10.3 Fused Multiply Accumulate

A fused multiply accumulate operation does not perform rounding between the multiply and add operations. The two operations are *fused* into one. NEON provides the following fused multiply accumulate instructions:

vfma   Fused Multiply Accumulate
vfnma   Fused Negate Multiply Accumulate
vfms   Fused Multiply Subtract
vfnms   Fused Negate Multiply Subtract

Using the fused multiply accumulate can result in improved speed and accuracy for many computations that involve the accumulation of products.

Syntax

```
<op>{<cond>}.<prec> Fd, Fn, Fm
```

<op> is one of vfma, vfnma, vfms, or vfnms.
<cond> is an optional condition code.
<prec> may be either f32 or f64.

Operations

Name	Effect	Description
vfma	$Fd \leftarrow Fd + Fn \times Fm$	Multiply and accumulate
vnfnma	$Fd \leftarrow Fd + Fn \times -Fm$	Negate, multiply, and accumulate
vfms	$Fd \leftarrow Fd - Fn \times Fm$	Multiply and subtract
vnfms	$Fd \leftarrow Fd - Fn \times -Fm$	Negate multiply, and subtract

Examples

```
1    vfma.f64    d8, d0, d8
2    vfms.f32    s20, s24, s28
3    vfmsle.f32  s6, s0, s26
```

10.10.4 *Saturating Multiply and Double (Low)*

These instructions perform multiplication, double the results, and perform saturation:

vqdmull   Saturating Multiply Double (Low)
vqdmlal   Saturating Multiply Double Accumulate (Low)
vqdmlsl   Saturating Multiply Double Subtract (Low)

Syntax

```
vqd<op>l.<type>   Qd, Dn, Dm
vqd<op>l.<type>   Qd, Dn, Dm[x]
```

- <op> is either mul, mla. or mls.
- <type> must be either s16 or s32.

Operations

Name	Effect	Description
vqdmull	**if** second operand is scalar **then** $$Qd[] \leftarrow \lfloor Dn[] \times Dm[x] \times 2 \rceil$$ **else** $$Qd[] \leftarrow \lfloor Dn[] \times Dm[] \times 2 \rceil$$ **end if**	Multiply elements, double the results, and store in the destination vector with saturation
vqdmull	**if** second operand is scalar **then** $$Qd[] \leftarrow \lfloor Qd[] + Dn[] \times Dm[x] \times 2 \rceil$$ **else** $$Qd[] \leftarrow \lfloor Qd[] + Dn[] \times Dm[] \times 2 \rceil$$ **end if**	Multiply elements , double the results, and add to the destination vector with saturation
vqdmull	**if** second operand is scalar **then** $$Qd[] \leftarrow \lfloor Qd[] - Dn[] \times Dm[x] \times 2 \rceil$$ **else** $$Qd[] \leftarrow \lfloor Qd[] - Dn[] \times Dm[] \times 2 \rceil$$ **end if**	Multiply elements , double the results, and subtract from the destination vector with saturation

Examples

```
1    vqdmull.s16  q1,d6,d8    @ Multiply elements, double,
2                             @ saturate
3    vqdmlal.s32 q0,d4,d5[0] @ Multiply elements, double, round,
4                             @ saturate, accumulate
```

10.10.5 Saturating Multiply and Double (High)

These instructions perform multiplication, double the results, perform saturation, and store the high half of the results:

vqdmulh   Saturating Multiply Double (High)
vqrdmulh Saturating Multiply Double (High) and Round

Syntax

```
vq{r}dmulh.<type>  Vd, Vn, Vm
vq{r}dmulh.<type>  Vd, Vn, Dm[x]
```

• <type> must be either s16 or s32.

Operations

Name	Effect	Description

vqdmulh	$n \leftarrow$ size of <type>
	if second operand is scalar **then**
	    $Vd[] \leftarrow Vn[] \times Dm[x] \times 2 \gg n$
	else
	    $Vd[] \leftarrow Vn[] \times Vm[] \times 2 \gg n$
	end if
		Multiply elements, double the results and store the high half in the destination vector with saturation

vqrdmulh	$n \leftarrow$ size of <type>
	if second operand is scalar **then**
	    $Vd[] \leftarrow \|Vn[] \times Dm[x] \times 2\| \gg n$
	else
	    $Vd[] \leftarrow \|Vn[] \times Vm[] \times 2\| \gg n$
	end if
		Multiply elements, double the results, round, and store the high half in the destination vector with saturation

Examples

```
1    vqrdmulh.s16 q1,q6,q8   @ Multiply elements, double, round
2                            @ saturate, store high half
3    vqdmulh.s32 q0,q4,d5[0] @ Multiply elements, double,
4                            @ accumulate high half, saturate
```

10.10.6 Estimate Reciprocals

These instructions perform the initial estimates of the reciprocal values:

vrecpe   Reciprocal Estimate
vrsqrte   Reciprocal Square Root Estimate

These work on floating point and unsigned fixed point vectors. The estimates from this instruction are accurate to within about eight bits. If higher accuracy is desired, then the Newton-Raphson method can be used to improve the initial estimates. For more information, see the Reciprocal Step instruction.

Syntax

```
v<op>.<type>      Vd, Vm
```

- `<op>` is either `recpe` or `rsqrte`.
- `<type>` must be either `u32`, or `f32`.
- If <type> is `u32`, then the elements are assumed to be $U(1, 31)$ fixed point numbers, and the most significant fraction bit (bit 30) must be 1, and the integer part must be zero. The `vclz` and shift by variable instructions can be used to put the data in the correct format.
- The result elements are always `f32`.

Operations

Name	Effect	Description

`vrecpe`	$n \leftarrow$ # of elements
	for $0 \leq i < n$) **do**
	  $Vd[i] \leftarrow\approx (1 \div Vm[i])$
	end for
		Find an approximate reciprocal of each element in a vector

`vrsqrte`	$n \leftarrow$ # of elements
	for $0 \leq i < n$) **do**
	  $Vd[i] \leftarrow\approx (1 \div \sqrt{Vm[i]})$
	end for
		Find an approximate reciprocal square root of each element in a vector

Examples

```
1   vrecpe.u32 q1,q6  @ Get initial reciprocal estimates
2   vrecpe.f32 d4,d5  @ Get initial reciprocal estimates
```

10.10.7 Reciprocal Step

These instructions are used to perform one Newton-Raphson step for improving the reciprocal estimates:

```
vrecps    Reciprocal Step
vrsqrts   Reciprocal Square Root Step
```

For each element in the vector, the following equation can be used to improve the estimates of the reciprocals:

$$x_{n+1} = x_n(2 - dx_n),$$

where x_n is the estimated reciprocal from the previous step, and d is the number for which the reciprocal is desired. This equation converges to $\frac{1}{d}$ if x_0 is obtained using `vrecpe` on d. The `vrecps` instruction computes

$$x'_{n+1} = 2 - dx_n,$$

so one additional multiplication is required to complete the update step. The initial estimate x_0 must be obtained using the `vrecpe` instruction.

For each element in the vector, the following equation can be used to improve the estimates of the reciprocals of the square roots:

$$x_{n+1} = x_n \frac{3 - dx_n^2}{2},$$

where x_n is the estimated reciprocal from the previous step, and d is the number for which the reciprocal is desired. This equation converges to $\frac{1}{\sqrt{d}}$ if x_0 is obtained using `vrsqrte` on d. The `vrsqrts` instruction computes

$$x'_{n+1} = \frac{3 - dx_n}{2},$$

so two additional multiplications are required to complete the update step. The initial estimate x_0 must be obtained using the `vrsqrte` instruction.

Syntax

```
v<op>.<type>        Vd, Vn, Vm
```

- `<op>` is either `recps` or `rsqrts`.
- `<type>` must be either `u32`, or `f32`.

Operations

Name	Effect	Description

vrecpe	$n \leftarrow$ # of elements
	for $0 \leq i < n$ **do**
	  $Vd[i] \leftarrow 2 - Vn[i] \times Vm[i]$
	end for
		Perform most of the Newton-Raphson reciprocal improvement step.

vrsqrte	$n \leftarrow$ # of elements
	for $0 \leq i < n$ **do**
	  $Vd[i] \leftarrow (3 - Vn[i] \times Vm[i]) \div 2$
	end for
		Perform most of the Newton-Raphson reciprocal square root improvement step

Examples

```
1    @ Divide elements of q0 by elements of q1 and store in q3
2    @ Doing a loop and testing for convergence would be slow,
3    @ so we will just do two improvement steps and hope it is
4    @ close enough.
5    vrecpe.f32   q3,q1    @ Get initial reciprocal estimates
6    vrecps.f32   q4,q1,q3 @ Improve estimates
7    vmul.f32     q3,q3,q4 @ Finish improvement step
8    vrecps.f32   q4,q1,q3 @ Improve estimates
9    vmul.f32     q3,q3,q4 @ Finish improvement step
10   vmul.f32     q3,q3,q0 @ Perform division
```

10.11 Pseudo-Instructions

The GNU assembler supports five pseudo-instructions for NEON. Two of them are vcle and vclt, which were covered in Section 10.6.1. The other three are explained in the following sections.

10.11.1 Load Constant

This pseudo-instruction loads a constant value into every element of a NEON vector, or into a VFP single-precision or double-precision register:

vldr   Load Constant.

This pseudo-instruction will use vmov if possible. Otherwise, it will create an entry in the literal pool and use vldr.

Syntax

```
vldr{<cond>}.<type>      Vd, =<imm>
```

- <cond> is an optional condition code.
- <type> must be one of i8, i16, i32, i64, s8, s16, s32, s64, u8, u16, u32, u64, f32, or f64.
- <imm> is a value appropriate for the specified <type>.

Operations

Name	Effect	Description
vldr	$Vd \leftarrow$ <imm>	Load a constant

Examples

```
1   vldr.s8  d0,=45 @ load 45 into d0
2   vldr.s32 d0,=0xAAAAAAAA
```

10.11.2 Bitwise Logical Operations with Immediate Data

It is often useful to clear and/or set specific bits in a register. The following
pseudo-instructions can provide bitwise logical operations:

vand   Bitwise AND Immediate

vorn   Bitwise Complement and OR Immediate

Syntax

```
v<op>.<type>  Vd, #<imm>
```

- <op> must be either and, or orn.
- V must be either q or d to specify whether the operation involves quadwords or
  doublewords.
- <type> must be i8, i16, i32, or i64.
- <imm> is a 16-bit or 32-bit immediate value, which is interpreted as a pattern for filling the
  immediate operand. The following table shows acceptable patterns for <imm>, based on
  what was chosen for <type>:

<type>	
i8,i16	i32,i64
OxFFXY	OxFFFFFFXY
OxXYFF	OxFFFFXYFF
	OxFFXYFFFF
	OxXYFFFFFF

Operations

Name	Effect	Description
vand	$Vd \leftarrow Vd \land imm : imm\dots$	Logical OR
vorn	$Vd \leftarrow \neg(Vd \lor imm : imm\dots)$	Bit Clear

Examples

```
1    vand.i8   d0,#0x00FF
2    vorn.i32  d0,#0xAAFFFFFF
```

10.11.3 Vector Absolute Compare

The following pseudo-instructions perform comparisons between the absolute values of all of the corresponding elements of two vectors in parallel:

vacle   Absolute Compare Less Than or Equal
vaclt   Absolute Compare Less Than

The vector absolute compare instruction compares the absolute value of each element of a vector with the absolute value of the corresponding element in a second vector, and sets an element in the destination vector for each comparison. If the comparison is true, then all bits in the result element are set to one. Otherwise, all bits in the result element are set to zero. Note that summing the elements of the result vector (as signed integers) will give the two's complement of the number of comparisons which were true.

Syntax

```
vac<op>.f32    Vd, Vn, Vm
```

* <op> must be either lt or lt.
* V can be d or q.
* The operand element type must be f32.
* The result element type is i32.

Operations

Name	Effect	Description

vac<op>	**for** $i \in vector\_length$ **do**
	   if $|Fm[i]|<op>|Fn[i]|$
	   then
	     $Fd[i] \leftarrow 111\ldots$
	   else
	     $Fd[i] \leftarrow 000\ldots$
	   end if
	 end for
		Compare each scalar in Fn to the corresponding scalar in Fm. If the comparison is true, then set all bits in the corresponding scalar in Fd to one. Otherwise set all bits in the corresponding scalar in Fd to zero.

Examples

```
1    vacle.f32   d0,d4,d5
2    vaclt.f32   q2,q8,q9
```

10.12 Performance Mathematics: A Final Look at Sine

In Chapter 9, four versions of the sine function were given. Those implementations used scalar and VFP vector modes for single-precision and double-precision. Those previous implementations are already faster than the implementations provided by GCC, However, it may be possible to gain a little more performance by taking advantage of the NEON architecture. All versions of NEON are guaranteed to have a very large register set, and that fact can be used to attain better performance.

10.12.1 Single Precision

Listing 10.1 shows a single precision floating point implementation of the sine function, using the ARM NEON instruction set. It performs the same operations as the previous implementations of the sine function, but performs many of the calculations in parallel. This implementation is slightly faster than the previous version.

```
1    @@@ sin_N_f implements the sine function using NEON single
2    @@@ precision floating point. It computes sine by summing
3    @@@ the first 7 terms of the Taylor series.
4    @@@ ------------------------------------------------------------
5            .data
6            @@ The following is a table of constants used in the
7            @@ Taylor series approximation for sine
8            .align  8           @ Align to cache (256-byte boundary)
9    ctab:   .word   0x3F800000    @   1.000000000000000
10           .word   0xBE2AAAAB    @  -0.166666671633720
11           .word   0x3C088889    @   0.008333333767951
12           .word   0xB9500D01    @  -0.000198412701138
13           .word   0x3638EF1D    @   0.000002755731884
14           .word   0xB2D7322B    @  -0.000000025052108
15   @@ ------------------------------------------------------------
16           .text
17           .align 2
18           .global sin_N_f
19   sin_N_f:
20           @@ Load the entire table into d16-d18
21           ldr             r0,=ctab
22           vldmia          r0,{d16-d18}
```

```
23        @@ Calculate vectors holding powers of x as follows:
24        @@ d0 <- x, x^3
25        @@ d1 <- x^5, x^7
26        @@ d2 <- x^9, x^11
27        vmul.f32      s8,s0,s0    @ Put x^2 in s8 (d4[0])
28        vmul.f32      s9,s8,s8    @ Put x^4 in s9 (d4[1])
29        vmul.f32      s1,s8,s0    @ Put x^3 in s1 (d0[1])
30        vmov.f32      s8,s9       @ d4 <- 2 copies of x^4
31        vmul.f32      d1,d0,d4    @ Get x^5 and x^7
32        vmul.f32      d3,d0,d16   @ Do first 2 multiplies
33        vmul.f32      d2,d1,d4    @ Get x^9, x^11
34        vmla.f32      d3,d1,d17   @ Accumulate 2 multiplies
35        vmla.f32      d3,d2,d18   @ Accumulate last 2 multiplies
36        vadd.f32      s0,s6,s7    @ Final addition
37        mov           pc,lr       @ Return result in s0
```

Listing 10.1
NEON implementation of the $\sin x$ function using single precision.

10.12.2 Double Precision

Listing 10.2 shows a double precision floating point implementation of the sine function. This code is intended to run on ARMv7 and earlier NEON/VFP systems with the full set of 32 double-precision registers. NEON systems prior to ARMv8 do not have NEON SIMD instructions for double precision operations. This implementation is faster than Listing 9.4 because it uses a large number of registers, does not contain a loop, and is written carefully so that multiple instructions can be at different stages in the pipeline at the same time. This technique of gaining performance is known as *loop unrolling*.

```
1   @@@ sin_N_d implements the sine function using NEON double
2   @@@ precision floating point by summing the first ten terms of the
3   @@@ Taylor series.
4   @@@ Versions of NEON before ARMv8 do not support vectors of
5   @@@ double precision floating point, but we can use loop
6   @@@ unrolling and lots of registers to get good performance.
7   @@@ --------------------------------------------------------------
8         .data
9         @@ The following is a table of constants used in the
10        @@ Taylor series approximation for sine
11        .align  8              @ Align to cache (256-byte boundary)
12  ctab:  .word   0x55555555,0xBFC55555 @ -0.166666666666667
13         .word   0x11111111,0x3F811111 @  0.008333333333333
14         .word   0x1A01A01A,0xBF2A01A0 @ -0.000198412698413
15         .word   0xA556C734,0x3EC71DE3 @  0.000002755731922
```

```
16        .word    0x67F544E4,0xBE5AE645  @ -0.000000025052108
17        .word    0x13A86D09,0x3DE61246  @  0.000000000160590
18        .word    0xE733B81F,0xBD6AE7F3  @ -0.000000000000765
19        .word    0x7030AD4A,0x3CE952C7  @  0.000000000000003
20        .word    0x46814157,0xBC62F49B  @ -0.000000000000000
21  @@@ ----------------------------------------------------------------
22        .text
23        .align 2
24        .global sin_N_d
25  sin_N_d:
26        ldr      r0,=ctab       @ Load pointer to coefficients
27        vmul.f64 d5,d0,d0       @ Put x^2 in d5
28        vmov     d2,d0          @ Copy x to d2
29        vldmia   r0!,{d4}       @ load first coefficient
30        vmul.f64 d3,d5,d0       @ Put x^3 in d3
31        vmul.f64 d5,d5,d5       @ Put x^4 in d5
32        vldmia   r0!,{d24,d25}  @ load 2 more coefficients
33        vmla.f64 d0,d3,d4       @ d0 <- x - ((x^3)/3!) = t_1 + t_2
34        vmul.f64 d6,d5,d5       @ Put x^8 in d6
35        vmul.f64 d16,d2,d5      @ d16 <- x^5 (x*x^4)
36        vmul.f64 d17,d3,d5      @ d17 <- x^7 (x^3*x^4)
37        vldmia   r0!,{d26,d27}  @ load 2 more coefficients
38        vmul.f64 d18,d2,d6      @ d18 <- x^9 ( x*x^8)
39        vmul.f64 d19,d3,d6      @ d19 <- x^11 (x^3*x^8)
40        vmul.f64 d20,d16,d6     @ d20 <- x^13 (x^5*x^8)
41        vmul.f64 d21,d17,d6     @ d21 <- x^15 (x^7*x^8)
42        vldmia   r0!,{d28-d29}  @ load 2 more coefficients
43        vmul.f64 d22,d18,d6     @ d22 <- x^17 (x^9*x^8)
44        vmul.f64 d23,d19,d6     @ d23 <- x^19 (x^11*x^8)
45        @@ Calculate all of the remaining terms
46        vmul.f64 d16,d16,d24    @ d16 <- (x^5)/5! = t3
47        vmul.f64 d17,d17,d25    @ d17 <- -(x^7)/7! = t4
48        vldmia   r0!,{d30,d31}  @ load 2 more coefficients
49        vmul.f64 d18,d18,d26    @ d18 <- (x^9)/9! = t5
50        vmul.f64 d19,d19,d27    @ d19 <- -(x^11)/11! = t6
51        vmul.f64 d20,d20,d28    @ d20 <- (x^13)/13! = t7
52        vmul.f64 d21,d21,d29    @ d21 <- -(x^15)/15! = t8
53        vmul.f64 d22,d22,d30    @ d22 <- (x^17)/17! = t9
54        vmul.f64 d23,d23,d31    @ d23 <- -(x^19)/19! = t10
55        @@ Sum all of the terms
56        vadd.f64 d16,d16,d17    @ d16 <- t_3 + t_4
57        vadd.f64 d17,d18,d19    @ d17 <- t_5 + t_6
58        vadd.f64 d18,d20,d21    @ d18 <- t_7 + t_8
59        vadd.f64 d19,d22,d23    @ d19 <- t_9 + t_10
60        vadd.f64 d16,d16,d17    @ d16 <- t_3 + t_4 + t_5 + t_6
61        vadd.f64 d17,d18,d19    @ d17 <- t_7 + t_8 + t_9 + t_10
```

```
62    vadd.f64   d16,d16,d17   @ d16 <- sum of t_3 to t_10
63    vadd.f64   d0,d0,d16     @ final sum
64    mov        pc,lr
```

Listing 10.2
NEON implementation of the sin x function using double precision.

10.12.3 Performance Comparison

Table 10.4 compares the implementations from Listings 10.1 and 10.2 with the VFP vector implementations from Chapter 9 and the sine function provided by GCC. Notice that in every case, using vector mode VFP instructions is slower than the scalar VFP version. As mentioned previously, vector mode is deprecated on NEON processors. On NEON systems, vector mode is emulated in software. Although vector mode is supported, using it will result in reduced performance, because each vector instruction causes the operating system to take over and substitute a series of scalar floating point operations on-the-fly. A great deal of time was spent by the operating system software in emulating the VFP hardware vector mode.

When compiler optimization is not used, the single precision scalar VFP implementation achieves a speedup of about 2.51, and the NEON implementation achieves a speedup of about 3.30 compared to the GCC implementation. The double precision scalar VFP implementation achieves a speedup of about 1.62, and the loop-unrolled NEON implementation achieves a speedup of about 2.05 compared to the GCC implementation.

Table 10.4 Performance of sine function with various implementations

Optimization	Implementation	CPU seconds
None	Single Precision VFP scalar Assembly	1.74
	Single Precision VFP vector Assembly	27.09
	Single Precision NEON Assembly	1.32
	Single Precision C	4.36
	Double Precision VFP scalar Assembly	2.83
	Double Precision VFP vector Assembly	106.46
	Double Precision NEON Assembly	2.24
	Double Precision C	4.59
Full	Single Precision VFP scalar Assembly	1.11
	Single Precision VFP vector Assembly	27.15
	Single Precision NEON Assembly	0.96
	Single Precision C	1.69
	Double Precision VFP scalar Assembly	2.56
	Double Precision VFP vector Assembly	107.5.53
	Double Precision NEON Assembly	2.05
	Double Precision C	4.27

When the best possible compiler optimization is used (-Ofast), the single precision scalar VFP implementation achieves a speedup of about 1.52, and the NEON implementation achieves a speedup of about 1.76 compared to the GCC implementation. The double precision scalar VFP implementation achieves a speedup of about 1.67, and the loop-unrolled NEON implementation achieves a speedup of about 2.08 compared to the GCC implementation. The single precision NEON version was 1.16 times as fast as the VFP scalar version and the double precision NEON implementation was 1.25 times as fast as the VFP scalar implementation.

Although the VFP versions of the sine function ran without modification on the NEON processor, re-writing them for NEON resulted in significant performance improvement. Performance of the vectorized VFP code running on a NEON processor was abysmal. The take-away lesson is that a programmer can improve performance by writing some functions in assembly that are specifically targeted to run on an specific platform. However, assembly code which improves performance on one platform may actually result in very poor performance on a different platform. To achieve optimal or near-optimal performance, it is important for the programmer to be aware of exactly which hardware platform is being used.

10.13 *Alphabetized List of NEON Instructions*

Name	Page	Operation
vaba	339	Absolute Difference and Accumulate
vabal	339	Absolute Difference and Accumulate Long
vabd	339	Absolute Difference
vabdl	339	Absolute Difference Long
vabs	340	Absolute Value
vacge	324	Absolute Compare Greater Than or Equal
vacgt	324	Absolute Compare Greater Than
vacle	353	Absolute Compare Less Than or Equal
vaclt	353	Absolute Compare Less Than
vadd	335	Add
vaddhn	336	Add and Narrow
vaddl	335	Add Long
vaddw	335	Add Wide
vand	326	Bitwise AND
vand	352	Bitwise AND Immediate
vbic	326	Bit Clear
vbic	327	Bit Clear Immediate
vbif	328	Bitwise Insert if False
vbit	328	Bitwise Insert
vbsl	328	Bitwise Select
vceq	323	Compare Equal
vcge	323	Compare Greater Than or Equal
vcgt	323	Compare Greater Than
vcle	323	Compare Less Than or Equal
vcls	342	Count Leading Sign Bits

vclt	323	Compare Less Than
vclz	342	Count Leading Zero Bits
vcnt	342	Count Set Bits
vcvt	322	Convert Between Half and Single
vcvt	321	Convert Data Format
vdup	312	Duplicate Scalar
veor	326	Bitwise Exclusive-OR
vext	313	Extract Elements
vfma	346	Fused Multiply Accumulate
vfms	346	Fused Multiply Subtract
vfnma	346	Fused Negate Multiply Accumulate
vfnms	346	Fused Negate Multiply Subtract
vhadd	337	Halving Add
vhsub	337	Halving Subtract
vld<n>	305	Load Copies of Structured Data
vld<n>	307	Load Multiple Structured Data
vld<n>	303	Load Structured Data
vldr	351	Load Constant
vmax	341	Maximum
vmin	341	Minimum
vmla	343	Multiply Accumulate
vmla	345	Multiply Accumulate by Scalar
vmlal	344	Multiply Accumulate Long
vmlal	345	Multiply Accumulate Long by Scalar
vmls	343	Multiply Subtract
vmls	345	Multiply Subtract by Scalar
vmlsl	344	Multiply Subtract Long
vmlsl	345	Multiply Subtract Long by Scalar
vmov	310	Move Immediate
vmov	309	Move Between NEON and ARM
vmovl	311	Move and Lengthen
vmovn	311	Move and Narrow
vmul	343	Multiply
vmul	345	Multiply by Scalar
vmull	343	Multiply Long
vmull	345	Multiply Long by Scalar
vmvn	310	Move Immediate Negative
vneg	340	Negate
vorn	326	Bitwise Complement and OR
vorn	352	Bitwise Complement and OR Immediate
vorr	326	Bitwise OR
vorr	327	Bitwise OR Immediate
vpadal	338	Add Pairwises and Accumulate Long
vpadd	338	Add Pairwise
vpaddl	338	Add Pairwise Long
vpmax	341	Pairwise Maximum
vpmin	341	Pairwise Minimum
vqabs	340	Saturating Absolute Value
vqadd	335	Saturating Add
vqdmlal	347	Saturating Multiply Double Accumulate (Low)

10.14 Chapter Summary

NEON can dramatically improve performance of algorithms that can take advantage of data parallelism. However, compiler support for automatically vectorizing and using NEON instructions is still immature. NEON intrinsics allow C and C++ programmers to access NEON instructions, by making them look like C functions. It is usually just as easy and more concise to write NEON assembly code as it is to use the intrinsics functions. A careful assembly language programmer can usually beat the compiler, sometimes by a wide margin. The greatest gains usually come from converting an algorithm to avoid floating point, and taking advantage of data parallelism.

Exercises

10.1 What is the advantage of using IEEE half-precision? What is the disadvantage?

10.2 NEON achieved relatively modest performance gains on the sine function, when compared to VFP.

    (a) Why?

    (b) List some tasks for which NEON could significantly outperform VFP.

10.3 There are some limitations on the size of the structure that can be loaded or stored using the `vld<n>` and `vst<n>` instructions. What are the limitations?

10.4 The sine function in Listing 10.2 uses a technique known as "loop unrolling" to achieve higher performance. Name at least three reasons why this code is more efficient than using a loop?

10.5 Reimplement the fixed-point sine function from Listing 8.7 using NEON instructions. Hint: you should not need to use a loop. Compare the performance of your NEON implementation with the performance of the original implementation.

10.6 Reimplement Exercise 9.10 using NEON instructions.

10.7 Fixed point operations may be faster than floating point operations. Modify your code from the previous example so that it uses the following definitions for points and transformation matrices:

```
1  typedef int[3] point;    // Point is an array of S(15,16)
2  typedef int[4][4] matrix; // Matrix is a 2-D array of S(15,16)
```

Use saturating instructions and/or any other techniques necessary to prevent overflow. Compare the performance of the two implementations.

Accessing Devices

Devices

Chapter Outline

As mentioned in Chapter 1, a computer system consists of three main parts: the CPU, memory, and devices. The typical computing system has many devices of various types for performing specific functions. Some devices, such as data caches, are closely coupled to the CPU, and are typically controlled by executing special CPU instructions that can only be accessed in assembly language. However, most of the devices on a typical system are accessed and controlled through the system data bus. These devices appear to the programmer to be ordinary memory locations. The hardware in the system bus decodes the addresses coming from the CPU, and some addresses correspond to devices rather than memory. Fig. 11.1 shows the memory layout for a typical system. The exact locations of the devices and memory are chosen by the system hardware designers. From the programmer's standpoint, writing data to certain memory addresses results in the data being transferred to a device rather than stored in memory. The programmer must read documentation on the hardware design to determine exactly where the devices are in memory.

11.1 Accessing Devices Directly Under Linux

There are devices that allow data to be read or written from external sources, devices that can measure time, devices for moving data from one location in memory to another, devices for modifying the addresses of memory regions, and devices for even more esoteric purposes. Some devices are capable of sending signals to the CPU to indicate that they need attention, while others simply wait for the CPU to check on their status.

A modern computer system, such as the Raspberry Pi, has dozens or even hundreds of devices. Programmers write device driver software for each device. A device driver provides a

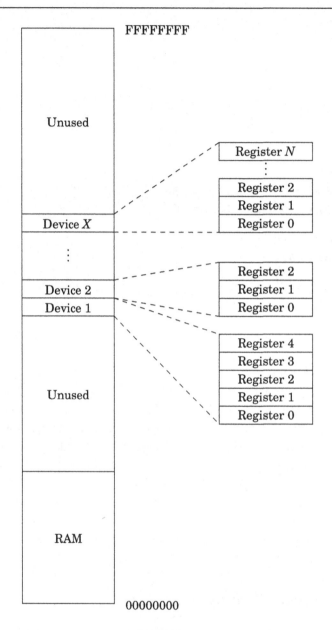

Figure 11.1
Typical hardware address mapping for memory and devices.

few standard function calls for each device, so that it can be used easily. The specific set of functions depends on the type of device and the design of the operating system. Operating system designers strive to define a small set of device types, and to define a standard software interface for each type in order to make devices interchangeable.

Devices are typically controlled by writing specific values to the device's internal *device registers*. For the ARM processor, access to most device registers is accomplished using the load and store instructions. Each device is assigned a *base address* in memory. This address corresponds with the first register inside the device. The device may also have other registers that are accessible at some pre-defined *offset address* from the base address. Some registers are read-only, some are write-only, and some are read-write. To use the device, the programmer must read from, and write appropriate data to, the correct device registers. For every device, there is a programmer's model and documentation explaining what each register in the device does. Some devices are well designed, easy to use, and well documented. Some devices are not, and the programmer must work harder to write software to use them.

Linux is a powerful, multiuser, multitasking operating system. The Linux kernel manages all of the devices and protects them from direct access by user programs. User programs are intended to access devices by making system calls. The kernel accesses the devices on behalf of the user programs, ensuring that an errant user program cannot misuse the devices and other resources on the system. Attempting to directly access the registers in any device will result in an exception. The kernel will take over and kill the offending process.

However, our programs will need direct access to the device registers. Linux allows user programs to gain direct access through the mmap() system call. Listing 11.1 shows how four devices can be mapped into the memory space of a user program on a Raspberry Pi. In most cases, the user program will need administrator privileges in order to perform the mapping. The operating system does not usually give permission for ordinary users to access devices directly. However Linux does provide the ability to change permissions on /dev/mem, or for user programs to run with elevated privileges.

```
1   @@@ Raspberry Pi devices
2   @@@ -----------------------------------------------------------
3   @@@ This file provides a function "IO_init" that will
4   @@@ map some devices into the user program's memory
5   @@@ space.  Pointers to the devices are stored in
6   @@@ global variables, and the user program can then
7   @@@ use those pointers to access the device registers.
8   @@@ -----------------------------------------------------------
9           .data
10  @@@ -----------------------------------------------------------
11  @@@ The following global variables will hold the addresses of
12  @@@ the devices that can be accessed directly after IO_init
13  @@@ has been called.
14          .global gpiobase
15  gpiobase:       .word   0
16          .global pwmbase
```

```
17   pwmbase :          .word   0
18          .global uart0base
19   uart0base:         .word   0
20          .global clkbase
21   clkbase :          .word   0
22
23   @@@ These are the addresses for the I/O devices (after
24   @@@ the firmware boot code has remapped them).
25          .equ   PERI_BASE,  0x20000000 @ start of all devices
26          @@ Base Physical Address of the GPIO registers
27          .equ   GPIO_BASE, (PERI_BASE + 0x200000)
28          @@ Base Physical Address of the PWM registers
29          .equ   PWM_BASE,  (PERI_BASE + 0x20C000)
30          @@ Base Physical Address of the UART 0 device
31          .equ   UART0_BASE,(PERI_BASE + 0x201000)
32          @@ Base Physical Address of the Clock/timer registers
33          .equ   CLK_BASE,  (PERI_BASE + 0x101000)
34
35          .equ   MAP_FAILED,-1
36          .equ   MAP_SHARED, 1
37          .equ   PROT_READ,  1
38          .equ   PROT_WRITE, 2
39          .equ   BLOCK_SIZE,(4*1024)
40
41          @@ some constants from fcntl.h
42          .equ   O_RDONLY,   00000000
43          .equ   O_WRONLY,   00000001
44          .equ   O_RDWR,     00000002
45          .equ   O_CREAT,    00000100
46          .equ   O_EXCL,     00000200
47          .equ   O_NOCTTY,   00000400
48          .equ   O_TRUNC,    00001000
49          .equ   O_APPEND,   00002000
50          .equ   O_NONBLOCK, 00004000
51          .equ   O_NDELAY,   O_NONBLOCK
52          .equ   O_SYNC,     00010000
53          .equ   O_FSYNC,    O_SYNC
54          .equ   O_ASYNC,    00020000
55
56   memdev:            .asciz  "/dev/mem"
57   successstr:        .asciz  "Successfully opened /dev/mem\n"
58   mappedstr:         .asciz  "Mapped %s device at 0x%08X\n"
59   openfailed:        .asciz  "IO_init: failed to open /dev/mem: "
60   mapfailedmsg:      .asciz  "IO_init: mmap of %s failed: "
61   gpiostr:           .asciz  "GPIO"
62   pwmstr:            .asciz  "PWM"
```

```
63   uart0str:        .asciz  "UART0"
64   clkstr:          .asciz  "CLK"
65
66
67         .text
68   @@@ ------------------------------------------------------------
69   @@@ IO_init() maps devices into memory space and stores their
70   @@@ addresses in global variables.
71   @@@ ------------------------------------------------------------
72         .global IO_init
73   IO_init:
74         stmfd   sp!,{r4,r5,lr}
75         @@ Try to open /dev/mem
76         ldr     r0,=memdev      @ load address of "/dev/mem"
77         ldr     r1,=(O_RDWR + O_SYNC) @ set up flags
78         bl      open            @ call the open syscall
79         cmp     r0,#0           @ check result
80         bge     init_opened     @ if open failed,
81         ldr     r0,=openfailed  @ print message and exit
82         bl      printf
83         bl      __errno_location
84         ldr     r0, [r0]
85         bl      strerror
86         bl      perror
87         mov     r0,#0           @ return 0 for failure
88         b       init_exit
89   init_opened:
90         @@ Open succeeded. Now map the devices
91         mov     r4,r0           @ move file descriptor to r4
92         ldr     r0,=successstr
93         bl      printf
94         @@ Map the GPIO device
95         mov     r0,r4           @ move file descriptor to r4
96         ldr     r1,=GPIO_BASE   @ address of device in memory
97         bl      trymap
98         cmp     r0,#MAP_FAILED
99         ldrne   r1,=gpiobase    @ if succeeded, load pointer
100        strne   r0,[r1]         @ if succeeded, store value
101        ldreq   r1,=gpiostr     @ if failed, load pointer to string
102        beq     map_failed_exit @ if failed, print message
103        mov     r2,r1
104        ldr     r2,[r2]
105        ldr     r0,=mappedstr   @ print success message
106        ldr     r1,=gpiostr
107        bl      printf
108        @@ Map the PWM device
```

```
109        mov     r0,r4           @ move file descriptor to r4
110        ldr     r1,=PWM_BASE    @ address of device in memory
111        bl      trymap
112        cmp     r0,#MAP_FAILED
113        ldrne   r1,=pwmbase     @ if succeeded, load pointer
114        strne   r0,[r1]         @ if succeeded, store value
115        ldreq   r1,=pwmstr      @ if failed, load pointer to string
116        beq     map_failed_exit @ if failed, print message
117        mov     r2,r1
118        ldr     r2,[r2]
119        ldr     r0,=mappedstr   @ print success message
120        ldr     r1,=pwmstr
121        bl      printf
122        @@ Map the UART0 device
123        mov     r0,r4           @ move file descriptor to r4
124        ldr     r1,=UART0_BASE  @ address of device in memory
125        bl      trymap
126        cmp     r0,#MAP_FAILED
127        ldrne   r1,=uart0base   @ if succeeded, load pointer
128        strne   r0,[r1]         @ if succeeded, store value
129        ldreq   r1,=uart0str    @ if failed, load pointer to string
130        beq     map_failed_exit @ if failed, print message
131        mov     r2,r1
132        ldr     r2,[r2]
133        ldr     r0,=mappedstr   @ print success message
134        ldr     r1,=uart0str
135        bl      printf
136        @@ Map the clock manager device
137        mov     r0,r4           @ move file descriptor to r4
138        ldr     r1,=CLK_BASE    @ address of device in memory
139        bl      trymap
140        cmp     r0,#MAP_FAILED
141        ldrne   r1,=clkbase     @ if succeeded, load pointer
142        strne   r0,[r1]         @ if succeeded, store value
143        ldreq   r1,=clkstr      @ if failed, load pointer to string
144        beq     map_failed_exit @ if failed, print message
145        mov     r2,r1
146        ldr     r2,[r2]
147        ldr     r0,=mappedstr   @ print success message
148        ldr     r1,=clkstr
149        bl      printf
150        @@ All mmaps have succeeded.
151        @@ Close file and return 1 for success
152        mov     r5,#1
153        b       init_close
154 map_failed_exit:
```

```
155         @@ At least one mmap failed. Print error,
156         @@ unmap everything and return
157         ldr    r0,=mapfailedmsg
158         bl     printf
159         bl     __errno_location
160         ldr    r0, [r0, #0]
161         bl     strerror
162         bl     perror
163         bl     IO_close
164         mov    r0,#0
165 init_close:
166         mov    r0,r4           @ close /dev/mem
167         bl     close
168 init_exit:
169         ldmfd  sp!,{r4,r5,pc}  @ return
170 @@@ ------------------------------------------------------------
171 @@@ trymap(int fd, unsigned offset) Calls mmap.
172 trymap: stmfd  sp!,{r5-r7,lr}
173         mov    r5,r1           @ copy address to r5
174         mov    r7,#0xFF         @ set up a mask for aligning
175         orr    r7,#0xF00
176         and    r6,r5,r7        @ get offset from page boundary
177         bic    r1,r5,r7        @ align phys addr to page boundary
178         stmfd  sp!,{r0,r1}     @ push last two params for mmap
179         mov    r0,#0           @ let kernel choose virt address
180         mov    r1,#BLOCK_SIZE
181         mov    r2,#(PROT_READ + PROT_WRITE)
182         mov    r3,#MAP_SHARED
183         bl     mmap
184         add    sp,sp,#8        @ pop params from stack
185         cmp    r0,#-1
186         addne  r0,r0,r6        @ add offset from page boundary
187         ldmfd  sp!,{r5-r7,pc}
188 @@@ ------------------------------------------------------------
189 @@@ IO_close unmaps all of the devices
190         .global IO_close
191 IO_close:
192         stmfd  sp!,{r4,r5,lr}
193         ldr    r4,=gpiobase    @ get address of first pointer
194         mov    r5,#4           @ there are 4 pointers
195 IO_closeloop:
196         ldr    r0,[r4]         @ load address of device
197         mov    r1,#BLOCK_SIZE
198         cmp    r0,#0
199         blgt   munmap          @ unmap it
200         mov    r0,#0
```

```
201        str     r0,[r4],#4        @ store and increment
202        subs    r5,r5,#1
203        bgt     IO_closeloop
204        ldmfd   sp!,{r4,r5,pc}
```

Listing 11.1
Function to map devices into the user program memory on a Raspberry Pi

Listing 11.2 shows how four devices can be mapped into the memory space of a user program on a pcDuino. The devices are equivalent to the devices mapped in Listing 11.1. Some of the devices are described in the following sections of this chapter. The pcDuino devices and Raspberry Pi devices operate differently, but provide similar functionality. Note that most of the code is the same for both listings. The only real differences between Listings 11.1 and 11.2 are the names of the devices and their hardware addresses.

```
 1
 2   @@@ pcDuino devices
 3   @@@ --------------------------------------------------------
 4   @@@ This file provides a function "IO_init" that will
 5   @@@ map some devices into the user program's memory
 6   @@@ space.  Pointers to the devices are stored in
 7   @@@ global variables, and the user program can then
 8   @@@ use those pointers to access the device registers.
 9   @@@ --------------------------------------------------------
10          .data
11   @@@ --------------------------------------------------------
12   @@@ The following global variables will hold the addresses of
13   @@@ the devices that can be accessed directly after IO_init
14   @@@ has been called.
15          .global gpiobase
16   gpiobase:       .word   0
17          .global pwmbase
18   pwmbase :       .word   0
19          .global uart2base
20   uart2base:      .word   0
21          .global ccubase
22   ccubase :       .word   0
23
24   @@@ These are the physical addresses for the I/O devices.
25          @@ Base Physical Address of the GPIO device
26          .equ    GPIO_BASE, 0x01C20800
27          @@ Base Physical Address of the PWM device
28          .equ    PWM_BASE,  0x01C20C00
29          @@ Base Physical Address of the UART2 device
```

```
30          .equ    UART2_BASE,0x01C28800
31          @@ Base Physical Address of the Clock Control Unit
32          .equ    CCU_BASE,  0x01C20000
33
34          .equ    MAP_FAILED,-1
35          .equ    MAP_SHARED, 1
36          .equ    PROT_READ,  1
37          .equ    PROT_WRITE, 2
38          .equ BLOCK_SIZE,(4*1024)
39
40          @@ some constants from fcntl.h
41          .equ    O_RDONLY,   00000000
42          .equ    O_WRONLY,   00000001
43          .equ    O_RDWR,     00000002
44          .equ    O_CREAT,    00000100
45          .equ    O_EXCL,     00000200
46          .equ    O_NOCTTY,   00000400
47          .equ    O_TRUNC,    00001000
48          .equ    O_APPEND,   00002000
49          .equ    O_NONBLOCK, 00004000
50          .equ    O_NDELAY,   O_NONBLOCK
51          .equ    O_SYNC,     00010000
52          .equ    O_FSYNC,    O_SYNC
53          .equ    O_ASYNC,    00020000
54
55  memdev:         .asciz  "/dev/mem"
56  successstr:     .asciz  "Successfully opened /dev/mem\n"
57  mappedstr:      .asciz  "Mapped %s device at 0x%08X\n"
58  openfailed:     .asciz  "IO_init: failed to open /dev/mem:\n "
59  mapfailedmsg:   .asciz  "IO_init: mmap of %s failed:\n "
60  gpiostr:        .asciz  "GPIO"
61  pwmstr:         .asciz  "PWM"
62  uart2str:       .asciz  "UART2"
63  ccustr:         .asciz  "CCU"
64
65          .text
66  @@@ ----------------------------------------------------------
67  @@@ IO_init() maps devices into memory space and stores their
68  @@@ addresses in global variables.
69  @@@ ----------------------------------------------------------
70          .global IO_init
71          .global IO_init
72  IO_init:
73          stmfd   sp!,{r4,r5,lr}
74          @@ Try to open /dev/mem
75          ldr     r0,=memdev      @ load address of "/dev/mem"
```

```
76         ldr      r1,=(O_RDWR + O_SYNC) @ set up flags
77         bl       open            @ call the open syscall
78         cmp      r0,#0           @ check result
79         bge      init_opened     @ jump if succeeded, else
80         ldr      r0,=openfailed  @ print message and exit
81         bl       printf
82         bl       __errno_location
83         ldr      r0, [r0]
84         bl       strerror
85         bl       perror
86         mov      r0,#0           @ return 0 for failure
87         b        init_exit
88  init_opened:
89         @@ Open succeeded. Print message and  map the devices
90         mov      r4,r0           @ move file descriptor to r4
91         ldr      r0,=successstr
92         bl       printf
93         @@ Map the GPIO device
94         mov      r0,r4           @ file descriptor for /dev/mem
95         ldr      r1,=GPIO_BASE   @ address of device in memory
96         bl       trymap
97         cmp      r0,#MAP_FAILED
98         ldrne    r1,=gpiobase    @ if succeeded, load pointer
99         strne    r0,[r1]         @ if succeeded, store value
100        ldreq    r1,=gpiostr     @ if failed, load pointer to string
101        beq      map_failed_exit @ if failed, print message
102        mov      r2,r1
103        ldr      r2,[r2]
104        ldr      r0,=mappedstr   @ print success message
105        ldr      r1,=gpiostr
106        bl       printf
107        @@ Map the PWM device
108        mov      r0,r4           @ file descriptor for /dev/mem
109        ldr      r1,=PWM_BASE    @ address of device in memory
110        bl       trymap
111        cmp      r0,#MAP_FAILED
112        ldrne    r1,=pwmbase     @ if succeeded, load pointer
113        strne    r0,[r1]         @ if succeeded, store value
114        ldreq    r1,=pwmstr      @ if failed, load pointer to string
115        beq      map_failed_exit @ if failed, print message
116        mov      r2,r1
117        ldr      r2,[r2]
118        ldr      r0,=mappedstr   @ print success message
119        ldr      r1,=pwmstr
120        bl       printf
121        @@ Map UART2
```

```
122        mov      r0,r4            @ file descriptor for /dev/mem
123        ldr      r1,=UART2_BASE   @ address of device in memory
124        bl       trymap
125        cmp      r0,#MAP_FAILED
126        ldrne    r1,=uart2base    @ if succeeded, load pointer
127        strne    r0,[r1]          @ if succeeded, store value
128        ldreq    r1,=uart2str     @ if failed, load pointer to string
129        beq      map_failed_exit  @ if failed, print message
130        mov      r2,r1
131        ldr      r2,[r2]
132        ldr      r0,=mappedstr    @ print success message
133        ldr      r1,=uart2str
134        bl       printf
135        @@ Map the clock control unit
136        mov      r0,r4            @ file descriptor for /dev/mem
137        ldr      r1,=CCU_BASE     @ address of device in memory
138        bl       trymap
139        cmp      r0,#MAP_FAILED
140        ldrne    r1,=ccubase      @ if succeeded, load pointer
141        strne    r0,[r1]          @ if succeeded, store value
142        ldreq    r1,=ccustr       @ if failed, load pointer to string
143        beq      map_failed_exit  @ if failed, print message
144        mov      r2,r1
145        ldr      r2,[r2]
146        ldr      r0,=mappedstr    @ print success message
147        ldr      r1,=ccustr
148        bl       printf
149        @@ All mmaps have succeeded.
150        @@ Close file and return 1 for success
151        mov      r5,#1
152        b        init_close
153 map_failed_exit:
154        @@ At least one mmap failed. Print error,
155        @@ unmap everything and return
156        ldr      r0,=mapfailedmsg
157        bl       printf
158        bl       __errno_location
159        ldr      r0, [r0, #0]
160        bl       strerror
161        bl       perror
162        bl       IO_close
163        mov      r0,#0
164 init_close:
165        mov      r0,r4            @ close /dev/mem
166        bl       close
167 init_exit:
```

```
168            ldmfd    sp!,{r4,r5,pc}   @ return
169    @@@ ------------------------------------------------------------
170    @@@ trymap(int fd, unsigned offset) Calls mmap.
171    trymap: stmfd    sp!,{r5-r7,lr}
172            mov      r5,r1            @ copy address to r5
173            mov      r7,#0xFF         @ set up a mask for aligning
174            orr      r7,#0xF00
175            and      r6,r5,r7         @ get offset from page boundary
176            bic      r1,r5,r7         @ align phys addr to page boundary
177            stmfd    sp!,{r0,r1}      @ push last two params for mmap
178            mov      r0,#0            @ let kernel choose virt address
179            mov      r1,#BLOCK_SIZE
180            mov      r2,#(PROT_READ + PROT_WRITE)
181            mov      r3,#MAP_SHARED
182            bl       mmap
183            add      sp,sp,#8         @ pop params from stack
184            cmp      r0,#-1
185            addne    r0,r0,r6         @ add offset from page boundary
186            ldmfd    sp!,{r5-r7,pc}
187    @@@ ------------------------------------------------------------
188    @@@ IO_close unmaps all of the devices
189            .global IO_close
190    IO_close:
191            stmfd    sp!,{r4,r5,lr}
192            ldr      r4,=gpiobase     @ get address of first pointer
193            mov      r5,#4            @ there are 4 pointers
194    IO_closeloop:
195            ldr      r0,[r4]          @ load address of device
196            mov      r1,#BLOCK_SIZE
197            cmp      r0,#0
198            blgt     munmap           @ unmap the device
199            mov      r0,#0
200            str      r0,[r4],#4       @ store and increment
201            subs     r5,r5,#1
202            bgt      IO_closeloop
203            ldmfd    sp!,{r4,r5,pc}
```

Listing 11.2
Function to map devices into the user program memory space on a pcDuino.

11.2 *General Purpose Digital Input/Output*

One type of device, commonly found on embedded systems, is the General Purpose I/O (GPIO) device. Although there are many variations on this device provided by different manufacturers, they all provide similar capabilities. The device provides a set of input and/or output bits, which allow signals to be transferred to or from the outside world. Each bit of input or output in a GPIO device is generally referred to as a pin, and a group of pins is referred to as a GPIO port. Ports commonly support 8 bits of input or output, but some devices have 16 or 32 bit ports. Some GPIO devices support multiple ports, and some systems have multiple GPIO devices in them.

A system with a GPIO device usually has some type of connector or wires that allow external inputs or outputs to be connected to the system. For example, the IBM PC has a type of GPIO device that was originally intended for communications with a parallel printer. On that platform, the GPIO device is commonly referred to as the parallel printer port.

Some GPIO devices, such as the one on the IBM PC, are arranged as sets of pins that can be switched as a group to either input or output. In many modern GPIO devices, each pin can be individually configured to accept or source different input and output voltages. On some devices, the amount of drive current available can be configured. Some include the ability to configure built-in pull-up and/or pull-down resistors. On most older GPIO devices, the input and output voltages are typically limited to the supply voltage of the GPIO device, and the device may be damaged by greater voltages. Newer GPIO devices generally can tolerate 5 V on inputs, regardless of the supply voltage of the device.

GPIO devices are very common in systems that are intended to be used for embedded applications. For most GPIO devices:

- individual pins or groups of pins can be configured,
- pins can be configured to be input or output,
- pins can be disabled so that they are neither input nor output,
- input values can be read by the CPU (typically high=1, low=0),
- output values can be read or written by the CPU, and
- input pins can be configured to generate interrupt requests.

Some GPIO devices may also have more advanced features, such as the ability to use Direct Memory Access (DMA) to send data without requiring the CPU to move each byte or word. Fig. 11.2 shows two common ways to use GPIO pins. Fig. 11.2A shows a GPIO pin that has been configured for input, and connected to a push-button switch. When the switch is open, the pull-up resistor pulls the voltage on the pin to a high state. When the switch is closed, the pin is pulled to a low state and some current flows through the pull-up resistor to ground. Typically, the pull-up resistor would be around $10\,k\Omega$. The specific value is not critical, but it must be high enough to limit the current to a small amount when the switch is closed. Fig. 11.2B shows a GPIO pin that is configured as an output and is being used to drive an LED. When a 1 is output on the pin, it is at the same voltage as Vcc (the power supply voltage), and no current flows. The LED is off. When a 0 is output on the pin, current is drawn through the resistor and the LED, and through the pin to ground. This causes the LED to be

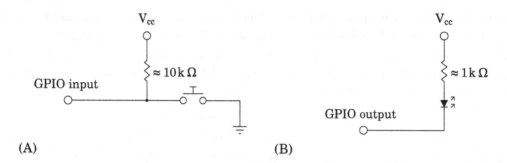

Figure 11.2

GPIO pins being used for input and output. (A) GPIO pin being used as input to read the state of a push-button switch. (B) GPIO pin being used as output to drive an LED.

illuminated. Selection of the resistor is not critical, but it must be small enough to light the LED without allowing enough current to destroy either the LED or the GPIO circuitry. This is typically around 1 kΩ. Note that, in general, GPIO pins can sink more current than they can source, so it is most common to connect LEDs and other devices in the way shown.

11.2.1 Raspberry Pi GPIO

The Broadcom BCM2835 system-on-chip contains 54 GPIO pins that are split into two banks. The GPIO pins are named using the following format: GPIOx, where x is a number between 0 and 53. The GPIO pins are highly configurable. Each pin can be used for general purpose I/O, or can be configured to serve up to six pre-defined alternate functions. Configuring a GPIO pin for an alternate function usually allows some other device within the BCM2835 to use the pin. For example, GPIO4 can be used

- for general purpose I/O,
- to send the signal generated by General Purpose Clock 0 to external devices,
- to send bit one of the Secondary Address Bus to external devices, or
- to receive JTAG data for programming the firmware of the device.

The last eight GPIO pins, GPIO46–GPIO53 have no alternate functions, and are used only for GPIO.

In addition to the alternate function, all GPIO pins can be configured individually as input or output. When configured as input, a pin can also be configured to detect when the signal changes, and to send an interrupt to the ARM CPU. Each input pin also has internal pull-up and pull-down resistors, which can be enabled or disabled by the programmer.

The GPIO pins on the BCM2835 SOC are very flexible and are quite complex, but are well designed and not difficult to program, once the programmer understands how the pins operate

Table 11.1 Raspberry Pi GPIO register map

Offset	Name	Description	Size	R/W
00_{16}	GPFSEL0	GPIO Function Select 0	32	R/W
04_{16}	GPFSEL1	GPIO Function Select 1	32	R/W
08_{16}	GPFSEL2	GPIO Function Select 2	32	R/W
$0C_{16}$	GPFSEL3	GPIO Function Select 3	32	R/W
10_{16}	GPFSEL4	GPIO Function Select 4	32	R/W
14_{16}	GPFSEL5	GPIO Function Select 5	32	R/W
$1C_{16}$	GPSET0	GPIO Pin Output Set 0	32	W
20_{16}	GPSET1	GPIO Pin Output Set 1	32	W
28_{16}	GPCLR0	GPIO Pin Output Clear 0	32	W
$2C_{16}$	GPCLR1	GPIO Pin Output Clear 1	32	W
34_{16}	GPLEV0	GPIO Pin Level 0	32	R
38_{16}	GPLEV1	GPIO Pin Level 1	32	R
40_{16}	GPEDS0	GPIO Pin Event Detect Status 0	32	R/W
44_{16}	GPEDS1	GPIO Pin Event Detect Status 1	32	R/W
$4C_{16}$	GPREN0	GPIO Pin Rising Edge Detect Enable 0	32	R/W
50_{16}	GPREN1	GPIO Pin Rising Edge Detect Enable 1	32	R/W
58_{16}	GPFEN0	GPIO Pin Falling Edge Detect Enable 0	32	R/W
$5C_{16}$	GPFEN1	GPIO Pin Falling Edge Detect Enable 1	32	R/W
64_{16}	GPHEN0	GPIO Pin High Detect Enable 0	32	R/W
68_{16}	GPHEN1	GPIO Pin High Detect Enable 1	32	R/W
70_{16}	GPLEN0	GPIO Pin Low Detect Enable 0	32	R/W
74_{16}	GPLEN1	GPIO Pin Low Detect Enable 1	32	R/W
$7C_{16}$	GPAREN0	GPIO Pin Async. Rising Edge Detect 0	32	R/W
80_{16}	GPAREN1	GPIO Pin Async. Rising Edge Detect 1	32	R/W
88_{16}	GPAFEN0	GPIO Pin Async. Falling Edge Detect 0	32	R/W
$8C_{16}$	GPAFEN1	GPIO Pin Async. Falling Edge Detect 1	32	R/W
94_{16}	GPPUD	GPIO Pin Pull-up/down Enable	32	R/W
98_{16}	GPPUDCLK0	GPIO Pin Pull-up/down Enable Clock 0	32	R/W
$9C_{16}$	GPPUDCLK1	GPIO Pin Pull-up/down Enable Clock 1	32	R/W

and what the various registers do. There are 41 registers that control the GPIO pins. The base address for the GPIO device is `20200000`. The 41 registers and their offsets from the base address are shown in Table 11.1.

Setting the GPIO pin function

The first six 32-bit registers in the device are used to select the function for each of the 54 GPIO pins. The function of each pin is controlled by a group of three bits in one of these registers. The mapping is very regular. Bits 0–2 of GPIOFSEL0 control the function of GPIO pin 0. Bits 3–5 of GPIOFSEL0 control the function of GPIO pin 1, and so on, up to bits 27–29 of GPIOFSEL0, which control the function of GPIO pin 9. The next pin, pin 10, is controlled by bits 0–2 of GPIOFSEL1. The pins are assigned in sequence through the remaining bits, until bits 27–29, which control GPIO pin 19. The remaining four GPIOFSEL registers control the remaining GPIO pins. Note that bits 30 and 31 of all of the GPIOFSEL registers are not used, and most of the bits in GPIOFSEL5 are not assigned to any pin. The meaning of each combination of the three bits is shown in Table 11.2. Note that the encoding is not as simple as one might expect.

The procedure for setting the function of a GPIO pin is as follows:

- Determine which GPIOFSEL register controls the desired pin.
- Determine which bits of the GPIOFSEL register are used.
- Determine what the bit pattern should be.
- Read the GPIOFSEL register.
- Clear the correct bits using the `bic` instruction.
- Set them to the correct pattern using the `orr` instruction.

For example, Listing 11.3 shows the sequence of code which would be used to set GPIO pin 26 to alternate function 1.

Setting GPIO output pins

To use a GPIO pin for output, the function select bits for that pin must be set to 001. Once that is done, the output can be driven high or low by using the GPSET and GPCLR registers. GPIO pin 0 is set to a high output by writing a 1 to bit 0 of GPSET0, and it is set to low output by writing a 1 to bit 0 of GPCLR0. GPIO pin 1 is similarly controlled by bit 1 in GPSET0 and GPCLR0. Each of the GPIO pins numbered 0 through 31 is assigned one bit in GPSET0 and one bit in GPCLR0. GPIO pin 32 is assigned to bit 0 of GPSET1 and GPCLR1, GPIO pin 33 is assigned to bit 1 of GPSET1 and GPCLR1, and so on. Since there are only 54 GPIO pins,

Table 11.2 GPIO pin function select bits

MSB-LSB	Function
000	Pin is an input
001	Pin is an output
100	Pin performs alternate function 0
101	Pin performs alternate function 1
110	Pin performs alternate function 2
111	Pin performs alternate function 3
011	Pin performs alternate function 4
010	Pin performs alternate function 5

```
1        .equ GPIOFSEL2, 0x7E200008
2          ⋮
3        ldr     r0,=GPIOFSEL2
4        ldr     r1,[r0]
5        bic     r1,r1,#0b111 lsl #(3*7)
6        orr     r1,r1,#0b101 lsl #(3*7)
7        str     r1,[r0]
```

Listing 11.3
ARM assembly code to set GPIO pin 26 to alternate function 1.

bits 22–31 of GPSET1 and GPCLR1 are not used. The programmer can set or clear several outputs simultaneously by writing the appropriate bits in the GPSET and GPCLR registers.

Reading GPIO input pins

To use a GPIO pin for input, the function select bits for that pin must be set to 000. Once that is done, the input can be read at any time by reading the appropriate GPLEV register and examining the bit that corresponds with the input pin. GPIO pin 0 is read as bit 0 of GPLEV0, GPIO pin 1 is similarly read as bit 1 of GPLEV1. Each of the GPIO pins numbered 0 through 31 is assigned one bit in GPLEV0. GPIO pin 32 is assigned to bit 0 of GPLEV1, GPIO pin 33 is assigned to bit 1 of GPLEV1, and so on. Since there are only 54 GPIO pins, bits 22–31 of GPLEV1 are not used. The programmer can read the status of several inputs simultaneously by reading one of the GPLEV registers and examining the bits corresponding to the appropriate pins.

Enabling internal pull-up or pull-down

Input pins can be configured with internal pull-up or pull-down resistors. This can simplify the design of the system. For instance, Fig. 11.2A, shows a push-button switch connected to an input, with an external pull-up resistor. That resistor is unnecessary if the internal pull-up for that pin is enabled.

Enabling the pull-up or pull-down is a two step process. The first step is to configure the type of change to be made, and the second step is to perform that change on the selected pin(s). The first step is accomplished by writing to the GPPUD register. The valid binary control codes are shown in Table 11.3.

Table 11.3 GPPUD control codes

Code	Function
00	Disable pull-up and pull-down
01	Enable pull-down
10	Enable pull-up

Once the GPPUD register is configured, the selected operation can be performed on multiple pins by writing to one or both of the GPPUDCLK registers. GPIO pins are assigned to bits in these two registers in the same way as the pins are assigned in the GPLEV, GPSET, and GPCLR registers. Writing 1 to bit 0 of GPPUDCLK0 will configure the pull-up or pull-down for GPIO pin 0, according to the control code that is currently in the GPPUD register.

Detecting GPIO events

The GPEDS registers are used for detecting events that have occurred on the GPIO pins. For instance a pin may have transitioned from low to high, and back to low. If the CPU does not read the GPLEV register often enough, then such an event could be missed. The GPEDS registers can be configured to capture such events so that the CPU can detect that they occurred.

GPIO pins are assigned to bits in these two registers in the same way as the pins are assigned in the GPLEV, GPSET, and GPCLR registers. If bit 1 of GPEDS0 is set, then that indicates that an event has occurred on GPIO pin 0. Writing a 0 to that bit will clear the bit and allow the event detector to detect another event. Each pin can be configured to detect specific types of events by writing to the GPREN, GPHEN, GPLEN, GPAREN, and GPAFEN registers. For more information, refer to the *BCM2835 ARM Peripherals* manual.

GPIO pins available on the Raspberry Pi

The Raspberry Pi provides access to several of the 54 GPIO pins through the expansion header. The expansion header is a group of physical pins located in the corner of the Raspberry Pi board. Fig. 11.3 shows where the header is located on the Raspberry Pi. Wires can be connected to these pins and then the GPIO device can be programmed to send and/or receive digital information. Fig. 11.4 shows which signals are attached to the various pins. Some of the pins are used to provide power and ground to the external devices.

Table 11.4 shows some useful alternate functions available on each pin of the Raspberry Pi expansion header. Many of the alternate functions available on these pins are not really useful. Those functions have been left out of the table. The most useful alternate functions are probably GPIO 14 and 15, which can be used for serial communication, and GPIO 18, which can be used for pulse width modulation. Pulse width modulation is covered in Section 12.2, and serial communication is covered in Section 13.2. The Serial Peripheral Interface (SPI) functions could also be useful for connecting the Raspberry Pi to other devices which support SPI. Also, the SDA and SCL functions could be used to communicate with I^2C devices.

11.2.2 pcDuino GPIO

The AllWinner A10/A20 system-on-chip contains 175 GPIO pins, which are arranged in seven ports. Each of the seven ports is identified by a letter between "A" and "I." The ports are part of the PIO device, which is mapped at address $01C20800_{16}$. The GPIO pins are named

Figure 11.3
The Raspberry Pi expansion header location.

using the following format: P*N*x, where *N* is a letter between "A" and "I" indicating the port, and *x* is a number indicating a pin on the given port. The assignment of pins to ports is somewhat irregular, as shown in Table 11.5. Some ports have as many as 28 physical pins, while others have as few as six. However, the layout of the registers in the device is very regular. Given any port and pin combination, finding the correct registers and sets of bits within the registers, is very straightforward.

Each of the 9 ports is controlled by a set of 9 registers, for a total of 81 registers. There are seven additional registers that can be used to configure pins as interrupt sources. Interrupt processing is explained in Section 14.2. All of the port and interrupt registers together make a total of 88 registers for the GPIO device. The complete register map with the offset of each register from the device base address is shown in Table 11.6.

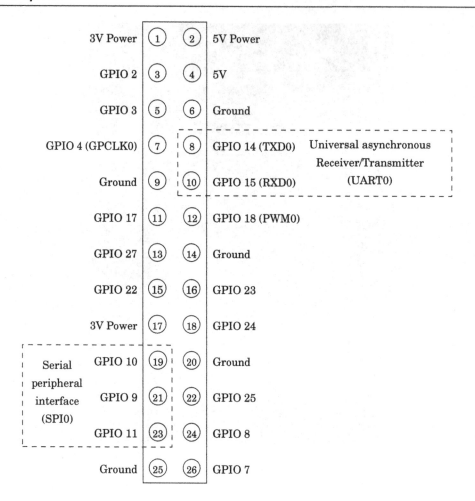

Figure 11.4
The Raspberry Pi expansion header pin assignments.

The GPIO pins are highly configurable. Each pin can be used either for general purpose I/O, or can be configured to serve one of up to six pre-defined alternate functions. Configuring a GPIO pin for an alternate function usually allows some other device within the A10/A20 SOC to use the pin. For example PB2 (pin 2 of port B) can be used for general purpose I/O, or can be used to output the signal from a Pulse Width Modulator (PWM) device (explained in Section 12.2). Each input pin also has internal pull-up and pull-down resistors, which can be enabled or disabled by the programmer.

Setting the GPIO pin function

The first four registers for each port are used to configure the functions for each of the pins. The function of each pin is controlled by three bits in one of the four configuration registers.

Table 11.4 Raspberry Pi expansion
header useful alternate functions

	Alternate Function	
Pin	0	5
GPIO 2	SDA1	
GPIO 3	SCL1	
GPIO 4	GPCLK0	
GPIO 7	SPI0_CE1_N	
GPIO 8	SPI0_CE0_N	
GPIO 9	SPI0_MISO	
GPIO 10	SPI0_MOSI	
GPIO 11	SPI0_SCLK	
GPIO 14	TXD0	TXD1
GPIO 15	RXD0	RXD1
GPIO 18	PCM_CLK	PWM0

Table 11.5 Number of pins available on
each of the AllWinner A10/A20 PIO
ports

Port	Pins
A	18
B	24
C	25
D	28
E	12
F	6
G	12
H	28
I	22

Pins 0–7 are controlled using configuration register 0. Pins 8–15 are controlled by configuration register 1, and so on. The assignment of pins to control bits is shown in Fig. 11.5. Note that eight pins are controlled by each register, and there is an unused bit between each group of three bits.

Each GPIO pin can be configured by writing a 3-bit code to the appropriate location in the correct port configuration register. The meanings of each possible code is shown in Table 11.7. For example, to configure port A, pin 10 (PA10) for output, the 3-bit code 001 must be written to bits 8–10 the PA_CFG1 register, without changing any other bit in the register. Listing 11.4 shows how this operation can be accomplished.

Table 11.6 Registers in the AllWinner GPIO device

Offset	Name	Description
000_{16}	PA_CFG0	Function select for Port A, Pins 0–7
004_{16}	PA_CFG1	Function select for Port A, Pins 8–15
008_{16}	PA_CFG2	Function select for Port A, Pins 16–17
$00C_{16}$	PA_CFG3	Not used
010_{16}	PA_DAT	Port A Data Register
014_{16}	PA_DRV0	Port A Multi-driving, Pins 0–15
018_{16}	PA_DRV1	Port A Multi-driving, Pins 16–17
$01C_{16}$	PA_PULL0	Port A Pull-Up/-Down, Pins 0–15
020_{16}	PA_PULL1	Port A Pull-Up/-Down, Pins 16–17
024_{16}	PB_CFG0	Function select for Port B, Pins 0–7
028_{16}	PB_CFG1	Function select for Port B, Pins 8–15
$02C_{16}$	PB_CFG2	Function select for Port B, Pins 16–23
030_{16}	PB_CFG3	Not used
034_{16}	PB_DAT	Port B Data Register
038_{16}	PB_DRV0	Port B Multi-driving, Pins 0–15
$03C_{16}$	PB_DRV1	Port B Multi-driving, Pins 16–23
040_{16}	PB_PULL0	Port B Pull-Up/-Down, Pins 0–15
044_{16}	PB_PULL1	Port B Pull-Up/-Down, Pins 16–23
048_{16}	PC_CFG0	Function select for Port C, Pins 0–7
$04C_{16}$	PC_CFG1	Function select for Port C, Pins 8–15
050_{16}	PC_CFG2	Function select for Port C, Pins 16–23
054_{16}	PC_CFG3	Function select for Port C, Pin 24
058_{16}	PC_DAT	Port C Data Register
$05C_{16}$	PC_DRV0	Port C Multi-driving, Pins 0–15
060_{16}	PC_DRV1	Port C Multi-driving, Pins 16–23
064_{16}	PC_PULL0	Port C Pull-Up/-Down, Pins 0–15
068_{16}	PC_PULL1	Port C Pull-Up/-Down, Pins 16–23
$06C_{16}$	PD_CFG0	Function select for Port D, Pins 0–7
070_{16}	PD_CFG1	Function select for Port D, Pins 8–15
074_{16}	PD_CFG2	Function select for Port D, Pins 16–23
078_{16}	PD_CFG3	Function select for Port D, Pins 24–27
$07C_{16}$	PD_DAT	Port D Data Register
080_{16}	PD_DRV0	Port D Multi-driving, Pins 0–15
084_{16}	PD_DRV1	Port D Multi-driving, Pins 16–27
088_{16}	PD_PULL0	Port D Pull-Up/-Down, Pins 0–15
$08C_{16}$	PD_PULL1	Port D Pull-Up/-Down, Pins 16–27
090_{16}	PE_CFG0	Function select for Port E, Pins 0–7
094_{16}	PE_CFG1	Function select for Port E, Pins 8–11
098_{16}	PE_CFG2	Not used
$09C_{16}$	PE_CFG3	Not used
$0A0_{16}$	PE_DAT	Port E Data Register
$0A4_{16}$	PE_DRV0	Port E Multi-driving, Pins 0–11
$0A8_{16}$	PE_DRV1	Not used
$0AC_{16}$	PE_PULL0	Port E Pull-Up/-Down, Pins 0–11
$0B0_{16}$	PE_PULL1	Not used
$0B4_{16}$	PF_CFG0	Function select for Port F, Pins 0–5

$0B8_{16}$	PF_CFG1	Not used
$0BC_{16}$	PF_CFG2	Not used
$0C0_{16}$	PF_CFG3	Not used
$0C4_{16}$	PF_DAT	Port F Data Register
$0C8_{16}$	PF_DRV0	Port F Multi-driving, Pins 0–5
$0CC_{16}$	PF_DRV1	Not used
$0D0_{16}$	PF_PULL0	Port F Pull-Up/-Down, Pins 0–5
$0D4_{16}$	PF_PULL1	Not used
$0D8_{16}$	PG_CFG0	Function select for Port G, Pins 0–7
$0DC_{16}$	PG_CFG1	Function select for Port G, Pins 8–11
$0E0_{16}$	PG_CFG2	Not used
$0E4_{16}$	PG_CFG3	Not used
$0E8_{16}$	PG_DAT	Port G Data Register
$0EC_{16}$	PG_DRV0	Port G Multi-driving, Pins 0–11
$0F0_{16}$	PG_DRV1	Not used
$0F4_{16}$	PG_PULL0	Port G Pull-Up/-Down, Pins 0–11
$0F8_{16}$	PG_PULL1	Not used
$0FC_{16}$	PH_CFG0	Function select for Port H, Pins 0–7
100_{16}	PH_CFG1	Function select for Port H, Pins 8–15
104_{16}	PH_CFG2	Function select for Port H, Pins 16–23
108_{16}	PH_CFG3	Function select for Port H, Pins 24–27
$10C_{16}$	PH_DAT	Port H Data Register
110_{16}	PH_DRV0	Port H Multi-driving, Pins 0–15
114_{16}	PH_DRV1	Port H Multi-driving, Pins 16–27
118_{16}	PH_PULL0	Port H Pull-Up/-Down, Pins 0–15
$11C_{16}$	PH_PULL1	Port H Pull-Up/-Down, Pins 16–27
120_{16}	PI_CFG0	Function select for Port I, Pins 0–7
124_{16}	PI_CFG1	Function select for Port I, Pins 8–15
128_{16}	PI_CFG2	Function select for Port I, Pins 16–21
$12C_{16}$	PI_CFG3	Not used
130_{16}	PI_DAT	Port I Data Register
134_{16}	PI_DRV0	Port I Multi-driving, Pins 0–15
138_{16}	PI_DRV1	Port I Multi-driving, Pins 16–21
$13C_{16}$	PI_PULL0	Port I Pull-Up/-Down, Pins 0–15
140_{16}	PI_PULL1	Port I Pull-Up/-Down, Pins 16–21
200_{16}	PIO_INT_CFG0	PIO Interrupt Configure Register 0
204_{16}	PIO_INT_CFG1	PIO Interrupt Configure Register 1
208_{16}	PIO_INT_CFG2	PIO Interrupt Configure Register 2
$20C_{16}$	PIO_INT_CFG3	PIO Interrupt Configure Register 3
210_{16}	PIO_INT_CTL	PIO Interrupt Control Register
214_{16}	PIO_INT_STATUS	PIO Interrupt Status Register
218_{16}	PIO_INT_DEB	PIO Interrupt Debounce Register

Table 11.7 Allwinner A10/A20 GPIO pin
function select bits

MSB-LSB	Function
000	Pin is an input
001	Pin is an output
010	Pin performs alternate function 0
011	Pin performs alternate function 1
100	Pin performs alternate function 2
101	Pin performs alternate function 3
110	Pin performs alternate function 4
111	Pin performs alternate function 5

```
1       .equ PA_CFG1, (0x01C20800 + 0x004)
2        ⋮
3       ldr     r0,=PA_CFG1
4       ldr     r1,[r0]
5       bic     r1,r1,#0b111 lsl #(2*4)
6       orr     r1,r1,#0b001 lsl #(2*4)
7       str     r1,[r0]
```

Listing 11.4
ARM assembly code to configure PA10 for output.

Reading and setting GPIO pins

An output pin can be set to a high state by setting the corresponding bit in the correct port data register. Likewise the pin can be set to a low state by clearing its corresponding bit. Care must be taken to avoid changing any other bits in the port data register. Listing 11.5 shows how this operation can be accomplished for setting a port to output a high state. To set the port output to a low state, the orr instruction would be replaced with a bic instruction.

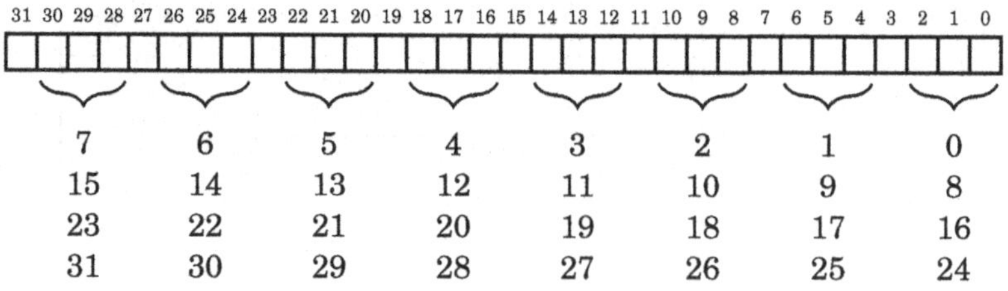

Figure 11.5
Bit-to-pin assignments for PIO control registers.

```
1        .equ PA_DAT, (0x01C20800 + 0x010)
2          :
3        ldr     r0,=PA_DAT
4        ldr     r1,[r0]
5        orr     r1,r1,#1 lsl #9
6        str     r1,[r0]
```

Listing 11.5
ARM assembly code to set PA10 to output a high state.

```
1        .equ PI_DAT, (0x01C20800 + 0x130)
2          :
3        ldr     r0,=PI_DAT
4        ldr     r1,[r0]
5        ands    r1,r1,#1 lsl #13 % set or clear Z flag
```

Listing 11.6
ARM assembly code to read the state of PI14 and set or clear the Z flag.

To determine the current state of an output pin or read an input pin, the programmer can read the contents of the correct port data register and use bitwise logical operations to isolate the appropriate bit. For example, to read the state of pin 14 of port I (PI14), the programmer would read the PI_DAT register and mask all bits except bit 14. Listing 11.6 shows how this operation can be accomplished. Another method would be to use the tst instruction, rather than the ands instruction, to set the CPSR flags.

Enabling internal pull-up or pull-down
Input pins can be configured with internal pull-up or pull-down resistors. This can simplify the design of the system. For instance, Fig. 11.2a, shows a push-button switch connected to an input with an external pull-up resistor. That resistor is unnecessary if the internal pull-up for that pin is enabled. Each pin is assigned two bits in one of the port pull-up/-down registers. The pull-up and pull-down resistors for pin 0 on port B are controlled using bits 0 and 1 of the

Table 11.8 Pull-up and pull-down
resistor control codes

Code	Function
00	Disable pull-up and pull-down
01	Enable pull-up
10	Enable pull-down
11	Reserved

PB_PULL0 register. Likewise the pull-up and pull-down resistors for pin 19 of port C are controlled using bits 6 and 7 of the PC_PULL1 register. Table 11.8 shows the bit patterns used to configure the pull-up and pull-down resisters for a pin.

Detecting GPIO events

When configured as an input, most of the pins on the pdDuino can be configured to generate an interrupt, which notifies the CPU than an event has occurred. Configuration of interrupts is beyond the scope of this chapter. It is accomplished using the PIO_INT registers.

GPIO pins available on the pcDuino

The pcDuino provides access to several of the 175 GPIO pins through the expansion headers. Fig. 11.6 shows where the headers are located on the pcDuino. Wires can be plugged into the holes in these headers and then the GPIO device can be programmed to send and/or receive digital and/or analog signals. The physical layout of the pcDuino header makes it compatible with a wide range of expansion modules designed for the Arduino family of microcontroller boards.

Figure 11.6
The pcDuino header locations.

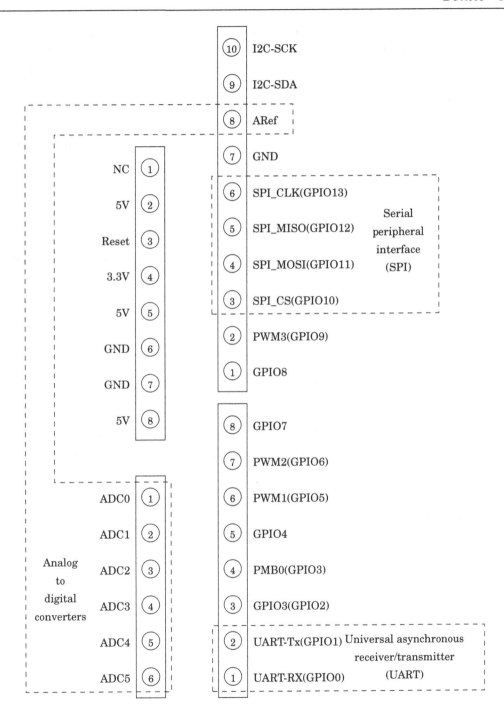

Figure 11.7
The pcDuino header pin assignments.

Some of the header holes can provide power and ground to the external devices. Analog signals can be read into the pcDuino using the ADC header connections. Fig. 11.7 shows the pcDuino names for the signals that are available on the headers. Table 11.9 shows how the pcDuino header signal names are mapped to the actual port pins on the AllWinner A10/A20 chip. It also shows the most useful alternate functions available on each of the pins. Many alternate functions are left out of the table because they are not really useful. Note that the pcDunio and the Raspberry Pi both provide pins to perform PWM, UART communications, and SPI.

11.3 Chapter Summary

All input and output are accomplished by using devices. There are many types of devices, and each device has its own set of registers which are used to control the device. The programmer must understand the operation of the device and the use of each register in order to use the device at a low level. Computer system manufacturers usually can provide documentation providing the necessary information for low-level programming. The quality of the documentation can vary greatly, and a general understanding of various types of devices can help in deciphering poor or incomplete documentation.

There are two major tasks where programming devices at the register level is required: operating system drivers and very small embedded systems. Operating systems provide an abstract view of each device and this allows programmers to use them more easily. However, someone must write that driver, and that person must have intimate knowledge of the device. On very small systems, there may not be a driver available. In that case, the device must be

Table 11.9 pcDuino GPIO pins and function select code assignments.

pcDuino Pin Name	Port	Pin	Function Select Code Assignment			
			010	011	100	110
UART-Rx(GPIO0)	I	19		UART2_RX		EINT31
UART-Tx(GPIO1)	I	18		UART2_TX		EINT30
GPIO3(GPIO2)	H	7			UART5_RX	EINT7
PWM0(GPIO3)	H	6			UART5_TX	EINT6
GPIO4	H	8				EINT8
PWM1(GPIO5)	B	2	PWM0			
PWM2(GPIO6)	I	3	PWM1			
GPIO7	H	9				EINT9
GPIO8	H	10				EINT10
PWM3(GPIO9)	H	5				EINT5
SPI_CS(GPIO10)	I	10	SPI0_CS0	UART5_TX		EINT22
SPI_MOSI(GPIO11)	I	12	SPI0_MOSI	UART6_TX	CLK_OUT_A	EINT24
SPI_MISO(GPIO12)	I	13	SPI0_MISO	UART6_RX	CLK_OUT_B	EINT25
SPI_CLK(GPIO13)	I	11	SPI0_CLK	UART5_RX		EINT23

accessed directly. Even when an operating system provides a driver, it is sometimes necessary or desirable for the programmer to access the device directly. For example, some devices may provide modes of operation or capabilities that are not supported by the operating system driver. Linux provides a mechanism which allows the programmer to map a physical device into the program's memory space, thereby gaining access to the raw device registers.

Exercises

11.1   Explain the relationships and differences between device registers, memory locations, and CPU registers.

11.2   Why is it necessary to map the device into user program memory before accessing it under Linux? Would this step be necessary under all operating systems or in the case where there is no operating system and our code is running on the "bare metal?"

11.3   What is the purpose of a GPIO device?

11.4   The Raspberry Pi and the PcDuino have very different GPIO devices.
   (a)   Are they functionally equivalent?
   (b)   Are they equally programmer-friendly?
   (c)   If you have answered no to either of the previous questions, then what are the differences?

11.5   Draw a circuit diagram showing how to connect:
   (a)   a pushbutton switch to GPIO 23 and an LED to GPIO 27 on the Raspberry Pi, and
   (b)   a pushbutton switch to GPIO12 and an LED to GPIO13 on the PcDuino.

11.6   Assuming the systems are wired according to the previous exercise, write two functions. One function must initialize the GPIO pins, and the other function must read the state of the switch and turn the LED on if the button is pressed, and off if the button is not pressed. Write the two functions for
   (a)   a Raspberry Pi, and
   (b)   a PcDuino.

11.7   Write the code necessary to route the output from PWM0 to GPIO 18 on a Raspberry Pi.

11.8   Write the code necessary to route the output from PWM0 to GPIO 5 on a PcDuino.

Pulse Modulation

Chapter Outline

The GPIO device provides a method for sending digital signals to external devices. This can be useful to control devices that have basically two states: on and off. In some situations, it is useful to have the ability to turn a device on at varying levels. For instance, it could be useful to control a motor at any required speed, or control the brightness of a light source. One way that this can be accomplished is through pulse modulation.

The basic idea is that the computer sends a stream of pulses to the device. The device acts as a low-pass filter, which averages the digital pulses into an analog voltage. By varying the percentage of time that the pulses are high, versus low, the computer can control how much average energy is sent to the device. The percentage of time that the pulses are high versus low is known as the *duty cycle*. Varying the duty cycle is referred to as *modulation*. There are two major types of pulse modulation: pulse density modulation (PDM) and pulse width modulation (PWM). Most pulse modulation devices are configured in three steps as follows:

1. The base frequency of the clock that drives the PWM device is configured. This step is usually optional.
2. The mode of operation for the pulse modulation device is configured by writing to one or more configuration registers in the pulse modulation device.
3. The cycle time is set by writing a "range" value into a register in the pulse modulation device. This value is usually set as a multiple of the base clock cycle time.

Once the device is configured, the duty cycle can be changed easily by writing to one or more registers in the pulse modulation device.

Modern Assembly Language Programming with the ARM Processor. http://dx.doi.org/10.1016/B978-0-12-803698-3.00012-7

12.1 Pulse Density Modulation

With PDM, also known as pulse frequency modulation (PFM), the duration of the positive pulses does not change, but the time between them (the pulse density) is modulated. When using PDM devices, the programmer typically sets the device cycle time t_c in a register, then uses another register to specify the number of pulses d that are to be sent during a device cycle. The number of pulses is typically referred to as the duty cycle and must be chosen such that $0 \leq d \leq t_c$. For instance, if $t_c = 1024$, then the device cycle time is 1024 times the cycle time of the clock that drives the device. If $d = 512$, then the device will send 512 pulses, evenly spaced, during the device cycle. Each pulse will have the same duration as the base clock. The device will continue to output this pulse pattern until d is changed.

Fig. 12.1 shows a signal that is being sent using PDM, and the resulting set of pulses. Each pulse transfers a fixed amount of energy to the device. When the pulses arrive at the device, they are effectively filtered using a low pass filter. The resulting received signal is also shown. Notice that the received signal has a delay, or phase shift, caused by the low-pass filtering. This approach is suitable for controlling certain types of devices, such as lights and speakers.

However, when driving such devices directly with the digital pulses, care must be taken that the minimum frequency of pulses remains above the threshold that can be detected by human senses. For instance, when driving a speaker, the minimum pulse frequency must be high enough that the individual pulses cannot be distinguished by the human ear. This minimum frequency is around 40 kHz. Likewise, when driving an LED directly, the minimum frequency must be high enough that the eye cannot detect the individual pulses, because they will be seen as a flickering effect. That minimum frequency is around 70 Hz. To reduce or alleviate this problem, designers may add a low-pass filter between the PWM device and the device that is being driven.

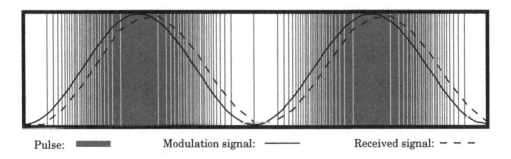

Pulse: ▬▬▬   Modulation signal: ———   Received signal: − − −

Figure 12.1
Pulse density modulation.

12.2 Pulse Width Modulation

In PWM, the frequency of the pulses remains fixed, but the duration of the positive pulse (the pulse width) is modulated. When using PWM devices, the programmer typically sets the device cycle time t_c in a register, then uses another register to specify the number of base clock cycles, d, for which the output should be high. The percentage $\frac{d}{t_c} \times 100$ is typically referred to as the duty cycle and d must be chosen such that $0 \leq d \leq t_c$. For instance, if $t_c = 1024$, then the device cycle time is 1024 times the cycle time of the clock that drives the device. If $d = 512$, then the device will output a high signal for 512 clock cycles, then output a low signal for 512 clock cycles. It will continue to repeat this pattern of pulses until d is changed.

Fig. 12.2 shows a signal that is being sent using PWM. The pulses are also shown. Each pulse transfers some energy to the device. The width of each pulse determines how much energy is transferred. When the pulses arrive at the device, they are effectively filtered using a low-pass filter. The resulting received signal is shown by the dashed line. As with PDM, the received signal has a delay, or phase shift, caused by the low-pass filtering.

One advantage of PWM over PDM is that the digital circuit is not as complex. Another advantage of PWM over PDM is that the frequency of the pulses does not vary, so it is easier for the programmer to set the base frequency high enough that the individual pulses cannot be detected by human senses. Also, when driving motors it is usually necessary to match the pulse frequency to the size and type of motor. Mismatching the frequency can cause loss of efficiency as well as overheating of the motor and drive electronics. In severe cases, this can cause premature failure of the motor and/or drive electronics. With PWM, it is easier for the programmer to control the base frequency, and thereby avoid those problems.

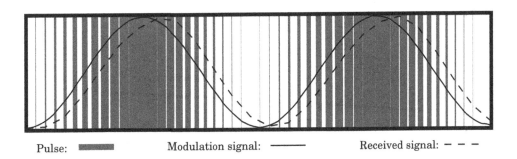

Pulse: ▬▬▬   Modulation signal: ———   Received signal: – – –

Figure 12.2
Pulse width modulation.

12.3 Raspberry Pi PWM Device

The Broadcom BCM2835 system-on-chip includes a device that can create two PWM signals. One of the signals (PWM0) can be routed through GPIO pin 18 (alternate function 5), where it is available on the Raspberry Pi expansion header at pin 12. PWM0 can also be routed through GPIO pin 40. On the Raspberry Pi, pin 40 it is sent through a low-pass filter, and then to the Raspberry Pi audio output port as the right stereo channel. The other signal (PWM1) can be routed through GPIO pin 45. From there, it is sent through a low-pass filter, and then to the Raspberry Pi audio output port as the left stereo channel. So, both PWM channels are accessible, but PWM1 is only accessible through the audio output port after it has been low-pass filtered. The raw PWM0 signal is available through the Raspberry Pi expansion header at pin 12.

There are three modes of operation for the BCM2835 PWM device:

1. PDM mode,
2. PWM mode, and
3. serial transmission mode.

The following paragraphs explain how the device can be used in basic PWM mode, which is the most simple and straightforward mode for this device. Information on how to use the PDM and serial transmission modes, the FIFO, and DMA is available in the *BCM2835 ARM Peripherals* manual.

The base address of the PWM device is $2020C000_{16}$ and it contains eight registers. Table 12.1 shows the offset, name, and a short description for each of the registers. The mode of operation is selected for each channel independently by writing appropriate bits in the PWMCTL register. The base clock frequency is controlled by the clock manager device, which is explained in Section 13.1. By default, the system startup code sets the base clock for the PWM device to 100 MHz.

Table 12.1 Raspberry Pi PWM register map

Offset	Name	Description	Size	R/W
00_{16}	PWMCTL	PWM Control	32	R/W
04_{16}	PWMSTA	PWM FIFO Status	32	R/W
08_{16}	PWMDMAC	PWM DMA Configuration	32	R/W
10_{16}	PWMRNG1	PWM Channel 1 Range	32	R/W
14_{16}	PWMDAT1	PWM Channel 1 Data	32	R/W
18_{16}	PWMFIF1	PWM FIFO Input	32	R/W
20_{16}	PWMRNG2	PWM Channel 2 Range	32	R/W
24_{16}	PWMDAT2	PWM Channel 2 Data	32	R/W

Table 12.2 Raspberry Pi PWM control register bits

Bit	Name	Description	Values
0	PWEN1	Channel 1 Enable	0: Channel is disabled 1: Channel is enabled
1	MODE1	Channel 1 Mode	0: PDM or PWM mode 1: Serial mode
2	RPTL1	Channel 1 Repeat Last	0: Transmission stops when FIFO empty 1: Last data are sent repeatedly
3	SBIT1	Channel 1 Silence Bit	0: Output goes low when not transmitting 1: Output goes high when not transmitting
4	POLA1	Channel 1 Polarity	0: 0 is low voltage and 1 is high voltage 1: 1 is low voltage and 0 is high voltage
5	USEF1	Channel 1 Use FIFO	0: Data register is used 1: FIFO is used
6	CLRF1	Channel 1 Clear FIFO	Write 0: No effect Write 1: Causes FIFO to be emptied
7	MSEN1	Channel 1 PWM Enable	0: PDM mode 1: PWM mode
8	PWEN2	Channel 2 Enable	0: Channel is disabled 1: Channel is enabled
9	MODE2	Channel 2 Mode	0: PDM or PWM mode 1: Serial mode
10	RPTL2	Channel 2 Repeat Last	0: Transmission stops when FIFO empty 1: Last data are sent repeatedly
11	SBIT2	Channel 2 Silence Bit	0: Output goes low when not transmitting 1: Output goes high when not transmitting
12	POLA2	Channel 2 Polarity	0: 0 is low voltage and 1 is high voltage 1: 1 is low voltage and 0 is high voltage
13	USEF2	Channel 2 Use FIFO	0: Data register is used 1: FIFO is used
14	Unused	Reserved	
16	MSEN2	Channel 2 PWM Enable	0: PDM mode 1: PWM mode
16–31	Unused	Reserved	

Table 12.2 shows the names and short descriptions of the bits in the PWMCTL register. There are 8 bits used for controlling channel 1 and 8 bits for controlling channel 2. PWENn is the master enable bit for channel n. Setting that bit to 0 disables the PWM channel, while setting it to 1 enables the channel. MODEn is used to select whether the channel is in serial transmission mode or in the PDM/PWM mode. If MODEn is set to 0, then MSENn is used to choose whether channel n is in PDM mode or PWM mode. If MODEn is set to 1, then RPTLn, SBITn, USEFn, and CLRFn are used to manage the operation of the FIFO for channel n. POLAn is used to enable or disable inversion of the output signal for channel n.

Example 12.1   Example of Determining Clock Values on the Raspberry Pi

Suppose we wish to use PWM0 to perform PWM with a base frequency of 100 kHz and the ability to control the duty cycle with a resolution of 0.1%. The steps would be as follows:

1. Verify that the clock manager device is configured to send a 100 MHz clock to the pulse modulator device through PWM_CLK.
2. To obtain a frequency of 100 kHz from a 100-MHz clock, it is necessary to divide by 1000. Therefore the second step is to store 1000 in the PWMRNG1 register.
3. Before enabling the PWM channel, it is prudent to initialize the duty cycle. The safest initial value is 0%, or completely off. This is accomplished by writing zero to the PWMDAT1 register.
4. Enable PWM channel 1 to operate in PWM mode by setting bit zero of PWMCTL to 1, bit one of PWMCTL to 0, bit five of PWMCTL to 0, and bit seven of PWMCTL to 1.

Once this initialization is performed, we can set or change the duty cycle at any time by writing a value between 0 and 1000 to the PWMDAT1 register.

The PWMRNGn registers are used to define the base period for the corresponding channel. In PDM mode, evenly distributed pulses are sent within a period of length defined by this register, and the number of pulses sent during the base period is controlled by writing to the corresponding PWMDATn register. In PWM mode, the PWMRNGn register defines the base frequency for the pulses, and the duty cycle is controlled by writing to the corresponding PWMDATn register. Example 12.1 gives an overview of the steps needed to configure PWM0 for use in PWM mode.

12.4 pcDuino PWM Device

The AllWinner A10/A20 SOCs have a hardware PWM device which is capable of generating two PWM signals. The PWM device is driven by the OSC24M signal, which is generated by the Clock Control Unit (CCU) in the AllWinner SOC. This base clock runs at 24 MHz by default, and changing the base frequency could affect many other devices in the system. The base clock can be divided by one of 11 predefined values using a prescaler built into the PWM device. Each of the two channels has its own prescaler. Table 12.3 shows the possible settings for the prescalers.

There are two modes of operation for the PWM device. In the first mode, the device operates like a standard PWM device as described in Section 12.2. In the second mode, it sends a single pulse and then waits until it is triggered again by the CPU. In this mode, it is a monostable multivibrator, also known as a one-shot multivibrator, or just one-shot. The duration of the pulse is controlled using the pre-scaler and the period register.

Table 12.3 Prescaler bits in the pcDuino PWM device

Value	Effect
0000	Base clock is divided by 120
0001	Base clock is divided by 180
0010	Base clock is divided by 240
0011	Base clock is divided by 360
0100	Base clock is divided by 480
0101,0110,0111	Not used
1000	Base clock is divided by 1200
1001	Base clock is divided by 2400
1010	Base clock is divided by 3600
1011	Base clock is divided by 4800
1100	Base clock is divided by 7200
1101,1110	Not used
1111	Base clock is divided by 1

Table 12.4 pcDuino PWM register map

Offset	Name	Description
200_{16}	PWMCTL	PWM Control
204_{16}	PWM_CH0_PERIOD	PWM Channel 0 Period
208_{16}	PWM_CH1_PERIOD	PWM Channel 1 Period

The PWM device is mapped at address $01C20C00_{16}$. Table 12.4 shows the registers and their offsets from the base address. All of the device configuration is done through a single control register, which can also be read in order to determine the status of the device. The bits in the control register are shown in Table 12.5.

Before enabling a PWM channel, the period register for that channel should be initialized. The two period registers are each organized as two 16-bit numbers. The upper 16 bits control the total number of clock cycles in one period. In other words, they control the base frequency of the PWM signal. The PWM frequency is calculated as

$$f = \frac{\frac{OSC24M}{PSC}}{N+1},$$

where $OSC24M$ is the frequency of the base clock (the default is 24 MHz), PSC is the prescale value set in the channel prescale bits in the PWM control register, and N is the value stored in the upper 16 bits of the channel period register.

Table 12.5 pcDuino PWM control register bits

Bit	Name	Description	Values
3-0	CH0_PRESCAL	Channel 0 Prescale	These bits must be set before PWM Channel 0 clock is enabled. See Table 12.3.
4	CH0_EN	Channel 0 Enable	0: Channel disabled 1: Channel enabled
5	CH0_ACT_STA	Channel 0 Polarity	0: Channel is active low 1: Channel is active high
6	SCLK_CH0_GATING	Channel 0 Clock	0: Clock disabled 1: Clock enabled
7	CH0_PUL_START	Start pulse	If configured for pulse mode, writing a 1 causes the PWM device to emit a single pulse.
8	PWM0_BYPASS	Bypass PWM	0: Output PWM device signal 1: Output base clock
9	SCLK_CH0_MODE	Select Mode	0: PWM mode 1: Pulse mode
10-14		Not Used	
18-15	CH1_PRESCAL	Channel 1 Prescale	These bits must be set before PWM Channel 1 clock is enabled. See Table 12.3.
19	CH1_EN	Channel 1 Enable	0: Channel disabled 1: Channel enabled
20	CH1_ACT_STA	Channel 1 Polarity	0: Channel is active low 1: Channel is active high
21	SCLK_CH1_GATING	Channel 1 Clock	0: Clock disabled 1: Clock enabled
22	CH1_PUL_START	Start pulse	If configured for pulse mode, writing a 1 causes the PWM device to emit a single pulse.
23	PWM1_BYPASS	Bypass PWM	0: Output PWM device signal 1: Output base clock
24	SCLK_CH1_MODE	Select Mode	0: PWM mode 1: Pulse mode
27-25		Not Used	
28	PWM0_RDY	CH0 Period Ready	0: PWM0 Period register is ready 1: PWM0 Period register is busy
29	PWM1_RDY	CH1 Period Ready	0: PWM1 Period register is ready 1: PWM1 Period register is busy
31–30		Not Used	

The lower 16 bits of the channel period register control the duty cycle. The duty cycle (expressed as % of full on) can be calculated as

$$d = \frac{D}{N} \times 100,$$

where N is the value stored in the upper 16 bits of the channel period register, and D is the value stored in the lower 16 bits of the channel period register. Note that the condition $D \leq N$

must always remain true. If the programmer allows D to become greater than N, the results are unpredictable.

The procedure for configuring the AllWinner A10/A20 PWM device is as follows:

1. Disable the desired channel:
   a. Read the PWM control register into x.
   b. Clear all of the bits in x for the desired PWM channel.
   c. Write x back to the PWM control register
2. Initialize the period register for the desired channel.
   a. Calculate the desired value for N.
   b. Let $D = 0$.
   c. Let $y = N \times 2^{16} + D$.
   d. Write y to the desired channel period register.
3. Set the prescaler.
   a. Select the four-bit code for the desired divisor from Table 12.3.
   b. Set the prescaler code bits in x.
   c. Write x back to the PWM control register.
4. Enable the PWM device.
   a. Set the appropriate bits in x to enable the desired channel, select the polarity, and enable the clock.
   b. Write x to the PWM control register.

Once the control register is configured, the duty cycle can be controlled by calculating a new value for D and then writing $y = N \times 2^{16} + D$ to the desired channel period register.

12.5 Chapter Summary

Pulse modulation is a group of methods for generating analog signals using digital equipment, and is commonly used in control systems to regulate the power sent to motors and other devices. Pulse modulation techniques can have very low power loss compared to other methods of controlling analog devices, and the circuitry required is relatively simple.

The cycle frequency must be programmed to match the application. Typically, 10 Hz is adequate for controlling an electric heating element, while 120 Hz would be more appropriate for controlling an incandescent light bulb. Large electric motors may be controlled with a cycle frequency as low as 100 Hz, while smaller motors may need frequencies around 10,000 Hz. It can take some experimentation to find the best frequency for any given application.

Exercises

12.1 Write ARM assembly programs to configure PWM0 and the GPIO device to send a signal out on Raspberry Pi header pin 12 with:
  (a) period of 1 ms and duty cycle of 25%, and
  (b) frequency of 150 Hz and duty cycle of 63%.

12.2 Write ARM assembly programs to configure PWM0 and the GPIO device to send a signal out on the pcDuino PWM1/GPIO5 pin with:
  (a) period of 1 ms and duty cycle of 25%, and
  (b) frequency of 150 Hz and duty cycle of 63%.

Common System Devices

Chapter Outline

There are some classes of devices that are found in almost every system, including the smallest embedded systems. Such common devices include hardware for managing the clock signals sent to other devices, and serial communications (typically RS232). Most mid-sized or large systems also include devices for managing virtual memory, managing the cache, driving a display, interfacing with keyboard and mouse, accessing disk and other storage devices, and networking. Small embedded systems may have devices for converting analog signals to digital and vice versa, pulse width modulation, and other purposes. Some systems, such as the Raspberry Pi and pcDuino, have all or most of the devices of large systems, as well as most of the devices found on embedded systems. In this chapter, we look at two devices found on almost every system.

13.1 Clock Management Device

Very simple computer systems can be driven by a single clock. Most devices, including the CPU, are designed as state machines. The clock device sends a square-wave signal at a fixed frequency to all devices that need it. The clock signal tells the devices when to transition to the next state. Without the clock signal, none of the devices would do anything.

More complex computers may contain devices which need to run at different rates. This requires the system to have separate clock signals for each device (or group of devices).

Modern Assembly Language Programming with the ARM Processor. http://dx.doi.org/10.1016/B978-0-12-803698-3.00013-9

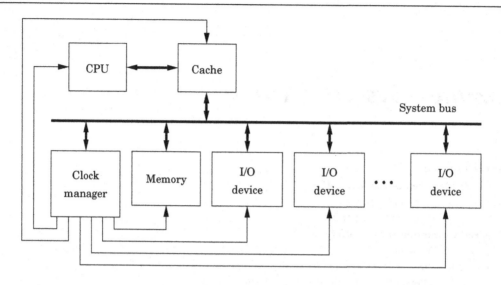

Figure 13.1
Typical system with a clock management device.

System designers often solve this problem by adding a clock manager device to the system. This device allows the programmer to configure the clock signals that are sent to the other devices in the system. Fig. 13.1 shows a typical system. The clock manager, just like any other device, is configured by the CPU writing data to its registers using the system bus.

13.1.1 Raspberry Pi Clock Manager

The BCM2835 system-on-chip contains an ARM CPU and several devices. Some of the devices need their own clock to drive their operation at the correct frequency. Some devices, such as serial communications receivers and transmitters, need configurable clocks so that the programmer has control over the speed of the device. To provide this flexibility and allow the programmer to have control over the clocks for each device, the BCM2835 includes a clock manager device, which can be used to configure the clock signals driving the other devices in the system.

The Raspberry Pi has a 19.2 MHz oscillator which can be used as a base frequency for any of the clocks. The BCM2835 also has three phase-locked-loop circuits that boost the oscillator to higher frequencies. Table 13.1 shows the frequencies that are available from various sources. Each device clock can be driven by one of the PLLs, the external 19.2 MHz oscillator, a signal from the HDMI port, or either of two test/debug inputs.

Among the clocks controlled by the clock manager device are the core clock (CM_VPU), the system timer clock (PM_TIME) which controls the speed of the system timer, the GPIO

Table 13.1 Clock sources available for the clocks provided by the clock manager

Number	Name	Frequency	Note
0	GND	0 Hz	Clock is stopped
1	oscillator	19.2 MHz	
2	testdebug0	Unknown	Used for system testing
3	testdebug1	Unknown	Used for system testing
4	PLLA	650 MHz	May not be available
5	PLLC	200 MHz	May not be available
6	PLLD	500 MHz	
7	HDMI auxiliary	Unknown	
8–15	GND	0 Hz	Clock is stopped

Table 13.2 Some registers in the clock manager device

Offset	Name	Description
070_{16}	CM_GP0_CTL	GPIO Clock 0 (GPCLK0) Control
074_{16}	CM_GP0_DIV	GPIO Clock 0 (GPCLK0) Divisor
078_{16}	CM_GP1_CTL	GPIO Clock 1 (GPCLK1) Control
$07c_{16}$	CM_GP1_DIV	GPIO Clock 1 (GPCLK1) Divisor
080_{16}	CM_GP2_CTL	GPIO Clock 2 (GPCLK2) Control
084_{16}	CM_GP2_DIV	GPIO Clock 2 (GPCLK2) Divisor
098_{16}	CM_PCM_CTL	Pulse Code Modulator Clock (PCM_CLK) Control
$09c_{16}$	CM_PCM_DIV	Pulse Code Modulator Clock (PCM_CLK) Divisor
$0a0_{16}$	CM_PWM_CTL	Pulse Modulator Device Clock (PWM_CLK) Control
$0a4_{16}$	CM_PWM_DIV	Pulse Modulator Device Clock (PWM_CLK) Divisor
$0f0_{16}$	CM_UART_CTL	Serial Communications Clock (UART_CLK) Control
$0f4_{16}$	CM_UART_DIV	Serial Communications Clock (UART_CLK) Divisor

clocks which are documented in the Raspberry Pi peripheral documentation, the pulse modulator device clocks, and the serial communications clocks. It is generally not a good idea to modify the settings of any of the clocks without good reason.

The base address of the clock manager device is 20101000_{16}. Some of the clock manager registers are shown in Table 13.2. Each clock is managed by two registers: a control register and a divisor. The control register is used to enable or disable a clock, to select which source oscillator drives the clock, and to select an optional multistage noise shaping (MASH) filter level. MASH filtering is useful for reducing the perceived noise when a clock is being used to generate an audio signal. In most cases, MASH filtering should not be used.

Table 13.3 shows the meaning of the bits in the control registers for each of the clocks, and Table 13.4 shows the fields in the clock manager divisor registers. The procedure for configuring one of the clocks is:

Table 13.3 Bit fields in the clock manager control registers

Bit	Name	Description
3–0	SRC	Clock source chosen from Table 13.1
4	ENAB	Writing a 0 causes the clock to shut down. The clock will not stop immediately. The BUSY bit will be 1 while the clock is shutting down. When the BUSY bit becomes 0, the clock has stopped and it is safe to reconfigure it. Writing a 1 to this bit causes the clock to start
5	KILL	Writing a 1 to this bit will stop and reset the clock. This does not shut down the clock cleanly, and could cause a glitch in the clock output
6	-	Unused
7	BUSY	A 1 in this bit indicates that the clock is running
8	FLIP	Writing a 1 to this bit will invert the clock output. Do not change this bit while the clock is running
10–9	MASH	Controls how the clock source is divided. 00: Integer division 01: 1-stage MASH division 10: 2-stage MASH division 11: 3-stage MASH division Do not change this while the clock is running.
23–11	–	Unused
31–24	PASSWD	This field must be set to $5A_{16}$ every time the clock control register is written to

Table 13.4 Bit fields in the clock manager divisor registers

Bit	Name	Description
11–0	DIVF	Fractional part of divisor. Do not change this while the clock is running
23–12	DIVI	Integer part of divisor. Do not change this while the clock is running
31–24	PASSWD	This field must be set to $5A_{16}$ every time the clock divisor register is written to

1. Read the desired clock control register.
2. Clear bit 4 in the word that was read, then OR it with $5A000000_{16}$ and store the result back to the desired clock control register.
3. Repeatedly read the desired clock control register, until bit 7 becomes 0.
4. Calculate the divisor required and store it into the desired clock divisor register.
5. Create a word to configure and start the clock. Begin with $5A000000_{16}$, and set bits 3–0 to select the desired clock source. Set bits 10–9 to select the type of division, and set bit 4 to 1 to enable the clock.
6. Store the control word into the desired clock control register.

Table 13.5 Clock signals in the AllWinner A10/A20 SOC

Clock Domain	Modules	Frequency	Description
OSC24M	Most modules	24 MHz	Main clock
CPU32_clk	CPU	2 kHz–1.2 GHz	Drives CPU
AHB_clk	AHB devices	8 kHz–276 MHz	Drives some devices
APB_clk	Peripheral bus	500 Hz–138 MHz	Drives some devices
SDRAM_clk	SDRAM	0 Hz–400 MHz	Drives SDRAM memory
USB_clk	USB	480 MHz	Drives USB devices

Selection of the divisor depends on which clock source is used, what type of division is selected, and the desired output of the clock being configured. For example, to set the PWM clock to 100 kHz, the 19.20 MHz clock can be used. Dividing that clock by 192 will provide a 100-KHz clock. To accomplish this, it is necessary to stop the PWM clock as described, store the value $5A0C0000_{16}$ in the PWM clock divisor register, and then start the clock by writing $5A000011_{16}$ into the PWM clock control register.

13.1.2 pcDuino Clock Control Unit

The AllWinner A10/A20 SOCs have a relatively simple clock manager, which is referred to as the Clock Control Unit. All of the clock signals in the system are driven by two crystal oscillators: the main oscillator runs at 24 MHz, and the real-time-clock oscillator, which runs at 32768 Hz. The real-time-clock oscillator is used only to provide a signal to the real-time-clock device.

The main clock oscillator drives many of the devices in the system, but there are seven phase-locked-loop circuits in the CCU which provide signals for devices which need clocks that are faster or slower than 24 MHz. Table 13.5 shows which devices are driven by the nine clock signals.

13.2 Serial Communications

There are basically two methods for transferring data between two digital devices: parallel and serial. Parallel connections use multiple wires to carry several bits at one time, typically including extra wires to carry timing information. Parallel communications are used for transferring large amounts of data over very short distances. However, this approach becomes very expensive when data must be transferred more than a few meters. Serial, on the other hand, uses a single wire to transfer the data bits one at a time. When compared to parallel transfer, the speed of serial transfer typically suffers. However, because it uses significantly fewer wires, the distance may be greatly extended, reliability improved, and cost vastly reduced.

13.2.1 UART

One of the oldest and most common devices for communications between computers and peripheral devices is the Universal Asynchronous Receiver/Transmitter, or UART. The word "universal" indicates that the device is highly configurable and flexible. UARTs allow a receiver and transmitter to communicate without a synchronizing signal.

The logic signal produced by the digital UART typically oscillates between zero volts for a low level and five volts for a high level, and the amount of current that the UART can supply is limited. For transmitting the data over long distances, the signals may go through a level-shifting or amplification stage. The circuit used to accomplish this is typically called a *line driver*. This circuit boosts the signal provided by the UART and also protects the delicate digital outputs from short circuits and signal spikes. Various standards, such as RS-232, RS-422, and RS-485 define the voltages that the line driver uses. For example, the RS-232 standard specifies that valid signals are in the range of $+3$ to $+15$ V, or -3 to -15 V. The standards also specify the maximum time that is allowable when shifting from a high signal to a low signal and vice versa, the amount of current that the device must be capable of sourcing and sinking, and other relevant design criteria.

The UART transmits data by sending each bit sequentially. The receiving UART re-assembles the bits into the original data. Fig. 13.2 shows how the transmitting UART converts a byte of data into a serial signal, and how the receiving UART samples the signal to recover the original data. Serializing the transmission and reassembly of the data are accomplished using shift registers. The receiver and transmitter each have their own clocks, and are configured so that the clocks run at the same speed (or close to the same speed). In this case, the receiver's clock is running slightly slower than the transmitter's clock, but the data are still received correctly.

To transfer a group of bits, called a *data frame*, the transmitter typically first sends a *start bit*. Most UARTs can be configured to transfer between four and eight data bits in each group. The transmitting and receiving UARTS must be configured to use the same number of data bits. After each group of data bits, the transmitter will return the signal to the low state and keep it there for some minimum period. This period is usually the time that it would take to send two bits of data, and is referred to as the two *stop bits*. The stop bits allow the receiver to have some time to process the received byte and prepare for the next start bit. Fig. 13.2A shows what a typical RS-232 signal would look like when transferring the value 56_{16} (the ASCII "V" character). The UART enters the idle state only if there is not another byte immediately ready to send. If the transmitter has another byte to send, then the start bit can begin at the end of the second stop bit.

Note that it is impossible to ensure that the receiver and transmitter have clocks which are running at exactly the same speed, unless they use the same clock signal. Fig. 13.2B shows

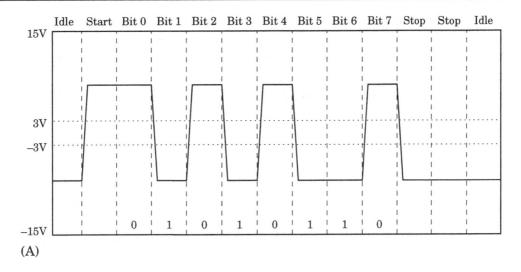

(A)

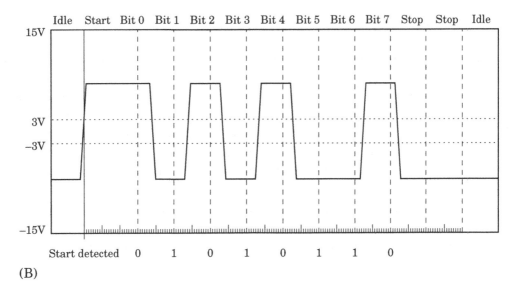

(B)

Figure 13.2

Transmitter and receiver timings for two UARTS. (A) Waveform of a UART transmitting a byte.
(B) Timing of UART receiving a byte.

how the receiver can reassemble the original data, even with a slightly different clock rate.
When the start bit is detected by the receiver, it prepares to receive the data bits, which will be
sent by the transmitter at an expected rate (within some tolerance). The receive circuitry of
most UARTS is driven by a clock that runs 16 times as fast as the baud rate. The receive

circuitry uses its faster clock to latch each bit in the middle of its expected time period. In Fig. 13.2B, the receiver clock is running slower than the transmitter clock. By the end of the data frame, the sample time is very far from the center of the bit, but the correct value is received. If the clocks differed by much more, or if more than eight data bits were sent, then it is very likely that incorrect data would be received. Thus, as long as their clocks are synchronized within some tolerance (which is dependent on the number of data bits and the baud rate), the data will be received correctly.

The RS-232 standard allows point-to-point communication between two devices for limited distances. With the RS-232 standard, simple one-way communications can be accomplished using only two wires: One to carry the serial bits, and another to provide a common ground. For bi-directional communication, three wires are required. In addition, the RS-232 standard specifies optional hand-shaking signals, which the UARTs can use to signal their readiness to transmit or receive data. The RS-422 and RS-485 standards allow multiple devices to be connected using only two wires.

The first UART device to enjoy widespread use was the 8250. The original version had 12 registers for configuration, sending, and receiving data. The most important registers are the ones that allow the programmer to set the transmit and receive bit rates, or baud. One baud is one bit per second. The baud is set by storing a 16 bit divisor in two of the registers in the UART. The chip is driven by an external clock, and the divisor is used to reduce the frequency of the external clock to a frequency that is appropriate for serial communication. For example, if the external clock runs at 1 MHz, and the required baud is 1200, then the divisor must be $833.\overline{3} \approx 833$. Note that the divisor can only be an integer, so the device cannot achieve exactly 1200 baud. However, as explained previously, the sending and receiving devices do not have to agree precisely on the baud. During the transmission and reception of a byte, 1200.48 baud is close enough that the bits will be received correctly even if the other end is running slightly below 1200 baud. In the 8250, there was only one 8-bit register for sending data and only one 8-bit register for receiving data. The UART could send an interrupt to the CPU after each byte was transmitted or received. When receiving, the CPU had to respond to the interrupt very quickly. If the current byte was not read quickly enough by the CPU, it would be overwritten by the subsequent incoming byte. When transmitting, the CPU needed to respond quickly to interrupts to provide the next byte to be sent, or the transmission rate would suffer.

The next generation of UART device was the 16550A. This device is the model for most UART devices today. It features 16-byte input and output buffers and the ability to trigger interrupts when a buffer is partially full or partially empty. This allows the CPU to move several bytes of data at a time and results in much lower CPU overhead and much higher data transmission and reception rates. The 16550A also supports much higher baud rates than the 8250.

13.2.2 Raspberry Pi UART0

The BCM2835 system-on-chip provides two UART devices: UART0 and UART1. UART 1 is part of the I^2C device, and is not recommended for use as a UART. UART0 is a PL011 UART, which is based on the industry standard 16550A UART. The major differences are that the PL011 allows greater flexibility in configuring the interrupt trigger levels, the registers appear in different locations, and the locations of bits in some of the registers is different. So, although it operates very much like a 16550A, things have been moved to different locations. The transmit and receive lines can be routed through GPIO pin 14 and GPIO pin 15, respectively. UART0 has 18 registers, starting at its base address of $2E20100_{16}$. Table 13.6 shows the name, location, and a brief description for each of the registers.

UART_DR: The UART Data Register is used to send and receive data. Data are sent or received one byte at a time. Writing to this register will add a byte to the transmit FIFO. Although the register is 32 bits, only the 8 least significant bits are used in transmission, and 12 least significant bits are used for reception. If the FIFO is empty, then the UART will begin transmitting the byte immediately. If the FIFO is full, then the last byte in the FIFO will be overwritten with the new byte that is written to the Data Register. When this register is read, it returns the byte at the top of the receive FIFO, along with four additional status bits to indicate if any errors were encountered. Table 13.7 specifies the names and use of the bits in the UART Data Register.

Table 13.6 Raspberry Pi UART0 register map

Offset	Name	Description
00_{16}	UART_DR	Data Register
04_{16}	UART_RSRECR	Receive Status Register/Error Clear Register
18_{16}	UART_ FR	Flag register
20_{16}	UART_ILPR	not in use
24_{16}	UART_IBRD	Integer Baud rate divisor
28_{16}	UART_FBRD	Fractional Baud rate divisor
$2c_{16}$	UART_LCRH	Line Control register
30_{16}	UART_CR	Control register
34_{16}	UART_IFLS	Interrupt FIFO Level Select Register
38_{16}	UART_IMSC	Interrupt Mask Set Clear Register
$3c_{16}$	UART_RIS	Raw Interrupt Status Register
40_{16}	UART_MIS	Masked Interrupt Status Register
44_{16}	UART_ICR	Interrupt Clear Register
48_{16}	UART_DMACR	DMA Control Register
80_{16}	UART_ITCR	Test Control register
84_{16}	UART_ITIP	Integration test input reg
88_{16}	UART_ITOP	Integration test output reg
$8c_{16}$	UART_TDR	Test Data reg

Table 13.7 Raspberry Pi UART data register

Bit	Name	Description	Values
7–0	DATA	Data	Read: Last data received Write: Data byte to transmit
8	FE	Framing error	0: No error 1: The received character did not have a valid stop bit
9	PE	Parity error	0: No error 1: The received character did not have the correct parity, as set in the EPS and SPS bits of the Line Control Register (UART_LCRH)
10	BE	Break error	0: No error 1: A break condition was detected. The data input line was held low for longer than the time it would take to receive a complete byte, including the start and stop bits.
11	OE	Overrun error	0: No error 1: Data was not read quickly enough, and one or more bytes were overwritten in the input buffer
31–12	-	Not used	Write as zero, read as don't care

UART_RSRECR: The UART Receive Status Register/Error Clear Register is used to check the status of the byte most recently read from the UART Data Register, and to check for overrun conditions at any time. The status information for overrun is set immediately when an overrun condition occurs. The Receive Status Register/Error Clear Register provides the same four status bits as the Data Register (but in bits 3–0 rather than bits 11–8). The received data character must be read first from the Data Register, before reading the error status associated with that data character from the RSRECR register. Since the Data Register also contains these 4 bits, this register may not be required, depending on how the software is written. Table 13.8 describes the bits in this register.

UART_FR: The UART Flag Register can be read to determine the status of the UART. The bits in this register are used mainly when sending and receiving data using the FIFOs. When several bytes need to be sent, the TXFF flag should be checked to ensure that the transmit FIFO is not full before each byte is written to the data register. When receiving data, the RXFE bit can be used to determine whether or not there is more data to be read from the FIFO. Table 13.9 describes the flags in this register.

UART_ILPR: This is the IrDA register, which is supported by some PL011 UARTs. IrDA stands for the Infrared Data Association, which is a group of companies that cooperate to provide specifications for a complete set of protocols for wireless infrared communications. The name "IrDA" also refers to that set of protocols. IrDA is not implemented on the Raspberry Pi UART. Writing to this register has no effect and reading returns 0.

UART_IBRD and UART_FBRD: UART_FBRD is the fractional part of the baud rate divisor value, and UART_IBRD is the integer part. The baud rate divisor is calculated as follows:

Table 13.8 Raspberry Pi UART receive status register/error clear register

Bit	Name	Description	Values
0	FE	Framing error	0: No error 1: The received character did not have a valid stop bit
1	PE	Parity error	0: No error 1: The received character did not have the correct parity, as set in the EPS and SPS bits of the Line Control Register (UART_LCRH)
2	BE	Break error	0: No error 1: A break condition was detected. The data input line was held low for longer than the time it would take to receive a complete byte, including the start and stop bits.
3	OE	Overrun error	0: No error 1: Data was not read quickly enough, and one or more bytes were overwritten in the input buffer
31–4		Not used	Write as zero, read as don't care

Table 13.9 Raspberry Pi UART flags register bits

Bit	Name	Description	Values
0	CTS	Clear To Send	0: Sender indicates they are ready to receive 1: Sender is NOT ready to receive
1	DSR	Data Set Ready	Not implemented: Write as zero, read as don't care
2	DCD	Data Carrier Detect	Not implemented: Write as zero, read as don't care
3	BUSY	UART is busy	0: UART is not transmitting data 1: UART is transmitting a byte
4	RXFE	Receive FIFO Empty	0: Receive FIFO contains bytes that have been received 1: Receive FIFO is empty
5	TXFF	Transmit FIFO is Full	0: There is room for at least one more byte in the transmit FIFO 1: Transmit FIFO is full – do not write to the data register at this time
6	RXFF	Receive FIFO is Full	0: There is no more room in the receive FIFO 1: There is still some space in the receive FIFO
7	TXFE	Transmit FIFO is Empty	0: There are no bytes waiting to be transmitted 1: There is at least one byte waiting to be transmitted
8	RI	Ring Indicator	Not implemented: Write as zero, read as don't care
31–9		Not used	Write as zero, read as don't care

$$BAUDDIV = \frac{UARTCLK}{16 \times Baudrate} \qquad (13.1)$$

where *UARTCLK* is the frequency of the UART_CLK that is configured in the Clock Manager device. The default value is 3 MHz. *BAUDDIV* is stored in two registers. UART_IBRD holds the integer part and UART_FBRD holds the fractional part. Thus *BAUDDIV* should be calculated as a U(16,6) fixed point number. The contents of the UART_IBRD and UART_FBRD registers may be written at any time, but the change will not have any effect until transmission or reception of the current character is complete.

Table 13.10 Raspberry Pi UART integer baud rate divisor

Bit	Name	Description	Values
15–0	IBRD	Integer Baud Rate Divisor	See Eq. (13.1)
31–16		Not used	Write as zero, read as don't care

Table 13.11 Raspberry Pi UART fractional baud rate divisor

Bit	Name	Description	Values
5–0	FBRD	Fractional Baud Rate Divisor	See Eq. (13.1)
31-6		Not used	Write as zero, read as don't care

Table 13.12 Raspberry Pi UART line control register bits

Bit	Name	Description	Values
0	BRK	Send Break	0: Normal operation 1: After the current character is sent, take the TXD output to a low level and keep it there
1	PEN	Parity Enable	0: Parity checking and generation is disabled 1: Generate and send parity bit and check parity on received data
2	EPS	Even Parity Select	0: Odd parity 1: Even parity
3	STP2	Two Stop Bits	0: Send one stop bit for each data word 1: Send two stop bits for each data word
4	FEN	FIFO Enable	0: Transmit and Receive FIFOs are disabled 1: Transmit and Receive FIFOs are enabled
6–5	WLEN	Word Length	00: 5 bits per data word 01: 6 bits per data word 10: 7 bits per data word 11: 8 bits per data word
31–7		Not used	Write as zero, read as don't care

Table 13.10 shows the arrangement of the integer baud rate divisor register, and Table 13.11 shows the arrangement of the fractional baud rate divisor register.

UART_LCRH: UART_LCRH is the line control register. It is used to configure the communication parameters. This register must not be changed until the UART is disabled by writing zero to bit 0 of UART_CR, and the BUSY flag in UART_FR is clear. Table 13.12 shows the layout of the line control register.

UART_CR: The UART Control Register is used for configuring, enabling, and disabling the UART. Table 13.13 shows the layout of the control register. To enable transmission, the TXE bit and UARTEN bit must be set to 1. To enable reception, the RXE bit and

Table 13.13 Raspberry Pi UART control register bits

Bit	Name	Description	Values
0	UARTEN	UART Enable	0: UART disabled 1: UART enabled.
1	SIREN	Not used	Write as zero, read as don't care
2	SIRLP	Not used	Write as zero, read as don't care
3–6		Not used	Write as zero, read as don't care
7	LBE	Loopback Enable	0: Loopback disabled 1: Loopback enabled. Transmitted data is also fed back to the receiver.
8	TXE	Transmit enable	0: Transmitter is disabled 1: Transmitter is enabled
9	RXE	Receive enable	0: Receiver is disabled 1: Receiver is enabled
10	DTR	Not used	Write as zero, read as don't care
11	RTS	Complement of nUARTRTS	
12	OUT1	Not used	Write as zero, read as don't care
13	OUT2	Not used	Write as zero, read as don't care
14	RTSEN	RTS Enable	0: Hardware RTS disabled. 1: Hardware RTS Enabled
15	CTSEN	CTS Enable	0: Hardware CTS disabled. 1: Hardware CTS Enabled
16–31		Not used	Write as zero, read as don't care

UARTEN bit must be set to 1. In general, the following steps should be used to configure or re-configure the UART:

1. Disable the UART.
2. Wait for the end of transmission or reception of the current character.
3. Flush the transmit FIFO by setting the FEN bit to 0 in the Line Control Register.
4. Reprogram the Control Register.
5. Enable the UART.

Interrupt Control: The UART can signal the CPU by asserting an interrupt when certain conditions occur. This will be covered in more detail in Chapter 14. For now, it is enough to know that there are five additional registers which are used to configure and use the interrupt mechanism.

UART_IFLS defines the FIFO level that triggers the assertion of the interrupt signal. One interrupt is generated when the FIFO reaches the specified level. The CPU must clear the interrupt before another can be generated.

UART_IMSC is the interrupt mask set/clear register. It is used to enable or disable specific interrupts. This register determines which of the possible interrupt conditions are allowed to generate an interrupt to the CPU.

UART_RIS is the raw interrupt status register. It can be read to raw status of interrupts conditions before any masking is performed.

UART_MIS is the masked interrupt status register. It contains the masked status of the interrupts. This is the register that the operating system should use to determine the cause of a UART interrupt.

UART_ICR is the interrupt clear register. writing to it clears the interrupt conditions. The operating system should use this register to clear interrupts before returning from the interrupt service routine.

UART_DMACR: The DMA control register is used to configure the UART to access memory directly, so that the CPU does not have to move each byte of data to or from the UART. DMA will be explained in more detail in Chapter 14.

Additional Registers: The remaining registers, UART_ITCR, UART_ITIP, and UART_ITOP, are either unimplemented or are used for testing the UART. These registers should not be used.

13.2.3 Basic Programming for the Raspberry Pi UART

Listing 13.1 shows four basic functions for initializing the UART, changing the baud rate, sending a character, and receiving a character using UART0 on the Raspberry Pi. Note that a large part of the code simply defines the location and offset for all of the registers (and bits) that can be used to control the UART.

```
 1
 2        @@ offsets to the UART registers
 3        .equ    UART_DR,      0x00 @ data register
 4        .equ    UART_RSRECR,  0x04 @ Receive Status/Error clear
 5        .equ    UART_FR,      0x18 @ flag register
 6        .equ    UART_ILPR,    0x20 @ not used
 7        .equ    UART_IBRD,    0x24 @ integer baud rate divisor
 8        .equ    UART_FBRD,    0x28 @ fractional baud rate divisor
 9        .equ    UART_LCRH,    0x2C @ line control register
10        .equ    UART_CR,      0x30 @ control register
11        .equ    UART_IFLS,    0x34 @ interrupt FIFO level select
12        .equ    UART_IMSC,    0x38 @ Interrupt mask set clear
13        .equ    UART_RIS,     0x3C @ raw interrupt status
14        .equ    UART_MIS,     0x40 @ masked interrupt status
15        .equ    UART_ICR,     0x44 @ interrupt clear register
16        .equ    UART_DMACR,   0x48 @ DMA control register
17        .equ    UART_ITCR,    0x80 @ test control register
18        .equ    UART_ITIP,    0x84 @ integration test input
```

```
19          .equ    UART_ITOP,      0x88 @ integration test output
20          .equ    UART_TDR,       0x8C @ test data register
21
22          @@ error condition bits when reading the DR (data register)
23          .equ    UART_OE,        (1<<11) @ overrun error bit
24          .equ    UART_BE,        (1<<10) @ break error bit
25          .equ    UART_PE,        (1<<9)  @ parity error bit
26          .equ    UART_FE,        (1<<8 ) @ framing error bit
27
28          @@ Bits for the FR (flags register)
29          .equ    UART_RI,        (1<<8) @ Unsupported
30          .equ    UART_TXFE,      (1<<7) @ Transmit FIFO empty
31          .equ    UART_RXFF,      (1<<6) @ Receive FIFO full
32          .equ    UART_TXFF,      (1<<5) @ Transmit FIFO full
33          .equ    UART_RXFE,      (1<<4) @ Receive FIFO empty
34          .equ    UART_BUSY,      (1<<3) @ UART is busy xmitting
35          .equ    UART_DCD,       (1<<2) @ Unsupported
36          .equ    UART_DSR,       (1<<1) @ Unsupported
37          .equ    UART_CTS,       (1<<0) @ Clear to send
38
39          @@ Bits for the LCRH (line control register)
40          .equ    UART_SPS,       (1<<7) @ enable stick parity
41          .equ    UART_WLEN1,     (1<<6) @ MSB of word length
42          .equ    UART_WLEN0,     (1<<5) @ LSB of word length
43          .equ    UART_FEN,       (1<<4) @ Enable FIFOs
44          .equ    UART_STP2,      (1<<3) @ Use 2 stop bits
45          .equ    UART_EPS,       (1<<2) @ Even parity select
46          .equ    UART_PEN,       (1<<1) @ Enable parity
47          .equ    UART_BRK,       (1<<0) @ Send break
48
49          @@ Bits for the CR (control register)
50          .equ    UART_CTSEN,     (1<<15) @ Enable CTS
51          .equ    UART_RTSEN,     (1<<14) @ Enable RTS
52          .equ    UART_OUT2,      (1<<13) @ Unsupported
53          .equ    UART_OUT1,      (1<<12) @ Unsupported
54          .equ    UART_RTS,       (1<<11) @ Request to send
55          .equ    UART_DTR,       (1<<10) @ Unsupported
56          .equ    UART_RXE,       (1<<9)  @ Enable receiver
57          .equ    UART_TXE,       (1<<8)  @ Enable transmitter
58          .equ    UART_LBE,       (1<<7)  @ Enable loopback
59          .equ    UART_SIRLP,     (1<<2)  @ Unsupported
60          .equ    UART_SIREN,     (1<<1)  @ Unsupported
61          .equ    UART_UARTEN,    (1<<0)  @ Enable UART
62
63
64          .text
```

```
65              .align  2
66      @@ ----------------------------------------------------------
67              .global UART_put_byte
68      UART_put_byte:
69              ldr     r1,=uartbase    @ load base address of UART
70              ldr     r1,[r1]         @ load base address of UART
71      putlp:  ldr     r2,[r1,#UART_FR] @ read the flag resister
72              tst     r2,#UART_TXFF   @ check if transmit FIFO is full
73              bne     putlp           @ loop while transmit FIFO is full
74              str     r0,[r1,#UART_DR] @ write the char to the FIFO
75              mov     pc,lr           @ return
76
77      @@@ ---------------------------------------------------------
78              .global UART_get_byte
79      UART_get_byte:
80              ldr     r1,=uartbase    @ load base address of UART
81              ldr     r1,[r1]         @ load base address of UART
82      getlp:  ldr     r2,[r1,#UART_FR] @ read the flag resister
83              tst     r2,#UART_RXFE   @ check if receive FIFO is empty
84              bne     getlp           @ loop while receive FIFO is empty
85              ldr     r0,[r1,#UART_DR] @ read the char from the FIFO
86              tst     r0,#UART_OE     @ check for overrun error
87              bne     get_ok1
88              @@ handle receive overrun error here - does nothing now
89      get_ok1:
90              tst     r0,#UART_BE     @ check for break error
91              bne     get_ok2
92              @@ handle receive break error here - does nothing now
93
94      get_ok2:
95              tst     r0,#UART_PE     @ check for parity error
96              bne     get_ok3
97              @@ handle receive parity error here - does nothing now
98
99      get_ok3:
100             tst     r0,#UART_FE     @ check for framing error
101             bne     get_ok4
102             @@ handle receive framing error here - does nothing now
103
104     get_ok4:
105             @@ return
106             mov     pc,lr           @ return the received character
107
108     @@@ ---------------------------------------------------------
109     @@@ UART init will set default values:
110     @@@ 115200 baud, no parity, 2 stop bits, 8 data bits
```

```
111            .global UART_init
112    UART_init:
113            ldr    r1,=uartbase      @ load base address of UART
114            ldr    r1,[r1]           @ load base address of UART
115            @@ set baud rate divisor
116            @@ (3MHz / ( 115200 * 16 )) = 1.62760416667
117            @@ = 1.101000 in binary
118            mov    r0,#1
119            str    r0,[r1,#UART_IBRD]
120            mov    r0,#0x28
121            str    r0,[r1,#UART_FBRD]
122            @@ set parity, word length, enable FIFOS
123            .equ BITS, (UART_WLEN1|UART_WLEN0|UART_FEN|UART_STP2)
124            mov    r0,#BITS
125            str    r0,[r1,#UART_LCRH]
126            @@ mask all UART interrupts
127            mov    r0,#0
128            str    r0,[r1,#UART_IMSC]
129            @@ enable receiver and transmitter and enable the uart
130            .equ FINALBITS, (UART_RXE|UART_TXE|UART_UARTEN)
131            ldr    r0,=FINALBITS
132            str    r0,[r1,#UART_CR]
133            @@ return
134            mov    pc,lr
135
136    @@ ----------------------------------------------------------
137    @@ UART_set_baud will change the baud rate to whatever is in r0
138    @@ The baud rate divisor is calculated as follows: Baud rate
139    @@ divisor BAUDDIV = (FUARTCLK/(16 Baud rate)) where FUARTCLK
140    @@ is the UART reference clock frequency. The BAUDDIV
141    @@ is comprised of the integer value IBRD and the
142    @@ fractional value FBRD. NOTE: The contents of the
143    @@ IBRD and FBRD registers are not updated until
144    @@ transmission or reception of the current character
145    @@ is complete.
146            .global UART_set_baud
147    UART_set_baud:
148            @@ set baud rate divisor using formula:
149            @@ (3000000.0 / ( R0 * 16 ))  ASSUMING 3Mhz clock
150            lsl    r1,r0,#4        @ r1 <- desired baud * 16
151            ldr    r0,=(3000000<<6)@ Load 3 MHz as a U(26,6) in r0
152            bl     divide          @ divide clk freq by (baud*16)
153            asr    r1,r0,#6        @ put integer divisor into r1
154            and    r0,r0,#0x3F     @ put fractional divisor into r0
155            ldr    r2,=uartbase    @ load base address of UART
156            ldr    r2,[r2]         @ load base address of UART
```

```
157    str    r1,[r2,#UART_IBRD] @ set integer divisor
158    str    r0,[r2,#UART_FBRD] @ set fractional divisor
159    mov    pc,lr
```

Listing 13.1
Assembly functions for using the Raspberry Pi UART.

13.2.4 pcDuino UART

The AllWinner A10/A20 SOC includes eight UART devices. They are all fully compatible with the 16550A UART, and also provide some enhancements. All of them provide transmit (TX) and receive (RX) signals. UART0 has the full set of RS232 signals, including RTS, CTS, DTR, DSR, DCD, and RING. UART1 has the RTS and CTS signals. The remaining six UARTs only provide the TX and RX signals. They can all be configured for serial IrDA. Table 13.14 shows the base address for each of the eight UART devices.

When the 16550 UART was designed, 8-bit processors were common, and most of them provided only 16 address bits. Memory was typically limited to 64 kB, and every byte of address space was important. Because of these considerations, the designers of the 16550 decided to limit the number of addresses used to 8, and to only use eight bits of data per address. There are 10 registers in the 16550 UART, but some of them share the same address. For example, there are three registers mapped to an offset address of zero, two registers mapped at offset four, and two registers mapped at offset eight. Bit seven in the Line Control Register is used to determine which of the registers is active for a given address.

Because they are meant to be fully backwards-compatible with the 16550, the AllWinner A10/A20 SOC UART devices also use only 8 bits for each register, and the first 12 registers correspond exactly with the 16550 UART. The only differences are that the pcDuino uses word addresses rather than byte addresses, and they provide four additional registers that are

Table 13.14 pcDuino UART addresses

Name	Address
UART0	0x01C28000
UART1	0x01C28400
UART2	0x01C28800
UART3	0x01C28C00
UART4	0x01C29000
UART5	0x01C29400
UART6	0x01C29800
UART7	0x01C29C00

Table 13.15 pcDuino UART register offsets

Register Name	Offset	Description
UART_RBR	0x00	UART Receive Buffer Register
UART_THR	0x00	UART Transmit Holding Register
UART_DLL	0x00	UART Divisor Latch Low Register
UART_DLH	0x04	UART Divisor Latch High Register
UART_IER	0x04	UART Interrupt Enable Register
UART_IIR	0x08	UART Interrupt Identity Register
UART_FCR	0x08	UART FIFO Control Register
UART_LCR	0x0C	UART Line Control Register
UART_MCR	0x10	UART Modem Control Register
UART_LSR	0x14	UART Line Status Register
UART_MSR	0x18	UART Modem Status Register
UART_SCH	0x1C	UART Scratch Register
UART_USR	0x7C	UART Status Register
UART_TFL	0x80	UART Transmit FIFO Level
UART_RFL	0x84	UART_RFL
UART_HALT	0xA4	UART Halt TX Register

used for IrDA mode. Table 13.15 shows the arrangement of the registers in each of the 8 UARTs on the pcDuino. The following sections will explain the registers.

The baud rate is set using a 16-bit Baud Rate Divisor, according to the following equation:

$$BAUDDIV = \frac{sclk}{16 \times Baudrate} \tag{13.2}$$

where *sclk* is the frequency of the UART serial clock, which is configured by the Clock Manager device. The default frequency of the clock is 24 MHz. *BAUDDIV* is stored in two registers. UART_DLL holds the least significant 8 bits, and UART_DLH holds the most significant 8 bits. Thus *BAUDDIV* should be calculated as a 16-bit unsigned integer. Note that for high baud rates, it may not be possible to get exactly the rate desired. For example, a baud rate of 115200 would require a divisor of $13.0208\overline{3}$. Since the baud rate divisor can only be given as an integer, the desired rate must be based on a divisor of 13, so the true baud rate will be $\frac{24000000}{16 \times 13} = 115384.615385$, or about 0.16% faster than desired. Although slightly fast, it is well within the tolerance for RS232 communication.

UART_RBR: The UART Receive Buffer Register is used to receive data, 1 byte at a time. If the receive FIFO is enabled, then as the UART receives data, it places the data into a receive FIFO. Reading from this address removes 1 byte from the receive FIFO. If the FIFO becomes full and another data byte arrives, then the new data are lost and an overrun error occurs. Table 13.16 shows the layout of the receive buffer register.

Table 13.16 pcDuno UART receive buffer register

Bit	Name	Description	Values
7–0	RBR	Data	Read only: One byte of received data. Bit 7 of LCR must be zero.
31–8		Unused	

Table 13.17 pcDuno UART transmit holding register

Bit	Name	Description	Values
7–0	THR	Data	Write only: One byte of data to transmit. Bit 7 of LCR must be zero.
31–8		Unused	

Table 13.18 pcDuno UART divisor latch low register

Bit	Name	Description	Values
7–0	DLL	Data	Write only: Least significant eight bits of the Baud Rate Divisor. Bit 7 of LCR must be one.
31–8		Unused	

UART_THR: Writing to the Transmit Holding Register will cause that byte to be transmitted by the UART. If the transmit FIFO is enabled, then the byte will be added to the end of the transmit FIFO. If the FIFO is empty, then the UART will begin transmitting the byte immediately. If the FIFO is full, then the new data byte will be lost. Table 13.17 shows the layout of the transmit holding register.

UART_DLL: The UART Divisor Latch Low register is used to set the least significant byte of the baud rate divisor. When bit 7 of the Line Control Register is set to one, writing to this address will access the DLL register. If bit 7 of the Line Control Register is set to zero, then writing to this address will access the transmit holding register. Table 13.18 shows the layout of the UART_DLL register.

UART_DLH: The UART Divisor Latch High register is used to set the most significant byte of the baud rate divisor. When bit 7 of the Line Control Register is set to one, writing to this address will access the DLH register. If bit 7 of the Line Control Register is set to zero, then writing to this address will access the Interrupt Enable Register rather than the Divisor Latch High register. Table 13.19 shows the layout of the UART_DLL register.

If the two Divisor Latch Registers (DLL and DLH) are set to zero, the baud clock is disabled and no serial communications occur. DLH should be set before DLL, and at least eight clock cycles of the UART clock should be allowed to pass before data are transmitted or received.

UART_FCR: is the UART FIFO control register. It is used to enable or disable the receive and transmit FIFOs (buffers), flush their contents, set the level at which the transmit and

Table 13.19 pcDuno UART divisor latch high register

Bit	Name	Description	Values
7–0	DLH	Data	Write only: Most significant eight bits of the Baud Rate Divisor. Bit 7 of LCR must be one.
31–8		Unused	

Table 13.20 pcDuno UART FIFO control register

Bit	Name	Description
0	FIFOE	FIFO Enable 0: transmit and receive FIFOs disabled 1: transmit and receive FIFOs enabled
1	RFIFOR	Receive FIFO Reset: writing a 1 to this bit causes the receive FIFO to be reset, and then continue normal operation
2	XFIFOR	Transmit FIFO Reset: writing a 1 to this bit causes the transmit FIFO to be reset, and then continue normal operation
3	DMAM	DMA Mode: 0: Mode 0 1: Mode 1
5–4	TET	Transmit Empty Trigger: These bits control the level at which the Transmit Holding Register Empty interrupt is triggered 00: FIFO is completely empty 01: There are two characters in the FIFO 10: The FIFO is 25% full 11: The FIFO is 50% full This setting has no effect if THRE_MODE_USER is disabled
7–6	RT	Receive Trigger: These bits control the level at which the Received Data Available interrupt is triggered. 00: There is one character in the FIFO 01: The FIFO is 25% full 10: The FIFO is 50% full 11: There is room for two more characters in the FIFO This setting has no effect if THRE_MODE_USER is disabled.
31–8	Unused	

receive FIFOs trigger an interrupt, and to control Direct Memory Access (DMA) Table 13.20 shows the layout of the UART_FCR register.

UART_LCR: The Line Control Register is used to control the parity, number of data bits, and number of stop bits for the serial port. Bit 7 also controls which registers are mapped at offsets 0, 4, and 8 from the device base address. Table 13.21 shows the layout of the UART_LCR register.

Table 13.21 pcDuno UART line control register

Bit	Name	Description
1–0	DLS	This field controls the number of data bits: 00: 5 data bits 01: 6 data bits 10: 7 data bits 11: 8 data bits
2	STOP	This bit controls the number of stop bits used for transmitting and receiving data. 0: 1 stop bit 1: If DLS is set to 00, then 1.5 stop bits, otherwise 2 stop bits
3	PEN	Parity Enable: 0: Parity disabled 1: Parity enabled
4	EPS	Even Parity Select: 0: Odd Parity 1: Even Parity
5	Unused	
6	BCB	Writing a one to this bit causes a break to be sent. This bit must be set to zero for normal operation.
7	DLAB	The Divisor Latch Access Bit controls the behavior of other registers: 0: The RBR, THR, and IER registers are accessible (RBR is used for read at offset 0, and THR for write at offset 0). 1: The DLL and DLM registers are accessible
31–8	Unused	

UART_LSR: The Line Status Register is used to read status information from the UART. Table 13.22 shows the layout of the UART_LSR register.

UART_USR: The UART Status Register is used to read information about the status of the transmit and receive FIFOs, and the current state of the receiver and transmitter. Table 13.23 shows the layout of the UART_USR register. This register contains essentially the same information as the status register in the Raspberry Pi UART.

UART_TFL: The UART Transmit FIFO Level register allows the programmer to determine exactly how many bytes are currently in the transmit FIFO. Table 13.24 shows the layout of the UART_TFL register.

UART_RFL: The UART Receive FIFO Level register allows the programmer to determine exactly how many bytes are currently in the receive FIFO. Table 13.25 shows the layout of the UART_RFL register.

Table 13.22 pcDuno UART line status register

Bit	Name	Description
0	DR	When the Data Ready bit is set to 1, it indicates that at least one byte is ready to be read from the receive FIFO or RBR.
1	OE	When the Overrun Error bit is set to 1, it indicates that an overrun error occurred for the byte at the top of the receive FIFO.
2	PE	When the Parity Error bit is set to 1, it indicates that a parity error occurred for the byte at the top of the receive FIFO.
3	FE	When the Framing Error bit is set to 1, it indicates that a framing error occurred for the byte at the top of the receive FIFO.
4	BI	When the Break Interrupt bit is set to 1, it indicates that a break has been received.
5	THRE	When the Transmit Holding Register Empty bit is 1, it indicates that there are there are no bytes waiting to be transmitted, but there may be a byte currently being transmitted.
6	TEMT	When the Transmitter Empty bit is 1, it indicates that there are no bytes waiting to be transmitted and no byte currently being transmitted.
7	FIFOERR	When this bit is 1, an error has occurred (PE, BE, or BI) in the receive FIFO. This bit is cleared when the Line Status Register is read.
31–8	Unused	

Table 13.23 pcDuno UART status register

Bit	Name	Description
0	BUSY	When the Busy bit is 1, it indicates that the UART is currently busy. When it is 0, the UART is idle or inactive.
1	TFNF	When the Transmit FIFO Not Full bit is 1, it indicates that at least one more byte can be safely written to the Transmit FIFO.
2	TFE	When the Transmit FIFO Empty bit is 1, it indicates that there are no bytes remaining in the transmit FIFO.
3	RFNE	When the Receive FIFO Not Empty bit is 1, it indicates that at least one more byte is waiting to be read from the receive FIFO.
4	RFF	When the Receive FIFO Full bit is 1, it indicates that there is no more room in the receive FIFO. If data is not read before the next character is received, an overrun error will occur.
31–5	Unused	

Table 13.24 pcDuno UART transmit FIFO level register

Bit	Name	Description
6–0	TFL	The Transmit FIFO level field contains an integer which indicates the number of bytes currently in the transmit FIFO.
31–7	Unused	

Table 13.25 pcDuno UART receive FIFO level register

Bit	Name	Description
6–0	RFL	The Receive FIFO level field contains an integer which indicates the number of bytes currently in the receive FIFO.
31–7	Unused	

Table 13.26 pcDuno UART transmit halt register

Bit	Name	Description
0	Unused	
1	CHCFG_AT_BUSY	Setting this bit to 1 causes the UART to allow changing the Line Control Register (except the DLAB bit) and allows setting the baud rate even when the UART is busy. When this bit is set to 0, changes can only occur when the BUSY bit in the UART Status Register is 0.
2	CHANGE_UPDATE	After writing 1 to CHCFG_AT_BUSY and performing the configuration, 1 should be written to this bit to signal that the UART should re-start with the new configuration. This bit will stay at 1 while the new configuration is loaded, and go back to 0 when the re-start is complete.
3	Unused	
4	SIR_TX_INVERT	This bit allows the polarity of the transmitter to be inverted. 0: Normal polarity 1: Polarity inverted
5	SIR_RX_INVERT	This bit allows the polarity of the receiver to be inverted. 0: Normal polarity 1: Polarity inverted
31–5	Unused	

UART_HALT: The UART transmit halt register is used to halt the UART so that it can be reconfigured. After the configuration is performed, it is then used to signal the UART to restart with the new settings. It can also be used to invert the receive and transmit polarity. Table 13.26 shows the layout of the UART_HALT register.

Interrupt Control: The UART can signal the CPU by asserting an interrupt when certain conditions occur. This will be covered in more detail in Chapter 14. For now, it is enough to know that there are five additional registers which are used to configure and use the interrupt mechanism.

UART_IFLS defines the FIFO level that triggers the assertion of the interrupt signal. One interrupt is generated when the FIFO reaches the specified level. The CPU must clear the interrupt before another can be generated.

UART_IER is the interrupt enable register. It is used to enable or disable the generation of interrupts for specific conditions.

UART_IIR is the Interrupt Identity Register. When an interrupt occurs, the CPU can read this register to determine what caused the interrupt.

Additional Registers There are several additional registers which are not needed for basic use of the UART.

UART_MCR is the Modem Control Register. It is used to configure the port for IrDA mode, enable Automatic Flow Control, and manage the RS-232 RTS and DTR hardware handshaking signals for the ports in which they are implemented. The default configuration disables these extra features.

UART_MSR is the Modem Status Register, which is used to read the state of the RS-232 modem control and status lines on ports that implement them. This register can be ignored unless a telephone modem is being used on the port.

UART_SCH is the Modem Scratch Register. It provides 8 bits of storage for temporary data values. In the days of 8 and 16-bit computers, when the 16550 UART was designed, this extra byte of storage was useful.

13.3 Chapter Summary

Most modern computer systems have some type of Universal Asynchronous Receiver/Transmitter. These are serial communications devices, and are meant to provide communications with other systems using RS-232 (most commonly) or some other standard serial protocol. Modern systems often have a large number of other devices as well. Each device may need its own clock source to drive it at the correct frequency for its operation. The clock sources for all of the devices are often controlled by yet another device: the clock manager.

Although two systems may have different UARTs, these devices perform the same basic functions. The specifics about how they are programmed will vary from one system to another. However, there is always enough similarity between devices of the same class that a programmer who is familiar with one specific device can easily learn to program another similar device. The more experience a programmer has, the less time it takes to learn how to control a new device.

Exercises

13.1 Write a function for setting the PWM clock on the Raspberry Pi to 2 MHz.

13.2 The UART_GET_BYTE function in Listing 13.1 contains skeleton code for handling errors, but does not actually do anything when errors occur. Describe at least two ways that the errors could be handled.

13.3 Listing 13.1 provides four functions for managing the UART on the Raspberry Pi. Write equivalent functions for the pcDuino UART.

CHAPTER 14

Running Without an Operating System

Chapter Outline

The previous chapters assumed that the software would be running in user mode under an operating system. Sometimes, it is necessary to write assembly code to run on "bare metal," which simply means: without an operating system. For example, when we write an operating system kernel, it must run on bare metal and a significant part of the code (especially during the boot process) must be written in assembly language. Coding on bare metal is useful to deeply understand how the hardware works and what happens in the lowest levels of an operating system. There are some significant differences between code that is meant to run under an operating system and code that is meant to run on bare metal.

The operating system takes care of many details for the programmer. For instance, it sets up the stack, text, and data sections, initializes static variables, provides an interface to input and output devices, and gives the programmer an abstracted view of the machine. When accessing data on a disk drive, the programmer uses the file abstraction. The underlying hardware only knows about blocks of data. The operating system provides the data structures and operations which allow the programmer to think of data in terms of files and streams of bytes. A user program may be scattered in physical memory, but the hardware memory management unit, managed by the operating system, allows the programmer to view memory as a simple memory map (such as shown in Fig. 1.7). The programmer uses system calls to access the abstractions provided by the operating system. On bare metal, there are no abstractions, unless the programmer creates them.

However, there are some software packages to help bare-metal programmers. For example, Newlib is a C standard library intended for use in bare-metal programs. Its major features are that:

- it implements the hardware-independent parts of the standard C library,
- for I/O, it relies on only a few low-level functions that must be implemented specifically for the target hardware, and
- many target machines are already supported in the Newlib source code.

To support a new machine, the programmer only has to write a few low-level functions in C and/or Assembly to initialize the system and perform low-level I/O on the target hardware.

14.1 ARM CPU Modes

Many early computers were not capable of protecting the operating system from user programs. That problem was solved mostly by building CPUs that support multiple "levels of privilege" for running programs. Almost all modern CPUs have the ability to operate in at least two modes:

User mode is the mode that normal user programs use when running under an operating system, and

Privileged mode is reserved for operating system code. There are operations that can be performed in privileged mode which cannot be performed in user mode.

The ARM processor provides six privileged modes and one user mode. Five of the privileged modes have their own stack pointer (r13) and link register (r14). When the processor mode is changed, the corresponding link register and stack pointer become active, "replacing" the user stack pointer and link register.

In any of the six privileged modes, the link registers and stack pointers of the other modes can be accessed. The privileged mode stack pointers and link registers are not accessible from user mode. One of the privileged modes, FIQ, has five additional registers which become active

31	30	29	28	27	...	24	...	19...16	...	9	8	7	6	5	0...4
N	Z	C	V	Q	...	J	...	GE[3:0]	...	E	A	I	F	T	M[4:0]

Figure 14.1
The ARM process status register.

when the processor enters FIQ mode. These registers "replace" registers r8 through r12. Additionally, five of the privileged modes have a Saved Process Status Register (SPSR). When entering those privileged modes, the CPSR is copied into the corresponding SPSR. This allows the CPSR to be restored to its original contents when the privileged code returns to the previously active mode. The full register set for all modes is shown in Table 14.1. Registers r0 through r7 and the program counter are shared by all modes. Some processors have an additional monitor mode, as part of the ARMv6-M and ARMv7-M security extensions.

All of the bits of the Program Status Register (PSR) are shown in Fig. 14.1. The processor mode is selected by writing a bit pattern into the mode bits (M[4:0]) of the PSR. The bit pattern assignment for each processor mode is shown in Table 14.2. Not all combinations of the mode bits define a valid processor mode. An illegal value programmed into M[4:0] causes the processor to enter an unrecoverable state. If this occurs, a hardware reset must be used to

Table 14.1 The ARM user and system registers

usr sys	svc	abt	und	irq	fiq
r0					
r1					
r2					
r3					
r4					
r5					
r6					
r7					
r8					r8_fiq
r9					r9_fiq
r10					r10_fiq
r11 (fp)					r11_fiq
r12 (ip)					r12_fiq
r13 (sp)	r13_svc	r13_abt	r13_und	r13_irq	r13_fiq
r14 (lr)	r14_svc	r14_abt	r14_und	r14_irq	r14_fiq
r15 (pc)					
CPSR	CPSR	CPSR	CPSR	CPSR	CPSR
	SPSR_svc	SPSR_abt	SPSR_und	SPSR_irq	SPSR_fiq

Table 14.2 Mode bits in the PSR

M[4:0]	Mode	Name	Register Set
10000	usr	User	R0-R14, CPSR, PC
10001	fiq	Fast Interrupt	R0-R7, R8_fiq-R14_fiq, CPSR, SPSR_fiq, PC
10010	irq	Interrupt Request	R0-R12, R13_irq, R14_irq, CPSR, SPSR_irq, PC
10011	svc	Supervisor	R0-R12, R13_svc R14_svc CPSR, SPSR_irq, PC
10111	abt	Abort	R0-R12, R13_abt R14_abt CPSR, SPSR_abt PC
11011	und	Undefined Instruction	R0-R12, R13_und R14_und, CPSR, SPSR_und PC
11111	sys	System	R0-R14, CPSR, PC

re-start the processor. Programs running in user mode cannot modify these bits directly. User programs can only change the processor mode by executing the software interrupt (swi) instruction (also known as the svc instruction), which automatically gives control to privileged code in the operating system. The hardware is carefully designed so that the user program cannot run its own code in privileged mode.

The swi instruction does not really cause an interrupt, but the hardware and operating system handle it in a very similar way. The software interrupt is used by user programs to request that the operating system perform some task on their behalf. Another general class of interrupt is the "hardware interrupt." This class of interrupt may occur at any time and is used by hardware devices to signal that they require service. Another type of interrupt may be generated within the CPU when certain conditions arise, such as attempting to execute an unknown instruction. These are generally known as "exceptions" to distinguish them from hardware interrupts. On the ARM processor, there are three bits in the CPSR which affect interrupt processing:

I: when set to one, normal hardware interrupts are disabled,

F: when set to one, fast hardware interrupts are disabled, and

A: (only on ARMv6 and later processors) when set to one, imprecise aborts are disabled (this is an abort on a memory write that has been held in a write buffer in the processor and not written to memory until later, perhaps after another abort).

Programs running in user mode cannot modify these bits. Therefore, the operating system gains control of the CPU whenever an interrupt occurs and the user program cannot disable interrupts and continue to run. Most operating systems use a hardware timer to generate periodic interrupts, thus they are able to regain control of the CPU every few milliseconds.

14.2 Exception Processing

Most of the privileged modes are entered automatically by the hardware when certain exceptional circumstances occur. For example, when a hardware device needs attention, it can signal the processor by causing an interrupt. When this occurs, the processor immediately enters IRQ mode and begins executing the IRQ exception handler function. Some

Table 14.3 ARM vector table

Address	Exception	Mode
0x00000000	Reset	svc
0x00000004	Undefined Instruction	und
0x00000008	Software Interrupt	svc
0x0000000C	Prefetch Abort	abt
0x00000010	Data Abort	abt
0x00000014	Reserved	
0x00000018	Interrupt Request	irq
0x0000001C	Fast Interrupt Request	fiq

devices can cause a fast interrupt, which causes the processor to immediately enter FIQ mode and begin executing the FIQ exception handler function. There are six possible exceptions that can occur, each one corresponding to one of the six privileged modes. Each exception must be handled by a dedicated function, with one additional function required to handle CPU reset events. The first instruction of each of these seven exception handlers is stored in a *vector table* at a known location in memory (usually address 0). When an exception occurs, the CPU automatically loads the appropriate instruction from the vector table and executes it. Table 14.3 shows the address, exception type, and the mode that the processor will be in, for each entry in ARM vector table. The vector table usually contains branch instructions. Each branch instruction will jump to the correct function for handling a specific exception type. Listing 14.1 shows a short section of assembly code which provides definitions for the ARM CPU modes.

```
1   @@@ FILE: modes.S
2   @@@ Definitions of mode field and interrupt bits in CPSR
3           .equ    I_BIT, 0x80     @ when I=1 IRQ is disabled
4           .equ    F_BIT, 0x40     @ when F=1 FIQ is disabled
5           .equ    USR_MODE, 0x10 @ shares sp,lr,CPSR with sys mode
6           .equ    FIQ_MODE, 0x11 @ fiq mode all interrupts masked
7           .equ    IRQ_MODE, 0x12 @ irq mode all interrupts masked
8           .equ    SVC_MODE, 0x13
9           .equ    ABT_MODE, 0x17
10          .equ    UND_MODE, 0x1B
11          .equ    SYS_MODE, 0x1F @ shares sp,lr,CPSR with usr mode
```

Listing 14.1
Definitions for ARM CPU modes.

Many bare-metal programs consist of a single thread of execution running in user mode to perform some task. This main program is occasionally interrupted by the occurrence of some exception. The exception is processed, and then control returns to the main thread. Fig. 14.2

Main process   Exception handler

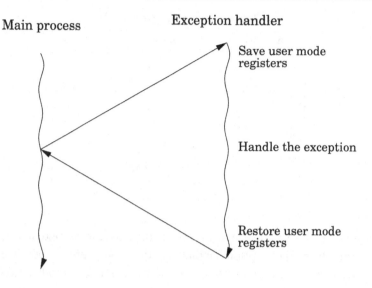

Save user mode
registers

Handle the exception

Restore user mode
registers

Figure 14.2
Basic exception processing.

shows the sequence of events when an exception occurs in such a system. The main program typically would be running with the CPU in user mode. When the exception occurs, the CPU executes the corresponding instruction in the vector table, which branches to the exception handler. The exception handler must save any registers that it is going to use, execute the code required to handle the exception, then restore the registers. When it returns to the user mode process, everything will be as it was before the exception occurred. The user mode program continues executing as if the exception never occurred.

More complex systems may have multiple tasks, threads of execution, or user processes running concurrently. In a single-processor system, only one task, thread, or user process can actually be executing at any given instant, but when an exception occurs, the exception handler may change the currently active task, thread, or user process. This is the basis for all modern multiprocessing systems. Fig. 14.3 shows how an exception may be processed on such a system. It is common on multi-processing systems for a timer device to be used to generate periodic interrupts, which allows the currently active task, thread, or user process to be changed at a fixed frequency.

When any exception occurs, it causes the ARM CPU hardware to perform a very well-defined sequence of actions:

1. The CPSR is copied into the SPSR for the mode corresponding to the type of exception that has occurred.
2. The CPSR mode bits are changed, switching the CPU into the appropriate privileged mode.

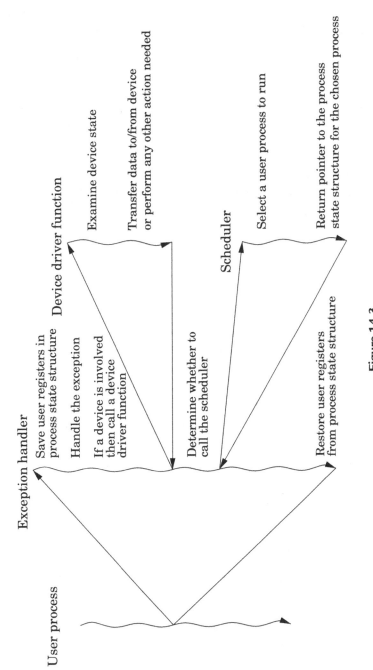

Figure 14.3

Exception processing with multiple user processes.

3. The banked registers for the new mode become active.
4. The I bit of the CPSR is cleared, which disables interrupts.
5. If the exception was an FIQ, or if a reset has occurred, then the FIQ bit is cleared, disabling fast interrupts.
6. The program counter is copied to the link register for the new mode.
7. The program counter is loaded with the address in the vector table corresponding with the exception that has occurred.
8. The processor then fetches the next instruction using the program counter as usual. However, the program counter has been set so that in loads an instruction from the vector table.

The instruction in the vector table should cause the CPU to branch to a function which handles the exception. At the end of that function, the program counter must be loaded with the address of the instruction where the exception occurred, and the SPSR must be copied back into the CPSR. That will cause the processor to branch back to where it was when the exception occurred, and return to the mode that it was in at that time.

14.2.1 Handling Exceptions

Listing 14.2 shows in detail how the vector table is initialized. The vector table contains eight identical instructions. These instructions load the program counter, which causes a branch. In each case, the program counter is loaded with a value at the memory location that is 32 bytes greater than the corresponding load instruction. An offset of 24 is used because the program counter will have advanced 8 bytes by the time the load instruction is executed. The addresses of the exception handlers have been stored in a second table, that begins at an address 32 bytes after the first load instruction. Thus, each instruction in the vector table loads a unique address into the program counter. Note that one of the slots in the vector table is not used and is reserved by ARM for future use. That slot is treated like all of the others, but it will never be used on any current ARM processor.

Listing 14.3 shows the stub functions for each of the exception handlers.

Note that the return sequence depends on the type of exception. For some exceptions, the return address must be adjusted. This is because the program counter may have been advanced past the instruction where the exception occurred. These stub functions simply return the processor to the mode and location at which the exception occurred. To be useful, they will need to be extended significantly. Note that these functions all return using a data processing instruction with the optional s specified and with the program counter as the destination register. This special form of data processing instruction indicates that the SPSR should be copied into the CPSR at the same time that the program counter is loaded with the return address. Thus, the function returns to the point where the exception occurred, and the processer switches back into the mode that it was in when the exception occurred.

```
1         .section .rodata      @ mark this data as read-only
2         .align      2
3         @@ All of the eight instructions in the vector table are
4         @@    ldr pc,[pc, #24]
5         @@ which loads the program counter with the program
6         @@ counter + 24. When the pc is used in this addressing
7         @@ mode, there is an 8-byte offset because of the
8         @@ pipeline (8+24=32). The address of the corresponding
9         @@ handler will be stored 32 bytes after each entry.
10  Vector_Table:
11        ldr pc,[pc, #24]
12        ldr pc,[pc, #24]
13        ldr pc,[pc, #24]
14        ldr pc,[pc, #24]
15        ldr pc,[pc, #24]
16        ldr pc,[pc, #24]
17        ldr pc,[pc, #24]
18        ldr pc,[pc, #24]
19  rh:   .word   reset_handler
20  uh:   .word   undef_handler
21  sh:   .word   swi_handler
22  ph:   .word   pAbort_handler
23  dh:   .word   dAbort_handler
24  vh:   .word   reserved_handler
25  ih:   .word   irq_handler
26  fh:   .word   fiq_handler
27        .equ    VT_SIZE, (. - Vector_Table)
28  @@@ -------------------------------------------------------------
29        .text
30        .align      2
31        .global setup_vector_table
32  setup_vector_table:
33        @@ Cortex-A and similar: set the vector base address
34        @@ to 0x0.  (The boot loader may have changed it.)
35        mov     r0,#0
36        MCR     p15,0,r0,c12,c0,0@ Write VBAR
37        @@ This section will copy Vector_Table to address 0x0
38        ldr     r0,=Vector_Table @ pointer to table of addresses
39        ldr     r1,=0x0
40        mov     r3,#VT_SIZE      @ stop after 64 bytes
41  movit: ldr     r2,[r0],#4
42        str     r2,[r1],#4
43        cmp     r1,r3
44        blt     movit
45        mov     pc,lr                   @ return
```

Listing 14.2
Function to set up the ARM exception table.

```
1    @@@ FILE: handlers.S
2            .text
3            .align  2
4    @@@ --------------------------------------------------------
5    @@@ On reset, jump to startup function. The CPU must not
6    @@@ be in usr or sys mode! (add some code to check it)
7            .global reset_handler
8    reset_handler:
9            b       _start
10
11   @@@ --------------------------------------------------------
12           .global irq_handler
13   irq_handler:                    @ must subtract 4 from lr
14           stmfd   sp!,{r0-r7, lr}
15           @@ handler body goes here
16           ldmfd   sp!,{r0-r7, lr}
17           subs    pc, lr, #4
18
19   @@@ --------------------------------------------------------
20           .global undef_handler
21   undef_handler:                  @ lr holds exact return address
22           stmfd   sp!,{r0-r7, lr}
23           @@ handler body goes here
24           ldmfd   sp!,{r0-r7, lr}
25           movs    pc,lr
26
27   @@@ --------------------------------------------------------
28           .global swi_handler
29   swi_handler:                    @ lr holds exact return address
30           stmfd   sp!,{r0-r7, lr}
31           @@ handler body goes here
32           ldmfd   sp!,{r0-r7, lr}
33           movs    pc,lr
34
35   @@@ --------------------------------------------------------
36           .global pAbort_handler
37   pAbort_handler:                 @ must subtract 4 from lr
38           stmfd   sp!,{r0-r7, lr}
39           @@ handler body goes here
40           ldmfd   sp!,{r0-r7, lr}
41           subs    pc,lr,#4
42
43   @@@ --------------------------------------------------------
44           .global dAbort_handler
45   dAbort_handler:                 @ must subtract 8 from lr
```

```
46          stmfd   sp!,{r0-r7, lr}
47          @@ handler body goes here
48          ldmfd   sp!,{r0-r7, lr}
49          subs    pc,lr,#8
50
51  @@@ -----------------------------------------------------------------
52          .global reserved_handler
53  reserved_handler:              @ this will never be called
54          stmfd   sp!,{r0-r7, lr}
55          @@ handler body goes here
56          ldmfd   sp!,{r0-r7, lr}
57          movs    pc,lr
58
59  @@@ -----------------------------------------------------------------
60          .global fiq_handler
61  fiq_handler:                   @ must subtract 4 from lr
62          stmfd   sp!,{r0-r7, lr}
63          @@ handler body goes here
64          ldmfd   sp!,{r0-r7, lr}
65          subs    pc,lr,#4
```

Listing 14.3
Stubs for the exception handlers.

A special form of the `ldm` instruction can also be used to return from an exception processing function. In order to use that method, the exception handler should start by adjusting the link register (depending on the type of exception) and then pushing it onto the stack. The handler should also push any other registers that it will need to use. At the end of the function, an `ldmfd` is used to restore the registers, but instead of restoring the link register, it loads the program counter. Also a carat (^) is added to the end of the instruction. Listing 14.4 shows the skeleton for an exception handler function using this method.

```
1          .global dAbort_handler
2  dAbort_handler: @ must subtract 8 from lr
3          sub     lr,lr,#8
4          stmfd   sp!,{r0-r7, lr}
5          @@ handler body goes here
6          ldmfd   sp!,{r0-r7, pc}^  @ return and restore the CPSR
```

Listing 14.4
Skeleton for an exception handler.

14.3 The Boot Process

In order to create a bare-metal program, we must understand what the processor does when power is first applied or after a reset. The ARM CPU begins to execute code at a predetermined address. Depending on the configuration of the ARM processor, the program counter starts either at address 0 or 0xFFFF0000. In order for the system to work, the startup code must be at the correct address when the system starts up.

On the Raspberry Pi, when power is first applied, the ARM CPU is disabled and the graphics processing unit (GPU) is enabled. The GPU runs a program that is stored in ROM. That program, called the first stage boot loader, reads the second stage boot loader from a file named (bootcode.bin) on the SD card. That program enables the SDRAM, and then loads the third stage bootloader, start.elf. At this point, some basic hardware configuration is performed, and then the kernel is loaded to address 0x8000 from the kernel.img file on the SD card. Once the kernel image file is loaded, a "b #0x8000" instruction is placed at address 0, and the ARM CPU is enabled. The ARM CPU executes the branch instruction at address 0, then immediately jumps to the kernel code at address 0x8000.

To run a bare-metal program on the Raspberry Pi, it is only necessary to build an executable image and store it as kernel.img on the SD card. Then, the boot process will load the bare-metal program instead of the Linux kernel image. Care must be taken to ensure that the linker prepares the program to run at address 0x8000 and places the first executable instruction at the beginning of the image file. It is also important to make a copy of the original kernel image so that it can be restored (using another computer). If the original kernel image is lost, then there will be no way to boot Linux until it is replaced.

The pcDuino uses u-boot, which is a highly configurable open-source boot loader. The boot loader is configured to attempt booting from the SD card. If a bootable SD card is detected, then it is used. Otherwise, the pcDuino boots from its internal NAND flash. In either case, u-boot finds the Linux kernel image file, named uImage, loads it at address 0x40008000, and then jumps to that location. The easiest way to run bare-metal code on the pcDuino is to create a duplicate of the operating system on an SD card, then replace the uImage file with another executable image. Care must be taken to ensure that the linker prepares the program to run at address 0x40008000 and places the first executable instruction at the beginning of the image file. If the SD card is inserted, then the bare-metal code will be loaded. Otherwise, it will boot normally from the NAND flash memory.

14.4 Writing a Bare-Metal Program

A bare-metal program should be divided into several files. Some of the code may be written in assembly, and other parts in C or some other language. The initial startup code, and the entry and exit from exception handlers, must be written in assembly. However, it may be much more productive to write the main program and the remainder of the exception handlers as C functions and have the assembly code call them.

14.4.1 Startup Code

Other than the code being loaded at different addresses, there is very little difference between getting bare-metal code running on the Raspberry Pi and the pcDuino. For either platform, the bare-metal program must include some start-up code. The startup code will:

- initialize the stack pointers for all of the modes,
- set up interrupt and exception handling,
- initialize the .bss section,
- configure the CPU and critical systems (optional),
- set up memory management (optional),
- set up process and/or thread management (optional),
- initialize devices (optional), and call the main function.

The startup code requires some knowledge of the target platform, and must be at least partly written in assembly language. Listing 14.5 shows a function named _start which sets up the stacks, initializes the .bss section, calls a function to set up the vector table, then calls the main function:

```
1   @@@ FILE:  start.S

2
3          .include "modes.S"

4
5   @@@ Stack locations
6          @@ uncomment one of the following two lines
7          @ .equ    stack_top, 0x10000000 @ Raspberry pi only
8          @ .equ    stack_top, 0x50000000 @ pcDuino only

9
10         .equ    fiq_stack_top, stack_top
11         .equ    irq_stack_top, stack_top - 0x1000
12         .equ    abt_stack_top, stack_top - 0x2000
13         .equ    und_stack_top, stack_top - 0x3000
14         .equ    mon_stack_top, stack_top - 0x4000
15         .equ    svc_stack_top, stack_top - 0x5000
16         .equ    sys_stack_top, stack_top - 0x6000

17
18  @@@ ------------------------------------------------------------
19  @@@ The startup code should be loaded by the boot loader.
20  @@@ The entry point is _start which performs initialization of
21  @@@ the hardware, then calls a C function.
```

```
22          .section .text.boot
23          .global _start
24          .func   _start
25  _start: @@ On reset, we should be in SVC mode.
26
27          @@ Switch to FIQ mode with interrupts disabled
28          msr     CPSR_c,#FIQ_MODE|I_BIT|F_BIT
29          ldr     sp,=fiq_stack_top @ set the FIQ stack pointer
30
31          @@ Switch to IRQ mode with interrupts disabled
32          msr     CPSR_c,#IRQ_MODE|I_BIT|F_BIT
33          ldr     sp,=irq_stack_top @ set the IRQ stack pointer
34
35          @@ Switch to ABT mode with interrupts disabled
36          msr     CPSR_c,#ABT_MODE|I_BIT|F_BIT
37          ldr     sp,=abt_stack_top @ set the ABT stack pointer
38
39          @@ Switch to UND mode with interrupts disabled
40          msr     CPSR_c,#UND_MODE|I_BIT|F_BIT
41          ldr     sp,=und_stack_top @ set the UND stack pointer
42
43          @@ Switch to SYS mode with interrupts disabled
44          msr     CPSR_c,#SYS_MODE|I_BIT|F_BIT
45          ldr     sp,=sys_stack_top @ set SYS/USR stack pointer
46
47          @@ Switch to SVC mode with interrupts disabled
48          msr     CPSR_c,#SVC_MODE|I_BIT|F_BIT
49          ldr     sp,=svc_stack_top @ set SVC stack pointer
50
51          @@ Clear the .bss segment to all zeros
52          @@ The __bss_start__ and __bss_end__ symbols are
53          @@ defined by the linker.
54          ldr     r1,=__bss_start__ @ load pointer to bss and
55          ldr     r2,=__bss_end__   @ to byte following bss
56          mov     r3,#0             @ load fill value (zero)
57  bssloop:cmp     r1,r2             @ Start filling
58          bge     bssdone
59          str     r3,[r1],#4
60          b       bssloop           @ loop until done
61  bssdone:
62          @@ Set up the vector table
63          bl      setup_vector_table
64
65          @@ Call the Main function
66          bl      main
67
```

```
68          @@ If main ever returns, cause an exception
69          swi    0xFFFFFF        @ this should never happen
70          .size  _start, . - _start
71          .endfunc
```

Listing 14.5
ARM startup code.

The first task for the startup code is to ensure that the stack pointer for each processor mode is initialized. When an exception or interrupt occurs, the processor will automatically change into the appropriate mode and begin executing an exception handler, using the stack pointer for that mode. Hardware interrupts can be disabled, but some exceptions cannot be disabled. In order to guarantee correct operation, a stack must be set up for each processor mode, and an exception handler must be provided. The exception handler does not actually have to do anything.

On the Raspberry Pi, memory is mapped to begin at address 0, and all models have at least 256 MB of memory. Therefore, it is safe to assume that the last valid memory address is 0x0FFFFFFF. If each mode is given 4 kB of stack space, then all of the stacks together will consume 32 kB, and the initial stack addresses can be easily calculated. Since the C compiler uses a full descending stack, the initial stack pointers can be assigned addresses 0x10000000, 0x0FFFF000, 0x0FFFE000, etc.

For the pcDuino, there is a small amount of memory mapped at address 0, but most of the available memory is in the region between 0x40000000 and 0xBFFFFFFF. The pcDuino has at least 1 GB of memory. One possible way to assign the stack locations is: 0x50000000, 0x4FFFF000, 0x4FFFE000, etc. This assignment of addresses will make it easy to write one piece of code to set up the stacks for either the Raspberry Pi or the pcDuino.

After initializing the stacks, the startup code must set all bytes in the .bss section to zero. Recall that the .bss section is used to hold data that is initialized to zero, but the program file does not actually contain all of the zeros. Programs running under an operating system can rely on the C standard library to initialize the .bss section. If it is not linked to a C library, then a bare-metal program must set all of the bytes in the .bss section to zero for itself.

14.4.2 Main Program

The final part of this bare-metal program is the main function. Listing 14.6 shows a very simple main program which reads from three GPIO pins which have pushbuttons connected to them, and controls three other pins that have LEDs connected to them. When a button is pressed the LED associated with it is illuminated. The only real difference between the pcDuino and Raspberry Pi versions of this program is in the functions which drive the GPIO device. Therefore, those functions have been removed from the main program file. This makes

```
 1   @@@ FILE: main.S
 2   @@@ This program reads from three buttons connected to GPIO3-5, and
 3   @@@ controls three leds connected to GPIO0-2. The main loop runs
 4   @@@ continuously.
 5           .global main
 6   main:   stmfd   sp!,{lr}
 7           @@ Set the GPIO pins
 8           mov     r0,#0           @ Port 0
 9           bl      GPIO_dir_output @ set for output
10           mov     r0,#1           @ Port 1
11           bl      GPIO_dir_output @ set for output
12           mov     r0,#2           @ Port 2
13           bl      GPIO_dir_output @ set for output
14
15           mov     r0,#3           @ Port 3
16           bl      GPIO_dir_input  @ set for input
17           mov     r0,#4           @ Port 4
18           bl      GPIO_dir_input  @ set for input
19           mov     r0,#5           @ Port 5
20           bl      GPIO_dir_input  @ set for input
21   @@@ Main loop just reads buttons and updates the LEDs.
22   loop:
23           @@ Read the state of the inputs and
24           @@ set the ouputs to the same state.
25           mov     r0,#3           @ Pin 3
26           bl      GPIO_get_pin    @ read it
27           mov     r1,r0           @ copy pin state to r1
28           mov     r0,#0           @ Pin 0
29           bl      GPIO_set_pin    @ write it
30
31           mov     r0,#4           @ Pin 4
32           bl      GPIO_get_pin    @ read it
33           mov     r1,r0           @ copy pin state to r1
34           mov     r0,#1           @ Pin 1
35           bl      GPIO_set_pin    @ write it
36
37           mov     r0,#5           @ Pin 5
38           bl      GPIO_get_pin    @ read it
39           mov     r1,r0           @ copy pin state to r1
40           mov     r0,#2           @ pin 2
41           bl      GPIO_set_pin    @ write it
42
43           b       loop
44           ldmfd   sp!,{pc}
```

Listing 14.6
A simple main program.

the main program *portable*; it can run on the pcDuino or the Raspberry Pi. It could also run on any other ARM system, with the addition of another file to implement the mappings and functions for using the GPIO device for that system.

14.4.3 The Linker Script

When compiling the program, it is necessary to perform a few extra steps to ensure that the program is ready to be loaded and run by the boot code. The last step in compiling a program is to link all of the object files together, possibly also including some object files from system libraries. A linker script is a file that tells the linker which sections to include in the output file, as well as which order to put them in, what type of file is to be produced, and what is to be the address of the first instruction. The default linker script used by GCC creates an ELF executable file, which includes startup code from the C library and also includes information which tells the loader where the various sections reside in memory. The default linker script creates a file that can be loaded by the operating system kernel, but which cannot be executed on bare metal.

For a bare-metal program, the linker must be configured to link the program so that the first instruction of the startup function is given the correct address in memory. This address depends on how the boot loader will load and execute the program. On the Raspberry Pi this address is 0x8000, and on the pcDuino this address is 0x40008000. The linker will automatically adjust any other addresses as it links the code together. The most efficient way to accomplish this is by providing a custom linker script to be used instead of the default system script. Additionally, either the linker must be instructed to create a flat binary file, rather than an ELF executable file, or a separate program (objcopy) must be used to convert the ELF executable into a flat binary file.

Listing 14.7 is an example of a linker script that can be used to create a bare-metal program. The first line is just a comment. The second line specifies the name of the function where the program begins execution. In this case, it specifies that a function named _start is where the program will begin execution. Next, the file specifies the sections that the output file will contain. For each output section, it lists the input sections that are to be used.

The first output section is the .text section, and it is composed of any sections whose names end in .text.boot followed by any sections whose names end in .text. In Listing 14.5, the _start function was placed in the .text.boot section, and it is the *only thing* in that section. Therefore the linker will put the _start function at the very beginning of the program. The remaining text sections will be appended, and then the remaining sections, in the order that they appear. After the sections are concatenated together, the linker will make a pass through the resulting file, correcting the addresses of branch and load instructions as necessary so that the program will execute correctly.

```
1   /* FILE: bare_metal.ld - linker script for bare metal */
2   ENTRY(_start)
3
4   SECTIONS
5   {
6       /* One of the following lines should be commented out! */
7       . = 0x8000; /* Raspbery Pi will load the image here */
8       . = 0x40008000; /* pcDuino will load the image here */
9
10      __text_start__ = .;
11      .text :
12      {
13          KEEP(*(.text.boot)) /* put the start function first ! */
14          *(.text)
15      }
16      . = ALIGN(4096); /* align to page size */
17      __text_end__ = .;
18
19      __rodata_start__ = .;
20      .rodata :
21      {
22          *(.rodata)
23      }
24      . = ALIGN(4096); /* align to page size */
25      __rodata_end__ = .;
26
27      __data_start__ = .;
28      .data :
29      {
30          *(.data)
31      }
32      . = ALIGN(4096); /* align to page size */
33      __data_end__ = .;
34
35      __bss_start__ = .;
36      .bss :
37      {
38          bss = .;
39          *(.bss)
40      }
41      . = ALIGN(4096); /* align to page size */
42      __bss_end__ = .;
43      _end = .;
44  }
```

Listing 14.7
A sample Gnu linker script.

14.4.4 *Putting it All Together*

Compiling a program that consists of multiple source files, a custom linker script, and special commands to create an executable image can become tedious. The `make` utility was created specifically to help in this situation. Listing 14.8 shows a make script that can be used to combine all of the elements of the program together and produce a `uImage` file for the pcDuino and a `kernel.img` file for the Raspberry Pi. Listing 14.9 shows how the program can be built by typing "make" at the command line.

14.5 *Using an Interrupt*

The main program shown in Listing 14.6 is extremely wasteful because it runs the CPU in a loop, repeatedly checking the status of the GPIO pins. It uses far more CPU time (and electrical power) than is necessary. In reality, the pins are unlikely to change state very often, and it is sufficient to check them a few times per second. It only takes a few nanoseconds to check the input pins and set the output pins so the CPU only needs to be running for a few nanoseconds at a time, a few times per second.

A much more efficient implementation would set up a timer to send interrupts at a fixed frequency. Then the main loop can check the buttons, set the outputs, and put the CPU to sleep. Listing 14.10 shows the main program, modified to put the processor to sleep after each iteration of the main loop. The only difference between this main function and the one in Listing 14.6 is the addition of a `wfi` instruction at line 43. The new implementation will consume far less electrical power and allow the CPU to run cooler, thereby extending its life. However, some additional work must be performed in order to set up the timer and interrupt system before the main function is called.

14.5.1 *Startup Code*

Some changes must be made to the startup code in Listing 14.5 so that after setting up the vector table, it calls a function to initialize the interrupt controller then calls another function to set up the timer. Listing 14.5 shows the modified startup function.

Lines 50 through 57 have been added to initialize the interrupt controller, enable the timer, and change the CPU into user mode before calling main. Of course, the hardware timers and interrupt controllers on the pcDuino and Raspberry Pi are very different.

14.5.2 *Interrupt Controllers*

The pcDuino has an ARM Generic Interrupt Controller (GIC-400) device to manage interrupts. The GIC device can handle a large number of interrupts. Each one is a separate

```
1   # source files
2   SOURCES_ASM := main.S pcDuino_GPIO.S start.S vectab.S \
3                 interrupts.S
4   SOURCES_C    :=
5
6   # object files
7   OBJS         := $(patsubst %.S,%.o,$(SOURCES_ASM))
8   OBJS         += $(patsubst %.c,%.o,$(SOURCES_C))
9
10  # Build flags
11  INCLUDES     := -I.
12  ASFLAGS      :=
13  CFLAGS       := $(INCLUDES)
14
15  # build targets
16  all: uImage kernel.img
17
18  # Build image for pcDuino
19  uImage: kernel.img
20          mkimage -A arm -T kernel -a 40008000 -C none \
21            -n "bare metal" -d kernel.img uImage
22
23  # Build image for Raspberry Pi
24  kernel.img: bare.elf
25          objcopy bare.elf -O binary kernel.img
26
27  # Build the ELF file
28  bare.elf: $(OBJS) bare_metal.ld
29          ld $(OBJS) -Tbare_metal.ld -o $@
30
31  # Compile C to object file
32  %.o: %.c
33          gcc $(CFLAGS) -c $< -o $@
34
35  # Compile Assembly to object file
36  %.o: %.S
37          gcc $(ASFLAGS) -c $< -o $@
38
39  # Clean up the build directory
40  clean:
41          $(RM) -f $(OBJS) kernel.elf kernel.img uImage
42
43  dist-clean: clean
44          $(RM) -f *~
```

Listing 14.8
A sample make file.

```
lpyeatt@pcDuino$ make
gcc  -c main.S -o main.o
gcc  -c pcDuino_GPIO.S -o pcDuino_GPIO.o
gcc  -c start.S -o start.o
gcc  -c vectab.S -o vectab.o
gcc  -c handlers.S -o handlers.o
ld main.o pcDuino_GPIO.o start.o vectab.o handlers.o -Tbare_metal.ld -o bare.elf
objcopy bare.elf -O binary kernel.img
mkimage -A arm -T kernel -a 40008000 -C none \
          -n "bare metal" -d kernel.img uImage
Image Name:    bare metal
Created:       Tue Oct 13 13:38:20 2015
Image Type:    ARM Linux Kernel Image (uncompressed)
Data Size:     4240 Bytes = 4.14 kB = 0.00 MB
Load Address: 40008000
Entry Point:  40008000
lpyeatt@pcDuino$
```

Listing 14.9
Running make to build the image.

input signal to the GIC. The GIC hardware prioritizes each input, and assigns each one a unique integer identifier. When the CPU receives an interrupt, it simply reads the GIC to determine which hardware device signaled the interrupt, calls the function which handles that device, then writes to one of the GIC registers to indicate that the interrupt has been processed. Listing 14.12 provides a few basic functions for managing this device.

The Raspberry Pi has a much simpler interrupt controller. It can enable and disable interrupt sources, and requires that the programmer read up to three registers to determine the source of an interrupt. For our purposes, we only need to manage the ARM timer interrupt. Listing 14.13 provides a few basic functions for using this device to enable the timer interrupt. Extending these functions to provide functionality equal to the GIC would not be very difficult, but would take some time. It would be necessary to set up a mapping from the interrupt bits in the interrupt register controller to integer values, so that each interrupt source has a unique identifier. Then the functions could be written to use those identifiers. The result would be a software implementation to provide capabilities equivalent to the GIC.

Note that although the devices are very different internally, they perform basically the same function. With the addition of a software driver layer, implemented in Listings 14.12 and 14.13 the devices become interchangeable and other parts of the bare-metal program do not have to be changed when porting from one platform to the other.

```
 1   @@@ FILE: main.S
 2   @@@ This program reads from three buttons connected to GPIO3-5, and
 3   @@@ controls three leds connected to GPIO0-2. The main loop puts
 4   @@@ the CPU to sleep after each iteration.  A timer interrupt wakes
 5   @@@ it up some time later.
 6           .global main
 7   main:   stmfd   sp!,{lr}
 8           @@ Set the GPIO pins
 9           mov     r0,#0           @ Port 0
10           bl      GPIO_dir_output @ set for output
11           mov     r0,#1           @ Port 1
12           bl      GPIO_dir_output @ set for output
13           mov     r0,#2           @ Port 2
14           bl      GPIO_dir_output @ set for output
15
16           mov     r0,#3           @ Port 3
17           bl      GPIO_dir_input  @ set for input
18           mov     r0,#4           @ Port 4
19           bl      GPIO_dir_input  @ set for input
20           mov     r0,#5           @ Port 5
21           bl      GPIO_dir_input  @ set for input
22   @@@ Main loop just reads buttons and updates the LEDs,
23   @@@ then puts the CPU to sleep until an interrupt occurs.
24   loop:   @@ Read the state of the inputs and
25           @@ set the ouputs to the same state.
26           mov     r0,#3           @ Pin 3
27           bl      GPIO_get_pin    @ read it
28           mov     r1,r0           @ copy pin state to r1
29           mov     r0,#0           @ Pin 0
30           bl      GPIO_set_pin    @ write it
31           mov     r0,#4           @ Pin 4
32           bl      GPIO_get_pin    @ read it
33           mov     r1,r0           @ copy pin state to r1
34           mov     r0,#1           @ Pin 1
35           bl      GPIO_set_pin    @ write it
36           mov     r0,#5           @ Pin 5
37           bl      GPIO_get_pin    @ read it
38           mov     r1,r0           @ copy pin state to r1
39           mov     r0,#2           @ pin 2
40           bl      GPIO_set_pin    @ write it
41           @@ Put CPU to sleep until an interrupt occurs
42           //wfi                   @ used on pcDunio
43           mov     r0,#0
44           mcr     p15,0,r0,c7,c0,4@ used on Raspberry Pi
45           b       loop
46           ldmfd   sp!,{pc}
```

Listing 14.10
An improved main program.

```
1          .include "modes.S"
2
3   @@@ Stack locations
4          @@ uncomment one of the following two lines
5          @ .equ    stack_top, 0x10000000 @ Raspberry pi only
6          @ .equ    stack_top, 0x50000000 @ pcDuino only
7
8          .equ    fiq_stack_top, stack_top
9          .equ    irq_stack_top, stack_top - 0x1000
10          .equ    abt_stack_top, stack_top - 0x2000
11          .equ    und_stack_top, stack_top - 0x3000
12          .equ    mon_stack_top, stack_top - 0x4000
13          .equ    svc_stack_top, stack_top - 0x5000
14          .equ    sys_stack_top, stack_top - 0x6000
15   @@@ -----------------------------------------------------------
16   @@@ The startup code should be loaded by the boot loader.
17   @@@ The entry point is _start which performs initialization of
18   @@@ the hardware, then calls a C function.
19          .section .start
20          .global _start
21          .func   _start
22   _start:
23
24          @@ On reset, we should be in SVC mode.
25          @@ Set up all stacks.
26          msr     CPSR_c,#FIQ_MODE|I_BIT|F_BIT  @ switch to FIQ mode
27          ldr     sp,=fiq_stack_top @ set the FIQ stack pointer
28
29          msr     CPSR_c,#IRQ_MODE|I_BIT|F_BIT  @ switch to IRQ mode
30          ldr     sp,=irq_stack_top @ set the IRQ stack pointer
31
32          msr     CPSR_c,#ABT_MODE|I_BIT|F_BIT  @ switch to ABT mode
33          ldr     sp,=abt_stack_top @ set the ABT stack pointer
34
35          msr     CPSR_c,#UND_MODE|I_BIT|F_BIT  @ switch to UND mode
36          ldr     sp,=und_stack_top @ set the UND stack pointer
37
38          msr     CPSR_c,#SYS_MODE|I_BIT|F_BIT  @ switch to SYS mode
39          ldr     sp,=sys_stack_top @ set SYS/USR stack pointer
40
41          msr     CPSR_c,#SVC_MODE|I_BIT|F_BIT  @ switch to SVC mode
42          ldr     sp,=svc_stack_top @ set svc mode stack pointer
43          @@ All stacks are initialized, and we are in SVC mode
44
45          @@ Clear the .bss segment to all zeros
```

```
46          @@ The __bss_start__ and __bss_end__ symbols are
47          @@ defined by the linker
48          ldr     r1,=__bss_start__ @ load pointer to bss and
49          ldr     r2,=__bss_end__   @ to byte following bss
50          mov     r3,#0             @ load fill value (zero)
51  bssloop:cmp     r1,r2             @ Start filling
52          bge     bssdone
53          str     r3,[r1],#4
54          b       bssloop           @ loop until done
55  bssdone:
56
57          @@ Set up the exception vector table
58          bl      setup_vector_table @ this function is in IVT.S
59          @@ The exception handlers are defined in interrupts.S
60
61          @@ Initialize the Interrupt Controller
62          bl      IC_init
63
64          @@ Set up the hardware timer to generate interrupts
65          @@ at a fixed frequency
66          bl      enable_timer
67
68          @@ switch to user mode with IRQ and FIQ enabled
69          msr     CPSR_c,#USR_MODE@ switch to USR mode
70
71          @@ Enter the C/C++ code at main
72          bl      main              @ call main function
73          @@ If main ever returns, cause an exception
74          swi     0xFFFFFF          @ this should never happen
75          .size   _start, . - _start
76          .endfunc
```

Listing 14.11
ARM startup code with timer interrupt.

```
1   @@@  Functions to manage the ARM Generic Interrupt Controller (GIC)
2          @@ offsets to GIC interfaces (GIC-400)
3          .equ    GIC_DIST, 0x1000
4          .equ    GIC_CPU, 0x2000
5          @@ Registers in the CPU interface. There are more
6          @@ registers, but I don't need them
7          .equ    ICCICR,  0x00
8          .equ    ICCPMR,  0x04
9          .equ    ICCEOIR, 0x10
10         .equ    ICCIAR,  0x0C
11         @@ Registers in the Distributor. There are more
```

```
12          @@ registers, but I don't need them
13          .equ    ICDDCR,   0x00
14          .equ    ICDISER,  0x100
15          .equ    ICDIPTR,  0x800
16  @@@ -----------------------------------------------------------------
17          .data
18          .align 2
19          @@ Addresses of the GIC Distributor and CPU interfaces
20  GIC_dist_base:  .word   0       @ address of GIC distributor
21  GIC_cpu_base:   .word   0       @ address of GIC CPU interface
22  @@@ -----------------------------------------------------------------
23          .text
24          .align 2
25  @@@ -----------------------------------------------------------------
26  @@@ Initialization of the Generic Interrupt Controller (GIC)
27          .global GIC_init
28  IC_init:
29          stmfd   sp!,{lr}
30          @@ Read GIC base from Configuration Base Address Register
31          @@ and use it to initialize GIC_dist_base and GIC_cpu_base
32          mrc     p15, 4, r0, c15, c0, 0
33          add     r2,r0,#GIC_DIST @ calculate address
34          ldr     r1,=GIC_dist_base
35          str     r2,[r1]         @ store address of GIC distributor
36          add     r2,r0,#GIC_CPU  @ calculate address
37          ldr     r1,=GIC_cpu_base
38          str     r2,[r1]         @ store address of GIC CPU iface
39          @@ Set the Interrupt CPU Control Priority Mask
40          @@ Register (ICCPMR) to enable interrupts
41          @@ of all priorities levels
42          ldr     r1,=0xFFFF
43          str     r1,[r2,#ICCPMR] @ r2 still has address of CPU iface
44          @@ Set the enable bit in the CPU Interface Control
45          @@ Register (ICCICR), allowing CPU(s) to receive interrupts
46          mov     r1, #1
47          str     r1, [r2,#ICCICR]
48          @@ Set the enable bit in the Distributor Control
49  @@@ FILE: GIC.S
50          @@ Register (ICDDCR), allowing interrupts to be generated
51          ldr     r2,=GIC_dist_base
52          ldr     r2,[r2] @ base address of Distributor Interface
53          mov     r1, #1
54          str     r1, [r2,#ICDDCR]
55          ldmfd sp!,{pc}
56  @@@ -----------------------------------------------------------------
57  @@@ config_interrupt (int ID, int CPU);
```

```
58   @@@ Configure one interrupt source to signal one or more CPU
59   @@@ CPU is an 8-bit bitmap, allowing up to eight CPUs to be
60   @@@ specified in the 8 bottom bits.
61          .global config_interrupt
62   config_interrupt:
63          stmfd   sp!,{r4-r5, lr}
64          @@ Configure the "Distributor Interrupt Set-Enable
65          @@ Registers" (ICDISERn). (enable the interrupt)
66          @@ reg_offset = (N / 32) * 4; (shift and clear some bits)
67          @@ value = 1 << (N mod 32);
68          ldr     r2,=GIC_dist_base
69          ldr     r2,[r2]        @ Read GIC distributor base address
70          add     r2,r2,#ICDISER @ r2 <- base address of ICDSER regs
71          lsr     r4,r0,#3       @ calculate reg_offset
72          bic     r4,r4,#3       @ r4 <- reg_offset
73          add     r4,r2,r4       @ r4 <- address of ICDISERn
74          @@ Create a bit mask
75          and     r2,r0,#0x1F    @ r2 <- N mod 32
76          mov     r5,#1          @ need to set one bit
77          lsl     r2,r5,r2       @ r2 <- value
78          @@ Using address in r4 and value in r2 set the correct bit
79          @@ in the GIC register
80          ldr     r3, [r4]       @ read ICDISERn
81          orr     r3, r3, r2     @ set the enable bit
82          str     r3, [r4]       @ store the new register value
83          @@ Configure the "Distributor Interrupt Processor Targets
84          @@ Register" (ICDIPTRn). (select target CPUs)
85          @@ reg_offset = (N / 4) * 4; (clear 2 bottom bits)
86          @@ index = N mod 4;
87          ldr     r2,=GIC_dist_base
88          ldr     r2,[r2]        @ Read GIC distributor base address
89          add     r2, r2, #ICDIPTR@ base address of ICDIPTR regs
90          bic     r4, r0, #3     @ r4 <- reg_offset
91          add     r4, r2, r4     @ r4 <- address of ICDIPTRn
92          @@ Get the address of the byte within ICDIPTRn
93          and     r2, r0, #0x3   @ r2 <- index
94          add     r4, r2, r4     @ r4 <- byte address to be set
95          @@ using address in r4 and value in r2, write to the
96          @@ appropriate byte
97          strb    r1, [r4]
98          ldmfd   sp!,{r4-r5, lr}
99   @@@ -------------------------------------------------------------
100  @@@ int get_interrupt_number();
101  @@@ Get the interrupt ID for the current interrupt. This should be
102  @@@ called at the beginning of interrupt processing. It also
103  @@@ changes the state of the interrupt from pending to active,
```

```
104  @@@ which helps to prevent other CPUs from trying to handle it.
105          .global get_interrupt_number
106  get_interrupt_number: @ Read the ICCIAR from the CPU Interface
107          ldr     r0,=GIC_cpu_base
108          ldr     r0,[r0]        @ Read GIC CPU interface address
109          ldr     r0,[r0,#ICCIAR] @ Read from ICCIAR
110          mov     pc,lr
111  @@@ --------------------------------------------------------------
112  @@@ void end_of_interrupt(int ID);
113  @@@ Notify the GIC that the interrupt has been processed.
114  @@@ The state goes from active to inactive, or it goes from
115  @@@ active and pending to pending.
116          .global end_of_interrupt
117  end_of_interrupt:
118          ldr     r1,=GIC_cpu_base
119          ldr     r1,[r1]        @ Read GIC CPU interface address
120          str     r0,[r1,#ICCEOIR]@ Write to ITTEOIR
121          mov     pc,lr
```

Listing 14.12
Functions to manage the pdDuino interrupt controller.

```
1   @@@ FILE: RasPiIC.S
2   @@@ Functions to manage the Interrupt Controller on the
3   @@@ Raspberry Pi
4           @@ Address of Interrupt Controller
5           .equ    IC,      0x7e00B000
6           @@ Register offsets
7           .equ    IRQBP,  0x200 @ IRQ basic pending
8           .equ    IRQP1,  0x204 @ IRQ pending 1
9           .equ    IRQP2,  0x208 @ IRQ pending 2
10          .equ    FIQC,   0x20C @ FIQ control
11          .equ    IRQEN1, 0x210 @ IRQ enable 1
12          .equ    IRQEN2, 0x214 @ IRQ enable 2
13          .equ    IRQBEN, 0x218 @ Enable basic IRQs
14          .equ    IRQDA1, 0x21C @ IRQ disable 1
15          .equ    IRQDA2, 0x220 @ IRQ disable 2
16          .equ    IRQBDA, 0x224 @ Disable basic IRQs
17  @@@ --------------------------------------------------------------
18          .text
19          .align 2
20  @@@ --------------------------------------------------------------
21  @@@ Initialization of the Interrupt Controller (IC)
22          .global IC_init
23  IC_init:
24          @@ disable all interrupts
```

```
25          ldr     r0,=IC
26          mov     r1,#0
27          str     r1,[r0,#IRQEN1]
28          str     r1,[r0,#IRQEN2]
29          str     r1,[r0,#IRQBEN]
30          mov     pc,lr
31  @@@ ---------------------------------------------------------------
32  @@@ config_interrupt (int ID, int CPU);
33  @@@ On Raspberry Pi, this just enables the timer interrupt
34          .global config_interrupt
35  config_interrupt:
36          ldr     r0,=IC
37          mov     r1,#1
38          str     r1,[r0,#IRQBEN]
39          mov     pc,lr
40  @@@ ---------------------------------------------------------------
41  @@@ int get_interrupt_number();
42  @@@ Get the interrupt ID for the current interrupt.
43  @@@ On Raspberry Pi, just read and return the pending register.
44          .global get_interrupt_number
45  get_interrupt_number: @ Read the ICCIAR from the CPU Interface
46          ldr     r0,=IC
47          ldr     r0,[r0,#IRQBP]
48          mov     pc,lr
49  @@@ ---------------------------------------------------------------
50  @@@ void end_of_interrupt(int ID);
51  @@@ Notify the IC that the interrupt has been processed.
52  @@@ On Raspberry Pi, this does nothing
53          .global end_of_interrupt
54  end_of_interrupt:
55          mov     pc,lr
```

Listing 14.13
Functions to manage the Raspberry Pi interrupt controller.

14.5.3 Timers

The pcDuino provides several timers that could be used, Timer0 was chosen arbitrarily. Listing 14.14 provides a few basic functions for managing this Device.

The Raspberry Pi also provides several timers that could be used, but the ARM timer is the easiest to configure. Listing 14.15 provides a few basic functions for managing this device:

```
1    @@@ FILE: pcDuino_timer.S
2            .equ    TIMER_BASE, 0x01C20C00 @ Allwinner A10/A20
3            .equ    TMR_IRQ_EN_REG,  0x0
4            .equ    TMR_IRQ_STA_REG, 0x4
5            .equ    TMR0_CTRL_REG,        0x10
6            .equ    TMR0_INTV_VALUE_REG, 0x14
7            .equ    TMR0_CUR_VALUE_REG,  0x18
8
9            .text
10           .align 2
11   @@@ ----------------------------------------------------------------
12   @@@ Configures and enables timer0 to generate interrupts at a
13   @@@ fixed frequency.  Also configures the Generic Interrupt
14   @@@ Controller (GIC) to send interrupts to CPU 0.
15           .global enable_timer
16   enable_timer:
17           stmfd   sp!,{lr}
18           ldr     r0,=TIMER_BASE
19           @@ Clear the control register and current count.
20           mov     r1,#0
21           str     r1,[r0,#TMR0_CTRL_REG]
22           str     r1,[r0,#TMR0_CUR_VALUE_REG]
23           @@ Set the interval value to 24000000/8 clocks
24           ldr     r1,=(24000000>>3) @ interrupt every 0.125 seconds
25           str     r1,[r0,#TMR0_INTV_VALUE_REG]
26           @@ Configure and start the timer.
27           @@ continuous mode -> 0
28           @@ prescale 1:1 -> 000
29           @@ use 24MHz oscillator -> 01
30           @@ reload counter -> 1
31           @@ start timer  -> 1
32           mov     r1,#0b00000111          @ load configuration word
33           str     r1,[r0,#TMR0_CTRL_REG]  @ configure and start timer
34           @@ Enable timer 0 to generate interrupts
35           ldr     r1,[r0,#TMR_IRQ_EN_REG]
36           orr     r1,r1,#1                @ Set the IRQ enable bit
37           str     r1,[r0,#TMR_IRQ_EN_REG]
38           @@ Configure GIC to allow interrupts from Timer 0, which
39           @@ is source 54 of the GIC (Generic Interrupt Controller)
40           @@ on the Allwinner A10/A20
41           mov     r0, #54 @ Timer 0 is SRC 54
42           mov     r1, #1  @ bit mask: this is cpu 0 only
43           bl      config_interrupt
44           ldmfd   sp!,{lr}
45           mov     pc,lr
```

```
46   @@@ --------------------------------------------------------------
47   @@@ int check_timer_interrupt()
48   @@@ Check and clear the timer 0 interrupt.  Returns 1 if the
49   @@@ interrupt was active.  Returns 0 otherwise.
50          .global check_timer_interrupt
51   check_timer_interrupt:
52          ldr     r0,=TIMER_BASE
53          ldr     r1,[r0,#TMR_IRQ_STA_REG]
54          ands    r2,r1,#1
55          movne   r1,#1
56          strne   r1,[r0,#TMR_IRQ_STA_REG]
57          mov     r0,r1
58          mov     pc,lr
```

Listing 14.14
Functions to manage the pdDuino timer0 device.

```
1    @@@ FILE: RasPi_timer.S
2    @@@ The timer runs off the 250MHz APB_clock source
3           .equ    TIMER_BASE, 0x7e008400 @ BCM2835
4           .equ    LOAD,    0x00  @ Load
5           .equ    VALUE,   0x04  @ Value (read only)
6           .equ    CONTROL,0x08   @ Control
7           .equ    IRQACK,  0x0C  @ IRQ Clear/Ack (write only)
8           .equ    RAWIRQ,  0x10  @ Raw IRQ (read only)
9           .equ    MSKIRQ,  0x14  @ Masked IRQ (read only)
10          .equ    RELOAD,  0x18  @ Reload
11          .equ    PREDIV,  0x1C  @ Pre-divider
12          .equ    COUNT,   0x20  @ Free-running counter
13
14          .text
15          .align 2
16   @@@ --------------------------------------------------------------
17   @@@ Configures and enables timer0 to generate interrupts at a
18   @@@ fixed frequency.  Also configures the Generic Interrupt
19   @@@ Controller (GIC) to send interrupts to CPU 0.
20          .global enable_timer
21   enable_timer:
22          ldr     r0,=TIMER_BASE
23          mov     r1,#0x7F        @ divide clock to 1,953,125Hz
24          str     r1,[r0,#PREDIV]
25          ldr     r1,=954         @ should give about 8Hz
26          str     r1,[r0,#LOAD]
27          ldr     r1,=0b11111000000000010101010
28          str     r1,[r0,#CONTROL]
29          mov     pc,lr
```

```
30   @@@ --------------------------------------------------------
31   @@@ int check_timer_interrupt()
32   @@@ Check and clear the timer 0 interrupt.  Returns 1 if the
33   @@@ interrupt was active.  Returns 0 otherwise.
34          .global check_timer_interrupt
35   check_timer_interrupt:
36          ldr    r1,=TIMER_BASE
37          ldr    r0,[r1,#MSKIRQ]
38          ands   r0,#1
39          strne  r0,[r1,#IRQACK]
40          mov    pc,lr
```

Listing 14.15
Functions to manage the Raspberry Pi timer0 device.

14.5.4 Exception Handling

The final step in writing the bare-metal code to operate in an interrupt-driven fashion is to modify the IRQ handler from Listing 14.3. Listing 14.16 shows a new version of the IRQ exception handler which checks and clears the timer interrupt, then returns to the location and CPU mode that were current when the interrupt occurred. This code works for both platforms.

14.5.5 Building the Interrupt-Driven Program

Finally, the make file must be modified to include the new source code that was added to the program. Listing 14.17 shows the modified make script. The only change is that two extra object files have been added. when make is run, those files will be compiled and linked with the program. Listing 14.9 shows how the program can be built by typing "make" at the command line.

14.6 ARM Processor Profiles

Since its introduction in 1982 as the flagship processor for Acorn RISC Machine, the ARM processor has gone through many changes. Throughout the years, ARM processors have always maintained a good balance of simplicity, performance, and efficiency. Although originally intended as a desktop processor, the ARM architecture has been more successful than any other architecture for use in embedded applications. That is at least partially because of good choices made by its original designers. The architectural decisions resulted in a processor that provides relatively high computing power with a relatively small number of transistors. This design also results in relatively low power consumption.

```
1    @@@ -------------------------------------------------------------------
2            .global irq_handler
3    irq_handler:
4            stmfd   sp!,{r0-r12,lr}
5
6            @@ find out which interrupt we are servicing
7            bl      get_interrupt_number @ returns in r0
8            stmfd   sp!,{r0}        @ save interrupt number
9
10           cmp     r0,#54          @ is it the timer interrupt?
11           bleq    check_timer_interrupt
12
13           ldmfd   sp!,{r0}        @ retrieve interrupt number
14           bl      end_of_interrupt@ tell GIC we are done
15
16           ldmfd   sp!,{r0-r12,lr}
17           subs    pc, lr, #4      @ must subtract 4 from lr
```

Listing 14.16
IRQ handler to clear the timer interrupt.

Today, there are almost 20 major versions of the ARMv7 architecture, targeted for everything from smart sensors to desktops and servers, and sales of ARM-based processors outnumber all other processor architectures combined. Historically, ARM has given numbers to various versions of the architecture. With the ARMv7, they introduced a simpler scheme to describe different versions of the processor. They divided their processor families into three major *profiles*:

ARMv7-A: Applications processors are capable of running a full, multiuser, virtual memory, multiprocessing operating system.

ARMv7-R: Real-time processors are for embedded systems that may need powerful processors, cache, and/or large amounts of memory.

ARMv7-M: Microcontroller processors only execute Thumb instructions and are intended for use in very small cost-sensitive embedded systems. They provide low cost, low power, and small size, and may not have hardware floating point or other high-performance features.

In 2014, ARM introduced the ARMv8 architecture. This is the first radical change in the ARM architecture in over 30 years. The new architecture extends the register set to thirty 64-bit general purpose registers, and has a completely new instruction set. Compatibility with ARMv7 and earlier code is supported by switching the processor into 32-bit mode, so that it

```
1   # source files
2   SOURCES_ASM := main.S pcDuino_IO.S start.S vectab.S \
3                 handlers.S pcDuino_GPIO.S
4   SOURCES_C   :=
5
6   # object files
7   OBJS        := $(patsubst %.S,%.o,$(SOURCES_ASM))
8   OBJS        += $(patsubst %.c,%.o,$(SOURCES_C))
9
10  # Build flags
11  INCLUDES    := -I.
12  ASFLAGS     :=
13  CFLAGS      := $(INCLUDES)
14
15  # build targets
16  all: uImage kernel.img
17
18  # Build image for pcDuino
19  uImage: kernel.img
20          mkimage -A arm -T kernel -a 40008000 -C none \
21           -n "bare metal" -d kernel.img uImage
22
23  # Build image for Raspberry Pi
24  kernel.img: bare.elf
25          objcopy bare.elf -O binary kernel.img
26
27  # Build the ELF file
28  bare.elf: $(OBJS) bare_metal.ld
29          ld $(OBJS) -Tbare_metal.ld -o $@
30
31  # Compile C to object file
32  %.o: %.c
33          gcc $(CFLAGS) -c $< -o $@
34
35  # Compile Assembly to object file
36  %.o: %.S
37          gcc $(ASFLAGS) -c $< -o $@
38
39  # Clean up the build directory
40  clean:
41          $(RM) -f $(OBJS) kernel.elf kernel.img uImage
42
43  dist-clean: clean
44          $(RM) -f *~
```

Listing 14.17
A sample make file.

```
lpyeatt@pcDuino$ make
gcc  -c main.S -o main.o
gcc  -c pcDuino_GPIO.S -o pcDuino_GPIO.o
gcc  -c start.S -o start.o
gcc  -c vectab.S -o vectab.o
gcc  -c handlers.S -o handlers.o
ld main.o pcDuino_GPIO.o start.o vectab.o handlers.o -Tbare_metal.ld -o bare.elf
objcopy bare.elf -O binary kernel.img
mkimage -A arm -T kernel -a 40008000 -C none \
          -n "bare metal" -d kernel.img uImage
Image Name:    bare metal
Created:       Tue Oct 13 13:38:20 2015
Image Type:    ARM Linux Kernel Image (uncompressed)
Data Size:     4240 Bytes = 4.14 kB = 0.00 MB
Load Address: 40008000
Entry Point:   40008000
lpyeatt@pcDuino$
```

Listing 14.18
Running make to build the image.

executes the 32-bit ARM instruction set. This is somewhat similar to the way that the Thumb instructions are supported on 32-bit ARM cores, but the change to 32-bit code can only be made when the processor is in privileged mode, and drops back to unprivileged mode.

14.7 Chapter Summary

Writing bare-metal programs can be a daunting task. However, that task can be made easier by writing and testing code under an operating system before attempting to run it bare metal. There are some functions which cannot be tested in this way. In those cases, it is best to keep those functions as simple as possible. Once the program works on bare metal, extra capabilities can be added.

Interrupt-driven processing is the basis for all modern operating systems. The system timer allows the O/S to take control periodically and select a different process to run on the CPU. Interrupts allow hardware devices to do their jobs independently and signal the CPU when they need service. The ability to restrict user access to devices and certain processor features provides the basis for a secure and robust system.

Exercises

14.1 What are the advantages of a CPU which supports user mode and privileged mode over a CPU which does not?

14.2 What are the six privileged modes supported by the ARM architecture?

14.3 The interrupt handling mechanism is somewhat complex and requires significant programming effort to use. Why is it preferred over simply having the processor poll I/O devices?

14.4 Where does program control transfer to when a hardware interrupt occurs?

14.5 What is the purpose of the Undefined Instruction exception? How can it be used to allow an older processor to run programs that have new instructions? What other uses does it have?

14.6 What is an `swi` instruction? What is its use in operating systems? What is the key difference between an `swi` instruction and an interrupt?

14.7 Which of the following operations should be allowed only in privileged mode? Briefly explain your decision for each one.
   (a) Execute an `swi` instruction.
   (b) Disable all interrupts.
   (c) Read the time-of-day clock.
   (d) Receive a packet of data from the network.
   (e) Shutdown the computer.

14.8 The main program in Listing 14.10 has two different methods to put the processor to sleep waiting for an interrupt. One method is for the Raspberry Pi, while the other is for the pcDuino. In order to compile the code, the correct lines must be uncommented and the unneeded lines must be commented out or removed. Explain two ways to change the code so that exactly the same main program can be used on both systems.

14.9 The programs in this chapter assumed the existence of libraries of functions for controlling the GPIO pins on the Raspberry Pi and the pcDuino. Both libraries provide the same high-level functions, but one operates on the Raspberry Pi GPIO device and the other operates on the pcDuino GPIO device. The C prototypes for the functions are: `int GPIO_get_pin(int pin)`, `void GPIO_set_pin(int pin,int state)`, `GPIO_dir_input(int pin)`, and `GPIO_dir_output(int pin)`. Write these libraries in ARM assembly language for both platforms.

14.10 Write an interrupt-driven program to read characters from the serial port on either the Raspberry Pi or the pcDuino. The UART on either system can be configured to send an interrupt when a character is received.
   When a character is received through the UART and an interrupt occurs, the character should be echoed by transmitting it back to the sender. The character should also be stored in a buffer. If the character received is newline (\n), or if the buffer becomes full, then the contents of the buffer should be transmitted through the UART. Then, the buffer cleared and prepared to receive more characters.

Index

Note: Page numbers followed by *b* indicate boxes, *f* indicate figures and *t* indicate tables.

Printed in the United States
By Bookmasters